DAVID O. MCKAY LIBRARY

3 1404 00818 6196

Surface Chemistry
Of Solid and Liquid Interfaces

D0796570

APR 2 5 2007

Surface Chemistry
Of Solid and Liquid Interfaces

Professor H. Yıldırım Erbil
Faculty of Engineering
Department of Chemical Engineering
Gebze Institute of Technology
Turkey

Blackwell
Publishing

© 2006 Yıldırım Erbil
Editorial offices:
Blackwell Publishing Ltd, 9600 Garsington Road, Oxford OX4 2DQ, UK
   Tel: +44 (0)1865 776868
Blackwell Publishing Inc., 350 Main Street, Malden, MA 02148-5020, USA
   Tel: +1 781 388 8250
Blackwell Publishing Asia Pty Ltd, 550 Swanston Street, Carlton, Victoria 3053, Australia
   Tel: +61 (0)3 8359 1011

The right of the Author to be identified as the Author of this Work has been asserted in
accordance with the Copyright, Designs and Patents Act 1988.

All rights reserved. No part of this publication may be reproduced, stored in a retrieval system,
or transmitted, in any form or by any means, electronic, mechanical, photocopying, recording
or otherwise, except as permitted by the UK Copyright, Designs and Patents Act 1988, without
the prior permission of the publisher.

First published 2006 by Blackwell Publishing Ltd

ISBN-10: 1-4051-1968-3
ISBN-13: 978-1-4051-1968-9

Library of Congress Cataloging-in-Publication Data

A catalogue record for this title is available from the British Library

Set in 10 on 12 pt Minion
by SNP Best-set Typesetter Ltd., Hong Kong
Printed and bound in India
by Replika Press Pvt Ltd

The publisher's policy is to use permanent paper from mills that operate a sustainable
forestry policy, and which has been manufactured from pulp processed using acid-free and
elementary chlorine-free practices. Furthermore, the publisher ensures that the text paper
and cover board used have met acceptable environmental accreditation standards.

For further information on Blackwell Publishing, visit our website:
www.blackwellpublishing.com

Contents

Preface

The main objective of this book is to provide a general physical chemistry background in a clear manner to students and scientists working with solid–liquid interfaces. The number of industries related to surface chemistry and physics have increased dramatically during the past decades through the rapid development of surface coatings, adhesives, textiles, oil recovery, cosmetics, pharmaceuticals, biomedical, biotechnology, agriculture, mineral flotation, lubrication industries and the introduction of newly emerged nanotechnology. A great number of publications on these subjects over the last decade have resulted in a better understanding of all of these scientific fields, which are mostly interdisciplinary in nature, where high-vacuum physicists, colloid and polymer chemists, biologists, and chemical, material, biomedical, environmental, and electronic engineers are all contributing.

Unfortunately, most standard textbooks on physical chemistry pay only limited attention to surface science at present. Consequently, many graduates in the fields of chemistry, physics, biology, engineering and materials science enter their careers ignorant of even the basic concepts of important surface interactions and are now under pressure to solve an increasing number of challenging new problems. Thus, knowledge of the fundamental principles of interfacial chemistry and physics is going to become a requisite for all the research and technical personnel in these industries. Although a number of excellent books are available in both surface and colloid sciences, and in polymer, solid physics and chemistry, these can seem daunting due to their *written for experts* nature. This book is intended to meet the needs of newcomers to the surface field from academia or industry who want to acquire a basic knowledge of the subject.

The text largely contains fundamental material and focuses on understanding the basic principles rather than learning factual information. Since it is impossible to include all branches of surface science in such an introductory book because of its wide and multidisciplinary scope, a specific and narrow topic, the *interfacial interactions between solids and liquids*, has been chosen for this book. For this reason, the ionic interactions, charged polymers, electrochemistry, electrokinetics and the colloid and particulate sciences cannot be included. Some fundamental physical chemistry subjects such as basic thermodynamics are covered, and many equations are derived from these basic concepts throughout the book in order to show the links between applied surface equations and the fundamental concepts. This is lacking in most textbooks and applied books in surface chemistry, and for this reason, this book can be used as a textbook for a course of 14–15 weeks.

The book is divided into three parts: (1) Principles, (2) Liquids and (3) Solids. In Principles, an introduction to surfaces and interfaces, molecular interactions and the thermodynamics of interfaces are presented in three chapters in their most basic form. In Liquids, the properties of pure liquid surfaces, liquid solution surfaces, the experimental determination of liquid surface tension and the potential energy of interaction between particles and surfaces through a liquid medium are covered in four chapters. In Solids, the properties of solid surfaces, contact angles of liquid drops on solids and some applications involving solid/liquid interfaces are presented in three chapters. The aim is that both final year undergraduate or graduate university students and also industrial researchers will find the necessary theory they seek in this book, and will be able to use and develop these concepts effectively. Considerable attention is devoted to experimental aspects throughout the book. At the end of each chapter are bibliographies to which the reader may turn for further details. Since much of the material in this book is drawn from existing treatments of the subject, our debt to earlier writers is considerable and, we hope, fully acknowledged. This introductory contribution cannot claim to be comprehensive; however it does hopefully provide a clear understanding of the field by assessing the importance of related factors. Some errors are unavoidable throughout the book and if found can be communicated by e-mail to (yerbil@gyte.edu.tr).

I am greatly indebted to my colleagues, who encouraged me to prepare this book, and my wife Ayse and my children, Ayberk, Billur, Beril and Onur for their patience and understanding throughout its preparation.

H. Yıldırım Erbil
2006

PART I

Principles

Chapter 1
Introduction to Surfaces and Interfaces

1.1 Definition of a Surface and an Interface

A *phase* of a substance is a form of matter that is uniform throughout in chemical composition and physical state. There are mainly three phases of matter namely solid, liquid and gas. The word *fluid* is used to describe both gas and liquid phases. We usually classify the phase of a material according to its state at the normal ambient temperature (20–25°C), which is well above the melting point of most fluids. We mostly deal with two or more phases, which coexist, in equilibrium or non-equilibrium conditions. Phase diagrams are used as a convenient method of representing the regions of stability of solid, liquid and gas phases under various conditions of temperature and pressure. An *interface* is the physical boundary between two adjacent bulk phases. The interface must be at least one molecular diameter in thickness for the purpose of constructing a molecular model. In some cases it may extend over several molecular thicknesses. We use the word *surface* in order to define the physical boundary of only one of these phases, such as *solid surface* and *liquid surface* etc. In reality, we deal with an *interface* in all cases other than absolute vacuum conditions for solids, since every single phase is in contact with another phase such as solid–air, liquid–air contacts. In many standard physical chemistry, surface chemistry and surface physics textbooks, the words *surface* and *interface* are used interchangeably because the authors neglect the small differences between the air phase and absolute vacuum conditions. On the other hand, some authors define and use the word *surface* for only the solid–gas and liquid–gas phase contacts, and the word *interface* for all of other phase contacts, but this has no scientific basis and should be abandoned.

The molecules that are located at the phase boundaries (that is between solid–gas, solid–liquid, liquid–gas and liquid$_1$–liquid$_2$) behave differently to those in the bulk phase. This rule generally does not apply for solid–solid and gas–gas interfaces where atomic and molecular bonding in the solid structure restrict the reorientation of interfacial molecules for the former, and the ease of miscibility of different gas molecules in free space does not allow any interface formation for the latter.

There is an orientation effect for molecules at fluid surfaces: the molecules at or near the fluid surface will be orientated differently with respect to each other than the molecules in the bulk phase. Any molecule at the fluid surface would be under an asymmetrical force field, resulting in surface or interfacial tension. The nearer the molecule is to the interfacial boundary, the greater the magnitude of the force due to asymmetry. As the distance from the interfacial boundary increases, the effect of the force field decreases and vanishes after a certain length. Thus, in reality there is no very sharp interfacial boundary between phases, rather there is a molecular gradient giving a change in the magnitudes of both density and orientation of interfacial molecules. We must remember that this layer is also very thin, usually between one and five monomolecular layers for liquid–vapour interfaces. The location of the so-called *mathematically dividing surface* (see Section 3.2.5) is only a theoretical concept to enable scientists to apply thermodynamics and statistical mechanics.

For solid surfaces, we should define the depth of the molecular layers, which are regarded as *surface*. We may say that only the top monolayer of surface atoms and molecules, which is the immediate interface with the other phases impinging on it, can be called the *surface*. However, in practice, it depends on the spectroscopic technique that we apply: some sophisticated spectroscopic instruments can determine the chemical structure of the top 2–10 atomic layers (0.5–3.0 nm), thus taking this depth as the solid surface layer. In many older books on surface chemistry or metallurgy, the surface is regarded as the top 100 nm or so of the solid, because some older technologies can only determine the chemical structure in the range of 10–100 nm and the *surface* was therefore assumed to be in this depth range. However, at present, this range may be called an *intermediate layer* between the bulk and surface structures, and only after 100 nm is it appropriate to regard such a layer in terms of its bulk solid-state properties. Thus, in the recent scientific literature, it has become necessary to report the solid surface layer with its thickness, such as the "top surface monolayer", the "surface film of 20 nm", the "first ten layers", or the "surface film less than 100 nm" etc.

1.2   Liquids and Liquid Surfaces

A liquid is defined as a medium which takes the shape of its container without necessarily filling it. Water is the most abundant liquid on earth and is essential for living systems, and non-aqueous liquids are widely used as solvents in synthetic, analytical, electrochemistry, polymer chemistry, spectroscopy, chromatography and crystallography. Molten metals are also considered to be liquid solvents. We may also consider another property that distinguishes liquids: the response of a liquid to an applied force is mainly inelastic; that is a liquid does not return to its original shape following the application, then removal, of a force.

The densities of liquids under normal pressures are not too dissimilar to those of their solids between the melting and boiling points. Generally a liquid is less dense than its solid at the melting point, but there are a few exceptions of which water is one; ice floats on water. These less dense solids have rather open crystal structures. Silicon, germanium and tin are other examples. Another similarity between normal liquids and solids is that they have a low compressibility due to there not being a great deal of space between the molecules in a liquid.

If we were able to observe the molecules in a liquid, we could see an extremely violent agitation in the liquid surface. The extent of this agitation may be calculated by considering the number of molecules that must evaporate each second from the surface, in order to maintain the measurable equilibrium vapor pressure of the liquid. When a liquid is in contact with its saturated vapor, the rate of evaporation of molecules is equal to the rate of condensation in the equilibrium conditions. For example, it is possible to calculate from kinetic theory that 0.25 g sec^{-1} water evaporates from each cm^2, which corresponds to 8.5×10^{21} molecules sec^{-1} at 20°C having a saturated water vapor pressure of 17.5 mmHg. However, the size of the water molecules permits only about $\approx 10^{15}$ molecules to be present, closely packed in each cm^2 of the surface layer at any particular moment, and we can conclude that the average lifetime of each water molecule is very short, $\approx 10^{-7}$ sec. This tremendous interchange between liquid and vapor is no doubt accompanied by a similar interchange between the interior of the liquid and its surface. However, although the thermal agitation of water molecules is so violent that they jump in and out of the surface very rapidly, the attractive forces between them are able to maintain the surface to within one to three molecules thickness. The evidence for this can be derived from the nature of light reflection from the liquid surfaces which results in completely plane polarized light from very clean liquid surfaces, indicating an abrupt transition between air and the liquid surface. On the other hand, the surface molecules of a liquid are mobile, that is they are also in constant motion parallel to the surface, diffusing long distances. Other properties of liquid surfaces will be described in Chapter 4.

Liquid surfaces tend to contract to the smallest possible surface area for a given volume resulting in a spherical drop in equilibrium with its vapor. It is clear that work must be done on the liquid drop to increase its surface area. This means that the surface molecules are in a state of higher free energy than those in the bulk liquid. This is due to the fact that a molecule in the bulk liquid is surrounded by others on all sides and is thus attracted in all directions and in a physical (vectorial) balance. However for a surface molecule there is no outward attraction to balance the inward pull because there are very few molecules on the vapor side. Consequently, every surface molecule is subject to a strong inward attraction perpendicular to the surface resulting in a physical force per length, which is known as *surface tension* (see Section 3.2.4). In addition, the tendency to minimize the surface area of a fluid surface can also be explained in terms of a thermodynamic free energy concept (see Chapter 3), since the surface molecules have fewer nearest neighbors and, as a consequence, fewer intermolecular interactions than the molecules in the bulk liquid. There is therefore a free energy change associated with the formation of a liquid surface reversibly and isothermally, and this is called the *surface free energy* or, more correctly, the *excess surface free energy*. The *surface free energy* term is preferred in the literature and it is equivalent to the work done to create a surface of unit area (Joules m^{-2} in the SI system). It must be remembered that this is not the *total free energy* of the surface molecules but the excess free energy that the molecules possess by virtue of their being located on the surface.

1.3 Surface Area to Volume Ratio

All liquid droplets spontaneously assume the form of a sphere in order to minimize their surface free energy by minimizing their surface area to volume ratio, as we will see in the thermodynamic treatment in Chapter 3. The surface area to volume ratio can easily be calculated for materials having exact geometric shapes by using well-known geometric formulas. For example, for a sphere and a cube this ratio is

$$\left(\frac{A}{V}\right)_{sph} = \frac{\left(4\pi r_{sph}^2\right)}{\left(4/3\,\pi r_{sph}^3\right)} = \frac{3}{r_{sph}}, \quad \left(\frac{A}{V}\right)_{cube} = \frac{6a_c^2}{a_c^3} = \frac{6}{a_c} \tag{1}$$

where r_{sph} is the radius of the sphere and a_c is the edge length of the cube. A comparison between the surface area to volume ratio of any geometrically shaped objects can be made by equating their volumes and then calculating their respective surface areas. For example, the edge length of a cube in terms of the radius of a sphere having the same volume is, $a_c = r_{sph}(3/4\pi)^{1/3}$ thus giving an area ratio of

$$\frac{A_{sph}}{A_{cube}} = \left(\frac{\pi}{6}\right)^{1/3} \cong 0.806 \tag{2}$$

The same approach may be applied to any geometric shape; the sphere has the minimum surface area for a given volume due to the fact that it has a completely curved profile without sharp edges (see Section 4.3 for the definition of *curvature* and its applications in surface science). For materials having irregular geometric shapes, well-known integration techniques from calculus are applied to calculate the surface area to volume ratio.

1.4   Solids and Solid Surface Roughness

An ideal solid surface is atomically flat and chemically homogenous. However, in reality there is no such ideal surface, all real solid surfaces have a *surface roughness* over varying length scales and they are also chemically heterogeneous to a degree due to the presence of impurities or polycrystallinity differences. Surface roughness is defined as the ratio of real surface area to the plan area and can be determined by Atomic Force Microscopy (AFM), Scanning Electron Microscopy (SEM) and other spectroscopic methods, and also by measuring the advancing and receding contact angles of liquid drops on substrates.

Solid surfaces are very complex: they have numerous small cracks in their surfaces and are not single crystals, but often aggregates of small crystals and broken pieces in all possible orientations, with some amorphous materials in the interstices. Even if the solid is wholly crystalline, and consists of a single crystal, there may be many different types of surface on it; the faces, edges and corners will all be different. A metallic, covalent and ionic crystal is regarded as a single giant molecule. The atoms of a solid crystal (or molecules in a molecular crystal) are practically fixed in position. They stay where they are placed when the surface is formed, and this may result in no two adjacent atoms (or molecules) having the same properties on a solid surface. Crystal defects affect the crystal's density and heat capacity slightly, but they profoundly alter the mechanical strength, electrical conductivity and catalytic activity. In a simple metal surface, there are many possible types of position and linkage of the surface atoms, which confer different chemical properties in different regions. For materials other than metals, the possibilities of variety of surface structure from atom to atom must be far greater through the formation of elevations and cavities. The presence of cracks in a surface decreases the strength of the crystals from that deduced from theoretical considerations. The loss of mechanical strength may be due to surface imperfections or internal cracks.

Wettability and *repellency* are important properties of solid surfaces from both fundamental and practical aspects. Wettability is the ability of a substrate to be covered with water or other liquids. Wettability can be seen when a liquid is brought into contact with a solid surface initially in contact with a gas or another liquid (see Chapters 9, 10). Many processes may happen: this liquid may spread on the substrate without any limit, displacing the entire original fluid surface from the entire solid surface area available, corresponding to a contact angle of 0° (see Section 9.1 on the definition of contact angles). Alternatively, the liquid may move out over the solid displacing the original fluid; however it halts when the contact angle between the liquid–fluid and solid–liquid interfaces reaches a certain value. This shows the magnitude of the wettability of the substrate with this liquid. Lastly, there is no wettability such that the substrate repels the liquid forcing it to form a spherical drop corresponding to a contact angle of 180°. Thus, wettability can be determined by measuring the contact angle of a water drop resting on a solid substrate. When the effects of surface stains or adsorption of other materials are ignored, the wettability of the solid surface is a characteristic material property and strongly depends on both the surface energy arising from the surface chemical structure and the surface roughness. Wenzel and later Cassie and Baxter provided different expressions showing the relationship between the contact angle and the roughness of a solid surface. The determination of wettability is important in adhesion, detergency, lubrication, friction, coating operations, flotation, catalysis and many other processes in chemical, mechanical, mineral, metallurgy, microelectronics, biomedical and biological industries.

1.5   Chemical Heterogeneity of Solid Surfaces

The chemical structure of the top surface layers of a solid determines its surface free energy. If these top surface layers consist of the same chemical groups, it is called *chemically homogeneous*, and if they consist of different chemical groups, it is called a *chemically heterogeneous* surface. The presence of two or more chemically different solid substances in a surface layer enormously multiplies the possibility of variety in the type of surface. Copolymer surfaces and catalysts having many different atoms at the surface are good examples for this. If desired, it is possible to impart chemical heterogeneity locally, for example concentric cylinders, stripes, patches etc. for various special applications. Measuring the contact angle hysteresis of liquid drops on a surface or measuring the heat of adsorption can check the presence of chemical heterogeneity (see also Chapter 9). The surface free energy is different for each region (or patch) for chemically heterogeneous surfaces and during an adsorption process, the higher free-energy spots (or patches) are filled first. Some dynamic experiments, for example adsorption chromatography, give information on surface heterogeneity.

Chemical heterogeneity of a surface is an important property affecting adhesion, adsorption, wettability, biocompatibility, printability and lubrication behavior of a surface. It seriously affects gas and liquid adsorption capacity of a substrate and also the extent of a catalysis reaction. As an example, the partial oxidation of carbon black surfaces has an important, influence on their adsorptive behavior. In a chemically heterogeneous catalyst, the composition and the chemical (valence) state of the surface atoms or molecules are very important, and such a catalyst may only have the power to catalyze a specific chemical reaction if the heterogeneity of its surface structure can be controlled and reproduced during the synthesis. Thus in many instances, it is necessary to determine the chemical composition, structural and electronic states and bonding properties of molecules at the solid surface. Such a complete surface investigation can be done by applying several advanced spectroscopic surface analysis techniques.

References

1. Adam, N.K. (1968). *The Physics and Chemistry of Surfaces.* Dover, New York.
2. Adamson, A.W. and Gast, A.P. (1997). *Physical Chemistry of Surfaces* (6th edn). Wiley, New York.
3. Lyklema, L. (1991). *Fundamentals of Interface and Colloid Science* (vol I). Academic Press, London.
4. Atkins, P.W. (1998). *Physical Chemistry* (6th edn). Oxford University Press, Oxford.
5. Murrell, J.N. and Jenkins, A.D. (1994). *Properties of Liquids and Solutions* (2nd edn). Wiley, Chichester.
6. Aveyard, R. and Haydon, D.A. (1973). *An Introduction to the Principles of Surface Chemistry.* Cambridge University Press, Cambridge.

Chapter 2
Molecular Interactions

Molecules in gas, liquid, solid and colloidal particles in a sol and biological macromolecules in living systems interact with each other. Knowledge of these interactions is mandatory since they determine all of the static and also the dynamic properties of the system. First of all, we should discriminate between the *chemical* and *physical* interactions. Chemical interatomic forces form chemical bonds within a molecule. However, the intermolecular forces between molecules are different from chemical interatomic forces because they are physical in nature. In the first part of this chapter (Section 2.1) we will examine the chemical bonding within a molecule and also the effects of geometry and dipole moments in molecules. We will consider the physical interactions between molecules in the rest of the chapter (Sections 2.2 to 2.9).

2.1 Intramolecular Forces: Formation of a Molecule by Chemical Bonding

2.1.1 Interatomic forces, bonds

The interatomic forces between atoms result in an atomic aggregate with sufficient stability to form chemical bonds within a molecule. According to the *valence bond* theory, a chemical bond is formed when an electron in one atomic orbital pairs its spin with that of an electron supplied by another atomic orbital, these electrons are then shared between two or more atoms so that the discrete nature of the atom is lost. Three main types of chemical bond are considered: *covalent, electrostatic (ionic)* and *metallic* bonds.

In the covalent bond type, two atoms share one or more pairs of electrons between them to attain a more stable electronic grouping. Sharing of pairs of electrons is a simple way of enabling two atoms to complete octets, which lack only a few electrons. Covalent bonding is the opposite extreme from pure ionic bonding because the bond is formed between like atoms and there is no excess net charge on one atom over that on the other. The electrons are shared by the two atoms and there is a physical accumulation of electrons in the space by the two nuclei. An electron outside two nuclei would tend to pull them apart, by pulling on the nearby nucleus more strongly than on the far one. On the other hand, an electron between two nuclei tends to draw them together. That shared electron acts almost like a glue in bonding atoms together. Examples of covalent bonds are provided by the diatomic molecules, H_2, N_2, O_2, Cl_2 etc. and the carbon compounds in the whole of organic chemistry. Covalent bonds can only be understood by wave mechanics. Wave and quantum mechanics explain that atoms are bonded covalently because an increase in electron probability in the internuclear region leads to lowering of the electrostatic potential energy (it becomes more negative) relative to the separated atoms.

Covalent bonds are very short-ranged, highly directional and strong. They have directionality so they are oriented at well-defined angles relative to each other. In crystalline solids, the covalent bonds determine the way they will coordinate themselves to form an ordered three-dimensional lattice. Covalent bonds operate over very short interatomic separations (0.1–0.2 nm). Typically covalent bond formation energies range from 80 to 400 kT per molecule, which corresponds to 200 to 1000 kJ mol^{-1} at room temperature (where $k = 1.381 \times 10^{-23}$ JK^{-1} is Boltzmann's constant) and they decrease in strength with increasing bond length. The covalent bond is so strong that extreme temperatures or strong electrostatic energy sources are required to disassociate the covalently bound molecules into their constituent atoms.

In an electrostatic (ionic) bond type, one atom gives one or more electrons to another atom or atoms; then ions are produced which are said to form an electrostatic bond between them to attain a more stable electronic grouping in a molecule. The most important electrostatic bond is the ionic bond resulting from the Coulomb attraction of opposite charges. The atoms of metallic elements such as sodium, potassium etc. lose their outer electrons easily, whereas those of nonmetallic elements tend to add additional electrons; in this way stable cations and anions may be formed which essentially retain their electronic structures as they approach one another to form a stable molecule or crystal. Ionic crystals have been found to be the most suited to simple theoretical treatment. The simplicity of the theory is due in part to the importance in the electrostatic interactions of the well-understood Coulomb terms (see Section 2.3) and in part to the spherical symmetry of the electron distributions of the ions with noble-gas configurations whose electronic groupings are the most stable, having complete octet electrons in their outer shell. The resultant ionic bonds are long-ranged, non-directional and not so strong. For example, the transfer of electrons from sodium metal atoms to chlorine gas atoms generates ionically bonded sodium chloride molecules for which the bond formation energy is 185 kT per molecule at room temperature.

In a metallic bond, the valence electrons of the atoms are common to the entire metal aggregate, so that a kind of gas of free electrons pervades it. The interaction between this electron gas and the positive metal ions leads to a strong interaction force. An ideal metal crystal consists of a regular array of *ion-cores* with valence electrons somewhat free to move throughout the whole mass resulting in exceptionally high electrical and thermal conductivity. Nevertheless, no electrons in any metal are able to move throughout its interior with total freedom. All of them are influenced to some extent by the other atomic particles present. Metallic bonds are sometimes as strong as ionic bonds.

Since the valence bond theory is insufficient to explain the structure and behavior of polyatomic molecules, the *molecular orbital* theory was developed. In this theory, it is accepted that electrons in a polyatomic molecule should not be regarded as belonging to particular bonds but should be treated as spreading throughout the entire molecule; every electron contributes to the strength of every bond. A molecular orbital is considered to be a linear combination of all the atomic orbitals of all the atoms in the molecule. Quantum mechanics has been successfully applied to explain the structures of simple molecules in this theory. For example, for N_2 and NaCl, we can solve the Schrödinger equation to obtain electron density maps, and we can calculate atomic dissociation energies, internuclear distances, bond lengths and angles of molecules from first principles. However the application of quantum mechanics to molecules containing many atoms was not so successful, and obtaining electron density maps and intramolecular energies and bond lengths of more complex polyatomic molecules requires a combination of theoretical and experimental information.

2.1.2   Molecular geometry

The bond length and bond strength are sufficient to define the molecular structure of a diatomic molecule. However, a triatomic molecule requires three parameters for similar definition: two bond lengths and a bond angle, θ. The presence of the bond angles introduces the molecular geometry concept. Triatomic molecules can be illustrated using plane geometry. We can still define molecules comprising more than three atoms, in terms of bond length and bond angles. However, in this case, three-dimensional geometry should be used. Although the bond angles and bond lengths remain constant, the variation of the twist between the inner atoms would allow the molecule to take on an infinite number of structures between these limits. We need to state a dihedral angle, ϕ, to define the structure, showing the angle of the twist in relation to a fixed position of the molecule. A molecule with five atoms in the chain will require two dihedral angles for its full description, as well as some agreement as to how one angle is measured with respect to the other. When we describe molecules with their dihedral angles, we reveal their conformations. Conformation is especially important for molecules containing chains of carbon atoms as found in organic and polymer chemistry.

2.1.3   Dipole moments

A molecule is built up from positively and negatively charged particles – atomic nuclei and electrons; and in a neutral molecule like H_2 the numbers of units of positive and negative charges are exactly equal. When the molecule is centro-symmetric, as it is for a diatomic molecule consisting of two atoms of the same kind, such as H_2, N_2, O_2 or Cl_2, the centers of gravity of the positive and negative charges will coincide, their electron distribution is symmetrical, and both atoms have the same electron density. However, when the atoms are different in a molecule, known as a *heteronuclear diatomic molecule* – such as hydrogen fluoride, HF, the centers of gravity of the positive and negative charges will not coincide. Such molecules carry no net charge but possess an *electric dipole* and are called *dipolar* or simply *polar* molecules. In the HF molecule, the fluorine atom tends to draw the hydrogen's electron towards itself so that it has greater *electronegativity* than the hydrogen atom. The resultant electron distribution in the covalent bond between these atoms is not symmetrical, and the electron pair is closer to the F-atom than to the H-atom. Thus, the F-end of the bond is slightly more negative, and the H-end is slightly more positive; this imbalance results in a *permanent dipole moment*, μ. The reason for the greater electron attraction of the F-atom is the larger positive charge in its nucleus and the effect of lone-pair electrons in its orbits. However, if we compare the amount of the residual negative charge at the F-atom and the residual positive charge at the H-atom, it is less than the unit electronic charge that is carried by each ion of Na^+Cl^- ionic bonds. Therefore, the HF molecule, as it exists in the gaseous state, is not an ionic compound as the formula H^+F^- would imply; rather it is partially covalent which can be shown as $H^{\delta+}F^{\delta-}$. Such a polar molecule with this property of charge imbalance will tend to orient itself in an electric field. The extent of polarity in a dipolar molecule is measured by the permanent dipole moment, μ, which is defined by the equation:

$$\mu = \delta l \tag{3}$$

where δ is the effective charge (positive δ^+ or negative δ^-) at each end of the molecule, and l is the distance between the centers of these respective charges. We shall represent dipole moments by an arrow $(- \rightarrow +)$ pointing from the negative charge to the positive charge. The unit of dipole moment is *Debye* where $1\,D = 3.336 \times 10^{-30}\,Cm$ (Coulomb meter). For comparison, if we consider two unit electric charges, $\delta = \pm e$ (where e is elementary charge $= 1.602 \times 10^{-19}$ Coulomb) separated by $l = 0.1\,nm$, the dipole moment can be calculated as $\mu = (1.602 \times 10^{-19})(10^{-10}) = 1.6 \times 10^{-29}\,Cm = 4.8\,D$. In practice, most of the molecules are not very polar, and the experimental dipole moment values for many organic molecules are 1–2 D.

The polarity of a molecule arises both from the difference in the electronegativities of the member atoms and also from the lack of molecular symmetry. In polyatomic molecules the polarity of the molecule is broadly determined by the combined effect of all of the bond dipoles present. Polyatomic molecules consisting of the same type of atom can be polar if they have low symmetry because the atoms may be in different environments and hence carry different partial charges. For example ozone, O_3, is polar because the electron density in the central oxygen atom is different from that of the outer two oxygen atoms. However, polyatomic molecules having different atoms within them may be non-polar if they have high symmetry because individual bond dipoles may then cancel. For example the linear triatomic carbon dioxide molecule, CO_2 $(O=C=O)$ is non-polar for this reason. To a first approximation, it is possible to resolve the dipole moment of a polyatomic molecule into contributions from various bond moments (μ_{bond}) in the molecule by applying a vectorial summation,

$$\mu_{molecule} = 2\mu_{bond} \cos\left(\frac{\theta_b}{2}\right) \tag{4}$$

where θ_b is the bond angle. Permanent dipole moments of some molecules, bond moments and group moments are given in Table 2.1. The bond and group moments in this table are approximate only, but are useful for estimating the dipole moments of new molecules. For example, if we want to calculate the dipole moment of water vapor, we must consider the bond angle between hydrogen and oxygen, which is 104.5°, and by using Equation (4) we obtain $\mu_{water} = 2(1.51)(\cos 52.25°) = 1.85\,D$ which fits the experimental result well.

The dipole moment of a molecular can be measured experimentally in an applied electrical field, preferably in gaseous, but also in the liquid, state. In these fluids the molecules are randomly arranged and in constant movement. If the molecules are polar but there is

Table 2.1 Dipole moments of molecules, bonds and molecular groups in Debye units. (Values compiled from standard references and books especially from David R. Lide (ed.) (2003) *CRC Handbook of Chemistry and Physics* 83rd edn. CRC Press, Boca Raton; Smyth, C. P. (1955) *Dielectric Behavior and Structure*. McGraw-Hill, New York; Israelachvili, J. (1991) *Intermolecular & Surface Forces* 2nd edn. Academic Press, London.)

Dipole moments of molecules

Molecule name	D	Molecule name	D	Molecule name	D
C_6H_6 (benzene)	0	C_6H_5OH	1.50	H_2O	1.85
C_nH_{2n+2} (alkanes)	0	$C_6H_5NH_2$	1.50	CH_3Cl	1.87
CCl_4	0	SO_2	1.62	C_2H_4O (ethylene oxide)	1.90
CO_2	0	CH_3OH	1.70	CH_3COCH_3 (acetone)	2.90
CO	0.11	C_2H_5OH	1.70	$HCONH_2$ (formamide)	3.70
$CHCl_3$ (chloroform)	1.06	$C_6H_{11}OH$ (cyclohexanol)	1.70	$C_6H_5NO_2$	4.20
HCl	1.08	CH_3COOH	1.70	NaCl	8.50
NH_3	1.47	C_6H_5Cl	1.80	CsCl	10.4

Dipole moments of bonds

Bond	D	Bond	D	Bond	D
C—C	0	C—H$^+$	0.40	C$^+$—Cl	1.5–1.7
C=C	0	C$^+$—O	0.74	F—H$^+$	1.94
C$^+$—N	0.22	N—H$^+$	1.31	N$^+$=O	2.00
N$^+$—O	0.30	O—H$^+$	1.51	C$^+$=O	2.3–2.7

Dipole moments of molecular groups

Group	D	Group	D	Group	D
C—$^+$CH$_3$	0.40	C—$^+$NH$_2$	1.2–1.5	C—$^+$COOH	1.70
C—$^+$OCH$_3$	1.30	C—$^+$OH	1.65	C$^+$—NO$_2$	3.1–3.8

no applied electric field, the dipoles are randomly oriented and the *molecular polarizability*, α, which is the average dipole per unit volume, is zero. However if an electric field is applied between parallel metal plates or by introducing an ion in a polar solvent, the random orientations of the dipoles disappear as the dipolar molecules orient along the direction of the electric field (for example oriented in a parallel position between two metal plates where electricity is applied, or oriented around a single ion) resulting in an *orientation polarization*. Molecular polarizability is also defined as the response of the electron cloud to an external electric field or the ability of electrons to move in the presence of an electric field. This external electric field should be weak compared to the internal electric fields between the nucleus and electron cloud. Molecular polarizability increases with increasing number of electrons in a molecule and also as the electrons become less tightly held by the atomic nuclei.

The application of an electric field always causes some physical changes in the medium: even if the liquid molecules are non-polar, the electrons in the molecule will be affected by the electric field. The movement of electrons within the molecule results in an *induced dipole, μ^i*, and the alignment of the induced dipoles with the electric field gives *induced polarization*. For example, if a positive charge is placed above the plane of a neutral benzene molecule, the average positions of the electrons will shift upward, giving the benzene molecule a dipole moment whose direction is perpendicular to the molecular plane. In summary, when a non-polar molecule is subjected to an electric field, the electrons in the molecule are displaced from their ordinary positions so that the electron clouds and nuclei are attracted in opposite directions and a dipole is induced; thus the molecule temporarily has an induced dipole moment, μ^i.

As we know from general physics, to avoid the notion of action at a distance, the concept of an electric field was introduced. An *electric field* is present around each charge. An electric charge, q_1 is said to produce an electric field in the space around itself, and this field exerts a force on any charge, q_2, that is present in the space around q_1. The *electric field strength*, E, whose unit is N C^{-1} (Newton Coulomb^{-1}) at a point P in space is defined as the electrical force per unit charge experienced by a test charge at rest at point P and is given as,

$$E = \frac{F}{\delta} \tag{5}$$

where F is the force exerted to a point charge and δ is the charge. The force on the point charge has a direction as well as a magnitude and thus electric field is a vector. The direction of E is the same as the direction of F on the line from the first charge to the second charge at point P. For a positive second charge, the vector E points outwards if the first charge is positive, inwards if it is negative.

Another important aspect of an electric field is described by the *electric potential, ϕ*. This quantity represents the potential energy of a unit positive charge in an electric field. The *electrical potential difference*, $(\phi_2 - \phi_1)$ between points 2 and 1 in an electric field, E, is defined as the work per unit charge to move a test charge reversibly from 1 to 2; $(\phi_2 - \phi_1) \equiv (dW_{1 \to 2}/dq)$. By assigning a value to the electric potential ϕ_1 at point 1, we have then defined the electric potential ϕ_2 at any point 2. The usual convention is to choose a point charge, 1 at $r = \infty$ (infinity) and to define $\phi_1 = 0$. If a material has a larger charge than the unit charge, and if we do reversible work to bring this material from infinity to point 1, we change its total electric potential energy so that $V_E = \phi q$. Thus, the potential energy of a point charge increases as it is brought closer to the positive charge that generates the electric field so that

$$d\phi = -Edr \tag{6}$$

where dr is the change in center-to-center distance. The SI unit of electric potential is the volt, $1\,V = 1\,J\,C^{-1} = 1\,kg\,m^2\,s^{-2}\,C^{-1}$ and thus the unit of E is (V m^{-1} = N C^{-1}). When a non-polar molecule is subjected to an electric field, the electron cloud and nuclei are attracted in opposite directions to form an induced dipole having a moment of μ^i, which is proportional to the electric field E experienced by the molecule

$$\mu^i = \alpha E \tag{7}$$

where α is the molecular polarizability as defined above. The unit of polarizability, α, in the SI system is $C^2\,m\,N^{-1}$ or $C\,m^2\,V^{-1}$ from Equation (7). Formerly, α was expressed as cm^3 in cgs units, in the range of $10^{-24}\,cm^3 = 1$ $(A°)^3$. The conversion from cgs units to SI units is $\alpha(C\,m^2\,V^{-1}) = 4\pi\varepsilon_o \times 10^{-6}\,\alpha(cm^3) = 1.11265 \times 10^{-16}\alpha(cm^3)$.

On the other hand, for polar molecules having permanent dipole moments, when subjected to an electric field, these molecules may have their existing permanent dipole moments modified temporarily by the applied field, and the measured dipole moment is the total dipole moment. For this reason, it is important to discriminate between *induced dipole moment*, μ^i, and *permanent dipole moment*, μ, for such polar molecules.

Dipole moments of molecules are calculated from the *relative permittivity* (previously known as the *dielectric constants*) measurements (see also Section 2.3, Equation (22) and Section 2.5), which are carried out in parallel metal plate capacitors (or condensers). When a capacitor is connected to a battery, there will be a potential drop of V (volts) across the capacitor. According to Equation (6) the electric field strength between the plates is $E = \phi/d$ where d is the distance (m) between the plates. The *capacitance, C*, of a capacitor is defined as the ratio of the total charge to the applied voltage across the capacitor, $C = \delta/\phi$ in a vacuum. However, if there is a medium other than a vacuum present between the parallel plates, the relationships are different: the electric field strength of a parallel metal plate capacitor depends on the nature of the medium between the plates, since the applied electric field causes an orientation of molecular charges present in the medium, and permits a greater accumulation of charge for a given voltage than is observed when the capacitor is evacuated. Thus, the ratio of the electric field strength in a vacuum (or approximately in air), E_o, to the electric field strength with a medium, E, is called the *relative permittivity*, ε_r, or *dielectric constant*, of the medium,

$$\varepsilon_r = \frac{E_o}{E} = \frac{\phi_o}{\phi} \tag{8}$$

and is calculated by measuring the voltage differences between the plates in a vacuum and in the test medium. The parameter, ε_r, increase, as the molecular polarizability, α, and permanent dipole moment of the molecules, μ, increase, and ε_r values range from 2 (for alkanes) to over 100 (as shown in Table 2.2).

The relation between relative permittivity and molecular properties is an important subject and for non-polar molecules the problem is easier to solve. The relation which was first derived by Clausius–Mossotti dates back to the nineteenth century,

$$\left(\frac{\varepsilon_r - 1}{\varepsilon_r + 2}\right)V_m = \frac{N_A \alpha}{3\varepsilon_o} = P_m \tag{9}$$

where N_A is the Avogadro's number, ε_o is the *vacuum permittivity*, a fundamental constant with the value $\varepsilon_o = 8.854 \times 10^{-12}\,C^2\,J^{-1}\,m^{-1}$ and V_m is the molar volume $(V_m = M/\rho)$, M is the molar mass, ρ is the density and P_m is the *molar polarization*; P_m varies with electrical frequency applied (see also Section 2.5). This equation may also be used for polar molecules for the cases where the frequency of the applied electric field is very high (larger than $\approx 10^{12}\,Hz$) so that the polar molecules cannot orient themselves quickly enough to follow the change in direction of the field and thus behave like non-polar molecules under this applied very high frequency.

Table 2.2 Dielectric constants, ε_r, of some gas, liquid and solid molecules (Values compiled from standard references and books especially from David R. Lide (ed.) (2003) *CRC Handbook of Chemistry and Physics* 83rd edn. CRC Press, Boca Raton; Israelachvili, J. (1991) *Intermolecular & Surface Forces* 2nd edn. Academic Press, London.)

Static dielectric constants of some gases at 25°C

Molecule	ε_r	Molecule	ε_r	Molecule	ε_r
Air (dry)	1.00054	Hydrogen	1.00025	Argon	1.00052
Nitrogen	1.00055	Methane	1.00081	Propane	1.00200
CF_4	1.00121	Dimethyl ether	1.00620	Carbon dioxide	1.00092
Chloromethane	1.01080	Water vapor (0°C)	1.00144	Water vapor (100°C)	1.00587

Static dielectric constants of some liquids at 25°C

n-C_6H_{14} (hexane)	1.9	Silicone oil	2.8	C_2H_5OH	24.3
n-C_8H_{18} (octane)	2.0	$CHCl_3$ (chloroform)	4.8	CH_3OH	32.6
C_6H_{12} (cyclohexane)	2.0	CH_3COOH	6.2	$C_6H_5NO_2$	34.8
$C_{12}H_{26}$ (dodecane)	2.0	NH_3 (ammonia)	16.9	HCOOH (16°C)	58.5
CCl_4	2.2	n-C_4H_9OH (butanol)	17.8	H_2O	78.5
Paraffin	2.2	n-C_3H_7OH (propanol)	20.2	HCN	115.0
C_6H_6 (benzene)	2.3	$(CH_3)_2CO$ (acetone)	20.7	$HCONH_2$ (formamide)	109.5

Static dielectric constants of some solids at 25°C

Paraffin wax	2.2	Nylon	≈4.0	NaCl	6.0
PTFE	2.0	Quartz glass	3.8	Soda glass	7.0
Polystyrene	2.4	SiO_2	4.5	Al_2O_3 (alumina)	8.5
Polycarbonate	3.0	Borosilicate glass	4.5	Water (ice, at 0°C)	92–106

However when a normal or low-frequency electric field is applied, for polar molecules, the ε_r of the medium is due to both *permanent* and *induced* dipoles in the molecule, and in order to measure the permanent dipole moment, the effect of the induced dipole moment must be evaluated. One molecular property helps us to solve this difficult problem: the induced dipole moment, μ^i is independent of the temperature since if the position of the molecule is disturbed by thermal collisions, the dipole is immediately induced again in the field direction. However, the contribution of permanent dipoles, μ, is temperature-dependent and decreases with increasing temperature because the random thermal collisions of the permanent dipole molecules oppose the tendency of their dipoles to line up in the electric field. In order to discriminate between μ^i and μ, it is necessary to calculate the average component of a permanent dipole in the field direction as a function of temperature. Debye modified the molar polarization for polar molecules by using Langevin function and Boltzmann distribution law ($e^{-E/kT}$) to give the relation between the relative permittivity and the electrical properties of a pure polar gas

$$P_m = \frac{N_A}{3\varepsilon_o}\left(\alpha + \frac{\mu^2}{3kT}\right) \tag{10}$$

where μ is the permanent dipole moment. By combining Equations (9) and (10), the well-known Debye–Langevin equation is obtained for polar molecules (see also Section 2.5),

$$\left(\frac{\varepsilon_r - 1}{\varepsilon_r + 2}\right)\frac{M}{\rho} = \frac{N_A}{3\varepsilon_o}\left(\alpha + \frac{\mu^2}{3kT}\right) = P_m \tag{11}$$

(in the absence of molecules having permanent dipole moments, the above equation reduces to the Clausius–Mossotti equation when $\mu = 0$ is applied). In practice, P_M is calculated from experimental relative permittivity values using the left side of Equation (11) and then it is possible to evaluate both α and μ from the intercept and slope of P_M versus $1/T$ plots using the middle part of Equation (11). The Debye–Langevin equation holds very well for gases, although ε_r needs to be measured very accurately. Unfortunately, when this equation is used for pure polar liquids the results are very unsatisfactory due to the presence of short-range interactions. However, Equation (11) can be used for dilute liquid solutions of polar molecules in non-polar solvents such as nitrobenzene dissolved in hexane etc. because of the removal of close packing conditions in dilute solutions. In order to determine permanent dipole moments of polar molecules, the refractive index data are also used. Maxwell found that the relative permittivity should be $\varepsilon_r = n^2$ for a transparent medium from electromagnetic theory, where n is the refractive index (n = speed of light in a vacuum/speed of light in the medium). Lorenz and Lorentz defined the *molar refraction, R_m*

$$\left(\frac{n^2 - 1}{n^2 + 2}\right)V_m = R_m \tag{12}$$

As can be seen from Equations (9), (11) and (12), both P_m and R_m are similar quantities, but there are differences too: only induced dipoles contribute to R_m in refractive index measurements which is independent of temperature and varies also with frequency. For non-polar molecules R_m and P_m are approximately equal over the whole frequency range provided that the wavelength at which n is determined is not close to an absorption band. For polar molecules, it is only at high frequencies (above $\approx 10^{12}$ Hz) $R_m = P_m$ that the Maxwell relation holds. At the normal frequency range, the use of refractive index data (Equation (12)) gives the electronic (basic) polarizabilities of the polar molecules; and the use of the Debye–Langevin equation (Equation (11)) gives the total polarizabilities of the polar molecules. Consequently, the measurement of a difference between R_m and P_m at a low frequency range indicates the presence of a polar molecule having a permanent dipole moment.

More recently, dipole moments have also been calculated from the effect of electric fields on molecular spectra (Stark effect) and from the electric resonance methods applied to molecular beams; these are beyond the scope of this book.

2.2   Intermolecular Forces and Potential Energies

After examining the formation of a molecule by chemical bonds between atoms, we may then ask the next question: can we calculate the interactions between un-bonded discrete molecules and predict their behavior in advance? It is a very significant question because these interactions are responsible for all the most important processes in nature and also in the chemical industry. It was found that molecules with complete valence shells are still able to interact with one another: they can exercise attractive forces over a range of several atomic diameters, and they can repel one another when pressed together. All the properties of a bulk material can be determined by the number and types of molecules it contains and their arrangement in space with respect to each other. They account, for instance, for the condensation of gases and the structures of molecular solids. However, these interactions arise from *physical forces* and they are different from *chemical forces* which give rise to chemical (or covalent) interatomic bonds to form a molecule. Physical binding between molecules is not a real *bond* because during chemical covalent bonding the electron charge distributions of the uniting atoms change completely and merge, whereas during physical binding they are only perturbed. Nevertheless, some authors call physical intermolecular interactions *non-covalent bonds* or *secondary bonds* in order to distinguish them from covalent bonds. Physical interactions lack the directionality and stoichiometry of chemical covalent bonds. In general, physical interactions can be as strong as covalent bonds to hold the molecules together in solids and liquids at room temperature, as well as in biological and colloidal systems. As a result of this physical binding the molecules in a liquid can rotate and move about, still remaining *bonded* to each other. These forces are also responsible for the structural organization of biological macromolecules since they twist the long polypeptide chains of proteins into characteristic shapes and then pin them together in the arrangement essential to their function. In summary, they are the regulating forces in all natural and technical phenomena that do not involve chemical reactions and thus are very important in all branches of science and industry.

In order to estimate the magnitude of physical intermolecular interactions, we can extend the molecular orbital approach and solve the Schrödinger equation to describe electron density between any two molecules and determine their intermolecular cohesion and adhesion potentials. However, this is a very difficult task analytically due to the fact that the electron charge distributions of the interacting molecules do not change completely and are only perturbed (and also it is a very difficult task numerically, in spite of the advance of computers, due to the lack of sufficient parameters); and in practice, it is more convenient and conceptually simpler to represent the total intermolecular potential energy as the sum of parts of several different interaction energies such as Coulomb, polar, induced-polarized, van der Waals, repulsive, hydrogen-bonding, hydrophobic and hydrophilic interactions, which will be explained in detail in this chapter. In this approach, investigation of the vapor phase interactions of the molecules is especially important because analyzing these relatively simple unfettered interactions helps to establish fundamental concepts that prevail with the more complex situations encountered in liquid and solid phases.

First of all, we must be aware of the fact that all physical forces acting between molecules are essentially electrostatic (Coulomb forces between charges and dipoles; induced dipole forces) or quantum mechanical (dispersion, and repulsive forces) in origin apart from a minor contribution due to mass attraction *gravitational forces*. Gravitational forces account for tidal motion and also determine the height of a liquid in a capillary when acting together with intermolecular forces. Two other terms are also used to indicate the physical forces between molecules, *cohesion* and *adhesion*: the term *cohesion* describes the physical interaction forces between the same type of molecules and the term *adhesion* between different types of molecules. All organic liquids and solids, ranging from small molecules like benzene to essentially infinitely large molecules like cellulose and synthetic polymers, are bound together by *cohesion* forces.

Second, we should consider the effect of the separation distance between molecules on the interaction forces. It was found that these fall off exponentially as the distance between atoms increases. The explanation for this behavior can be found in classical Newton mechanics: according to Newton's laws, a *force*, F, is a push or pull exerted on a body; it is a vector quantity with magnitude (*newton*, N = $kg\,m\,s^{-2}$, in the SI system) and direction. The *work*, W, done by the force acting on a body is given as

$$dW = F\,dl \tag{13}$$

where l is the displacement (m) of the body. Work is a scalar quantity and fully characterized by its magnitude (*joule* = newton-m = $kg\,m^2\,s^{-2}$) and can be calculated by integrating Equation (13). If the force is not constant, then it should be expressed in terms of displacement, $f(l)$, in order to carry out this integration, as shown below:

$$W = \int_{l_1}^{l_2} F\,dl = \int_{l_1}^{l_2} f(l)\,dl \tag{14}$$

Energy is given to an object when a force does work on it. The amount of energy given to an object equals the work done. Energy is also a scalar quantity and both energy and work have the same unit (joule). *Potential energy*, V, is defined as the energy possessed by an object because of its position in a gravitational and/or electrical field. It is the capacity to do work. Newton applied gravitational interaction energy potential, $V(r_g)$ to spherical bodies so that

$$V(r_g) = -G\frac{m_1 m_2}{r_g} \tag{15}$$

where m_1 and m_2 are the masses (kg) of the interacting bodies, r_g is the center to center separation distance between these bodies (m), $V(r_g)$ is also called the *pair potential* and G is the gravitation constant (G = 6.67×10^{-11} $N\,m^2\,kg^{-2}$). All interactions may be attractive or repulsive and the minus sign in Equation (15) is due to the international convention for attractive interactions because of the opposite vectors of displacement and attraction between the m_1 and m_2 bodies. In other words, we must specify a reference state in describing interactions between two bodies. Values of potential energies and free energies always refer to the difference between the actual state and a reference state. The reference state is taken to be at separation distance $r = \infty$. In Equation (15) this corresponds to $V(r_g) = 0$, and so a negative value of $V(r_g)$ corresponds to *attraction* and a positive value of $V(r_g)$ corresponds to *repulsion* in both the force and potential energy equations. (See also Figure 2.1 *a* and *b*.) Then Equation (13) can be rewritten for any interaction potential energy between two spherical bodies as

$$dV(r) = -F(r)\,dr \tag{16}$$

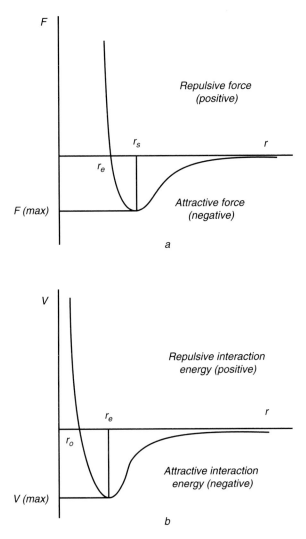

Figure 2.1 **a.** The change of physical interaction force versus separation distance between two spherical molecules across a vacuum. **b.** The change of interaction potential energy versus separation distance between two spherical molecules across a vacuum.

By combining Equation, (15) and (16) and differentiating, we can find for Newton's gravitational force between two spherical bodies

$$F(r_g) = -\frac{dV(r)}{dr_g} = -G\frac{m_1 m_2}{r_g^2} \tag{17}$$

However it should be kept in mind that Equation (17) is a special form of a general interaction potential equation. For interaction potential energies other than gravitational, a general form of Equation (15) must be used so that

$$V(r) = -C\frac{x_1 x_2}{r^n} \tag{18}$$

where C is an interaction constant, x is the amount of the molecular property involved in the bimolecular interaction, r is the center to center separation distance between these bodies and n is an integer believed to be greater than 3 for molecules (possibly 4 or 5) if intermolecular forces are not to extend over large distances. (In contrast, $n = 1$ for Newton's gravitational law where very big planets, suns etc. interact over very large distances.) It is instructive to show how we can conclude that $n > 3$ for molecules interacting over short distances only: for gas molecules, we may consider a region of space where the number density of these molecules is ρ (particle m^{-3}), their diameter is d (m) and the size of the system is L (i.e. one side length of a cube box containing the gas molecules, m). Let us add all the interaction energies of one particular gas molecule together with all the other gas molecules in the system. The number of gas molecules in a region of space between r and $(r + dr)$ away will be $\rho 4\pi r^2\,dr$ (since $4\pi r^2$ is the surface area of spherical shell and $4\pi r^2\,dr$ is the volume of a spherical shell of radius r and thickness dr). The total interaction energy of one molecule with all the other molecules in the system can therefore be given by summing (integrating between d and L) all the pair potentials as given in classical physics,

$$V_T = \int_d^L V(r)\rho 4\pi r^2 dr = -4\pi Cx_1x_2\rho \int_d^L r^{2-n} dr = \frac{-4\pi Cx_1x_2\rho}{(n-3)d^{n-3}}\left[1 - \left(\frac{d}{L}\right)^{n-3}\right] \qquad (19)$$

When $L \gg d$ and for $n > 3$, the term in the bracket can be taken as approximately equal to 1 and then Equation (19) becomes

$$V_T = -\frac{4\pi Cx_1x_2\rho}{(n-3)d^{n-3}} \qquad (20)$$

so that the size of the system (box), L, is not taken into account. Since d must be much smaller than L for gas molecules in a box, $1 > (d/L)$, large distance contributions to the interaction will disappear only for values of n greater than 3, and this fits the initial requirement so that the intermolecular forces are not to extend over large distances. Thus, the bulk properties of gases, liquids and solids do not depend on their volumes but depend only on the forces between molecules in close proximity to each other (*short-range forces*). Short-range forces operate over very short distances of the order of interatomic separations (0.1–1.0 nm) usually corresponding to very close to molecular contact. However, important long-range intermolecular forces also exist, especially between macroscopic particles and surfaces (see Chapter 7), but their effective range of action rarely exceeds 100 nm. If n is smaller than 3, the second term in Equation (19) will be large and the contribution from more distant molecules will dominate over that of nearby molecules. In such cases the size of the system must be taken into account. This condition is suitable for Newton's gravitational force where $n = 1$ and the distant objects (stars, planets etc) are still strongly interacting with each other.

The intermolecular potential energy is defined as the difference between the total energy of interacting molecules and the sum of their separate molecular energies. If we consider two spherical molecules interacting across a vacuum according to simple Newton mechanics, Figure 2.1 *a* shows typical force, F, and Figure 2.1 *b* shows potential energy, V curves. The negative parts at larger distances are caused by the attraction forces (mostly *van der Waals forces*; see Section 2.6) and the positive repulsive part at shorter separation is due to the outer electron shell overlap, called *Born repulsion forces* (see Section 2.7). As can be seen in Figure 2.1 the minimum and zero positions in the force curve do not coincide with those of the potential energy curve but these points can be interrelated. The force, $F = 0$ at $r = r_e$, and the interaction potential energy $V = 0$ at two points, one at $r = r_o$ and the other at $r = \infty$. At $r = r_s$ the force curve shows a minimum (a maximum for attractive force magnitude, F_{max}) which can be calculated mathematically from $dF/dr = 0$. At $r = r_e$ the potential energy curve shows a minimum (a maximum for attractive potential energy magnitude, V_{max}) which can be calculated from $dV/dr = 0$. We can calculate the force, $F(r)$ as the negative of the slope of the $V(r) - r$ curve as given in Equation (17) and, the minimum of $F(r)$, (F_{max}) at $r = r_s$ where the $V(r) - r$ curve has an inflection point, which can be calculated by $d^2V/dr^2 = 0$ at this point. If we examine the potential energy $V(r) - r$ curve, we see that the potential energy decreases (attraction increases) with the decrease in their separation distance due to the fact that work must be performed when they are drawn apart. If the attractive interactions are unopposed, V diminishes as the molecules approach one another and finally the two molecules would tend towards coalescence, the potential energy falling to $-\infty$, as r becomes 0. However, on the contrary, strong repulsive forces come into play when two molecules come close together as a result of the negative charges carried by the electrons belonging to the two molecules. The repulsion, rising dramatically as r decreases, causes the curve to rise very steeply on the left-hand side. The curve combines both attractive and repulsive interactions. At a certain distance, r_e, the forces are in equilibrium and this distance corresponds to the minimum potential energy (V_{max}), i.e. the stable state. This distance, r_e, represents the normal intermolecular distance, and the height, V_{max}, represents the energy that must be supplied to cause two molecules to separate to infinite distance.

In the early twentieth century, the purely Newtonian mechanistic view of intermolecular forces was abandoned and statistical thermodynamic concepts such as free energy and entropy were adopted by van der Waals, Boltzmann, Maxwell and Gibbs (see Chapter 3). Mie and later Lennard-Jones proposed semi-empirical *pair potential* equations which fit the $V(r) - r$ curve given in Figure 2.1 b (see Sections 2.6 and 2.7). On the other hand, we should be aware of the fact that the above description assumes all molecules are spherical and also rigid structures. However in reality, they are not; and also the atoms of a molecule are vibrating even in their lowest energy state (zero-point vibration state) so that the potential will change as atoms move. Nevertheless, intermolecular potentials can be considered to be averages over the vibration motions of the two molecules and can be used successfully in many applications.

Now we may ask, how can we relate the *pair potential energies* to the behavior of real gases, liquids and solids? First of all, historically, pair-wise interactions between molecules derived from experiments on dilute gases, where mainly pair-wise interactions take place since collisions in a dilute gas are sufficiently infrequent that they mostly occur between pairs of molecules, with much fewer involving three or more molecules colliding simultaneously. Under these conditions we can add up all the individual contributions to obtain the total interaction, following the principle of linear superposition of potentials; in other words, we can sum all the pair potentials. From the experimental data on the pressure–volume–temperature behavior of gases, it was soon found that many different semi-empirical potential energy equations with a wide range of adjustable parameters could satisfactorily be applied. Then analyzing these relatively simple, unfettered gas interactions, the fundamental concepts were established that also prevail with the more complex, condensed liquid and solid phase situations. However, in liquids and solids, unlike gases, the molecules have simultaneous contacts with several neighbors so that there is no fundamental reason why the bulk properties must depend only on the pair potentials. Thus, the intermolecular interaction theories applied especially to liquids are not as successful as for gases because it is apparent that there is a large difference between knowing the force or pair potential energy between two isolated molecules and understanding how an ensemble of such molecules will behave. Even today there is no ready recipe for deriving the properties of condensed phases from the intermolecular pair potentials, and vice versa. But there are some exceptions: for non-polar liquids, contrary to expectations, the semi-empirical pair potentials deduced from liquid state properties are very similar to their true pair potentials, and many-body potentials can be neglected. Unfortunately, when polar liquids are considered the many-body terms are probably not negligible, although every system of interest may be examined independently. *Effective pair potential* is a term used for pair potentials deduced semi-empirically from the properties of any liquid for which the many-body energies are not negligible. For metallic liquids many-body terms cannot be neglected because in the free-electron model, the electrons move in a potential that is constant over the whole metal phase and the total energy depends on the size and shape of the phase.

For fifty years, it was thought that the application of an *ab initio* method (from first principles) by solving the Schrödinger equation in wave mechanics could explain the behavior and interactions of a molecule; unfortunately, the form and exact solutions of the Schrödinger equation even for many simple molecules are not easy obtain. The Hamiltonian operator for any molecule must contain kinetic-energy operators for nuclei, kinetic-energy operators for electrons, potential energy of repulsion between the nuclei, potential energy of attraction between the electrons and potential energy of attraction between the electrons and the nuclei. Thus the molecular Schrödinger equation is extremely complex and it was almost hopeless to attempt an exact solution. Born and Oppenheimer showed that it is a good approximation to treat the electronic and nuclear motions separately. The nuclei move far more slowly than the electrons because of their much greater mass and the electrons "see" the slow-moving electrons as almost stationary point charges, whereas the nuclei "see" the fast-moving electrons as essentially a three-dimensional distribution of charge. For small molecules made up of light atoms such as H_2, He etc. the *Born–Oppenheimer Approximation* gives quite accurate descriptions of pair potentials by this means, however for more complex molecules we can only determine the general functional behavior of pair potentials such as how they depend on intermolecular distance and angles of orientation. Instead of an *ab initio* method we may apply semi-empirical methods by assuming a form of potential initially and testing its fit with experimental results. The difference in the fit is used to adjust the form of pair potential; such semi-empirical methods sometimes give satisfactory results (see Chapter 4 for liquids).

In summary, our understanding of intermolecular forces is far from complete and qualitative results have been obtained only for simple and idealized models of real matter. For this reason, it has been found useful to classify intermolecular interactions into a number of seemingly different categories even though they all have the same fundamental origin. Thus, such commonly encountered terms as Coulomb, polar, induced-polarized, van der Waals, repulsive, hydrogen-bonding, hydrophobic and hydrophilic interactions and solvation forces are a result of this classification, often accompanied by further divisions into short-range and long-range forces. Furthermore, the quantitative relations which link intermolecular forces to macroscopic thermodynamic properties by using statistical mechanics are also at present limited to simple and idealized cases. Therefore, the theory of intermolecular forces gives us no more than a semi-quantitative basis to interpret and generalize the existing experimental data.

More recently, the scope of intermolecular forces has broadened to include surface science, such as thin-film phenomena, biological macromolecules, self-assembling polymers, nanomaterial interactions etc. In addition to equilibrium (static) interactions, dynamic and time-dependent interactions were also included, together with the application of computer simulation techniques. We should keep in mind that in chemistry and biology the forces are generally short range and rarely extend over one or two atomic distances, whereas in colloid science the forces such as electrical double-layer and steric polymer interaction forces are generally long range.

2.3   Coulomb Interactions

Coulomb interactions between ions are responsible for the cohesion within some condensed phases such as ionic solids. These interactions are also operative in liquid solutions. Although such interactions are usually regarded as versions of chemical valence forces between atoms in the molecules, they also act as a physical force between molecules. The physical Coulomb force between two ionic molecules is by far the strongest force we see, stronger even than most chemical bonds. However, physical Coulomb force is long range up to 70 nm distance compared with the extremely short range chemical covalent bonds (0.1–0.2 nm)

In classical physics, *electrostatics* describes the interaction of stationary electric charges, and the Coulomb potential energy equation is the fundamental expression in electrostatics: if an ion (a point-like body) of charge q_1 (note that q is the ionic charge and is different from the residual charge, δ, which is present in the polar molecules) is at a center-to-center distance, r, away from another ion of charge, q_2, in a vacuum, then the Coulomb interaction potential energy, V, is given by

$$V(r) = \frac{q_1 q_2}{4\pi\varepsilon_o r} \tag{21}$$

The charge unit is the coulomb, C, and the constant, ε_o is the *vacuum permittivity*, a fundamental constant with the value $\varepsilon_o = 8.854 \times 10^{-12} \, C^2 J^{-1} m^{-1}$. The usual convention is that $V(r) = 0$ is assumed to be the reference state for infinite separations ($r = \infty$). The vacuum permittivity parameter, ε_o was derived from electromagnetic theory in classical physics. As we know from Maxwell's electromagnetic theory, light consists of electromagnetic waves, so that $c = (4\pi\varepsilon_o \times 10^{-7} \, Ns/C)^{-1/2}$, where c is the speed of light in a vacuum ($c \cong 3 \times 10^8 \, m/s$).

When the Coulomb interaction takes place in a medium other than a vacuum, the molecules of the medium will be oriented and polarized in the electric field created by these two charges, and the magnitude of $V(r)$ is reduced so that Equation (21) becomes,

$$V(r) = \frac{q_1 q_2}{4\pi\varepsilon_o\varepsilon_r r} \tag{22}$$

where ε_r is the *relative permittivity*, or *dielectric constant* of the medium as defined initially by Equation (8) in Section 2.1.3. Water has one of the highest ε_r values of all liquids ($\varepsilon_r = 78.4$ at

25°C) and this is the main reason for its ability to solubilize ionic solids, because the electrostatic attraction between anions and cations decreases 78.4 times in water. Polar liquids have much higher ε_r values than non-polar molecules, and hydrogen-bonded liquids have exceptionally high values. It should be noted that ε_r is a macroscopic parameter derived for a continuous dielectric medium and cannot be used for interactions over molecular distances.

The magnitude and sign of each ionic charge may be given in terms of the elementary charge ($e = 1.602 \times 10^{-19}$ C) multiplied by the ionic valency, z. For bi-ionic interactions, the two ions can be shown as $q_1 = z_1 e$ and $q_2 = z_2 e$ and then Equation (22) becomes

$$V(r) = \frac{z_1 z_2 e^2}{4\pi\varepsilon_o\varepsilon_r r} \tag{23}$$

For example, $z = +1$ for monovalent cations such as K^+, $z = -1$ for monovalent anions such as Cl^-, $z = +2$ for divalent cations such as Mg^{2+}, and so on. By combining Equations (17) and (23) and differentiating, one obtains the Coulomb force between two static ions,

$$F(r) = \frac{z_1 z_2 e^2}{4\pi\varepsilon_o\varepsilon_r r^2} \tag{24}$$

(Historically, the Coulomb force was found by Charles Coulomb in 1785, earlier than the potential energy concept.) When we apply Coulomb forces to condensed states, we should define the reference states, so that the ions in the lattice come together to form a condensed phase from the gaseous state, where $r = \infty$ and the electrostatic interaction takes place in a vacuum ($\varepsilon_r = 1$). If two ions are interacting in a liquid medium, the reference state is also at $r = \infty$, however, the dielectric constant has a value ($\varepsilon_r \neq 1$).

On the other hand, the Coulomb force between two static ions is the product of the electric field strength of one of the charges, q_1 acting on the other charge, q_2 (see Section 2.1.3). An electric charge, q_1 is said to produce an electric field in the space around itself, and this field exerts a force on any charge, q_2 that is present in the space around q_1. The electric field strength, E at a point P in space is defined as the electrical force per unit charge experienced by a test charge at rest at point P. Thus, the electric field strength, E_1 created by the first ion, q_1 on the second ion, q_2 at a distance r away from the center of q_1 is defined as, from Equation (5), as ($E_1 \equiv F(r)/q_2$), so that we may write

$$E_1 = \frac{q_1}{4\pi\varepsilon_o\varepsilon_r r^2} \tag{25}$$

and when this electrical field acts on a second ion having a charge of q_2 at distance r, this results in a force from Equation (5) of

$$F(r) = E_1 q_2 = \frac{q_1}{4\pi\varepsilon_o\varepsilon_r r^2} q_2 \tag{26}$$

which gives the same result as in Equation (24). The direction of E is the same as the direction of $F(r)$ on the line from the charge q_1 to the charge q_2 at point P. For a positive q_2, the vector E points outwards if q_1 is positive, inwards if q_1 is negative.

It is sometimes more convenient to use the *electric potential* terms, ϕ instead of the electric field strength terms (see also Section 2.1.3). We can define the *Coulomb potential, ϕ_1*, of a charge q_1 in the presence of another charge q_2 by using the electric field equation (Equation (6)),

$$\phi_1 = \frac{q_1}{4\pi\varepsilon_o\varepsilon_r r} \tag{27}$$

Electric potential, ϕ has units of joules per coulomb (J C^{-1}) which is equivalent to volts. Since, $V(r) = \phi_1 q_2$ from classical physics, we obtain Equation (22) again. If there are several charges, $q_1, q_2, q_3, \ldots$ present in the system, the total Coulomb potential experienced by the charge, q is the sum of the potential generated by each charge

$$\phi_T = \phi_1 + \phi_2 + \phi_3 + \ldots \tag{28}$$

and, $V(r) = \phi_T q$ is used for this case. Both $V(r)$ and $F(r)$ are positive for like charges ($++$, $--$) and they are negative for unlike charges ($+-$) in Equations (23) and (24). When either $F(r)$ or $V(r)$ is positive, the interaction is repulsive, and when $F(r)$ or $V(r)$ is negative the interaction is attractive. To illustrate this, we can calculate $V(r)$ for two isolated Na$^+$ and Cl$^-$ gas ions in contact in a vacuum: the separation distance is now the sum of two ionic radii, $r = 0.276$ nm, and thus

$$V(r) = \frac{(1)(-1)(1.602 \times 10^{-19})^2}{4\pi(8.854 \times 10^{-12})(1)(0.276 \times 10^{-9})} = -8.36 \times 10^{-19} \, J$$

If we want to use the thermal energy terms at a temperature of 300 K, $kT = (1.38 \times 10^{-23})$ $(300) = 4.1 \times 10^{-21}$ J, and the above $V(r)$ figure for two isolated Na$^+$ and Cl$^-$ ions turns out to be of the order of $200 \, kT$ in a vacuum, which is very large and similar to chemical covalent bond energies within molecules. In addition, we may calculate that around $r = 56$ nm, $V(r)$ falls below $1 \, kT$ in a vacuum. In summary, the physical Coulomb interaction energy is very strong and long-range. However, the above calculation of $V(r)$ for two isolated Na$^+$ and Cl$^-$ gas ions in contact in a vacuum is too simplistic for estimating the total interaction energy for a pair of Na$^+$ and Cl$^-$ ions in a solid ionic crystal lattice. The solids have a long range structural order and the Coulomb energy of an ion with all the other ions in the crystal lattice has to be summed for the accurate determination of the lattice energy, because the attractive part of the potential energy is not a vector, it is a scalar quantity and does not cancel, but is additive. Since the Coulomb interaction energy is effective for long range, all the possible interactions should be considered, and not only with its nearest neighbors. If we examine the NaCl crystal lattice carefully, each Na$^+$ ion has six nearest neighbor Cl$^-$ ions at $r = 0.276$ nm, 12 next-nearest neighbor Na$^+$ ions at $r_2 = \sqrt{2}r$, eight more Cl$^-$ ions at $r_3 = \sqrt{3}r$ and so on. Similar to the derivation of Equation (19), the total interaction energy for a pair of Na$^+$ and Cl$^-$ ions in the ionic crystal lattice is given by summing the pair potentials (integrating) as

$$V_T = -\frac{e^2}{4\pi\varepsilon_o r}\left(6 - \frac{12}{\sqrt{2}} + \frac{8}{\sqrt{3}} - \frac{6}{20} + \ldots\right) = -1.748\frac{e^2}{4\pi\varepsilon_o r} = -1.46 \times 10^{-18} \, J \tag{29}$$

and the above V_T figure for the crystal lattice energy of NaCl turns out to be of the order of $350 \, kT$, which is 1.75 times the two isolated Na$^+$ and Cl$^-$ ion energies in a vacuum. This is the result of the long-range interaction effects of distant ions. The constant 1.748 for an ionic NaCl crystal lattice is known as the *Madelung constant*, and varies numerically for other ionic crystal structures. In surface thermodynamics (Chapter 3) we will see that the total Coulomb potential energy, V_T is the cohesive energy of the ionic lattice. On the other hand, we should be careful to realize that the dielectric constant parameter, ε_r is absent in the denominator of Equation (29), although $\varepsilon_r \cong 6$ for NaCl. This is because of the reference state definitions: the ions in the lattice come together to form a condensed phase from the gaseous state where the reference state is at $r = \infty$ and the interaction occurs in a vacuum, so there is no room for ε_r in this calculation. However, if two ions are interacting in a liquid medium, say in water, the dielectric constant parameter, ε_r, should now appear in the denominator although the reference state is also at $r = \infty$, because the interaction is now taking place within the solvent medium.

In reality, we should also consider the *screening effect* of neighbor ions for Coulomb interactions. Since positive ions always have negative ions nearby, whether they are in a lattice or in solution, the electric field becomes screened and thus decays more rapidly away from them than from a truly isolated ion. This screening effect causes the decay of the electric field, exponentially with distance. Then the resulting Coulomb interactions between ionic crystals, dissolved ions, charged particles and charged surfaces are of a shorter range than the expected range from Equations (23) and (26). However, in general, the physical Coulomb forces are still very effective in ionic interactions and of a much longer range than chemical covalent forces.

2.4   Polar Interactions

As defined in Section 2.1.3, the permanent dipole moments of molecules, μ, are important because the partial charges in molecules may interact with the charges of ions (and contribute to orientation and solvation), and also with the partial charges in other polar molecules (and contribute to molecular orientation and cohesion). The permanent dipole moments of a dipolar molecule represent the *first moment* of charge distribution which is asymmetrical and has two equal separated residual charges δ^+ and δ^-. However, there are higher dipoles called *multipoles*, such as *quadrupole, octupole* etc. and the relevant charge distribution functions are called *second moment* for quadrupoles and *third moment* for octupoles etc. Quadrupoles also have an asymmetrical charge distribution where the distribution of charge can be represented by four point charges, two positive and two negative; for octupoles there are eight point charges, four positive and four negative. The presence of quadrupoles and octupoles can be best explained by giving an example: as shown in Section 2.1.3, the CO_2 molecule has a zero dipole moment because the two dipoles associated with the carbon–oxygen bonds exactly cancel out; nevertheless it is capable of interacting with an ionic charge, by its second moment (or quadrupole moment) which represents the effective behavior of the set of separated charges, $2\delta^+$ on the carbon atom and δ^- on each of the oxygen atoms. Similarly, the CH_4 molecule has a zero dipole and quadrupole moment but has a significant third moment, and SF_6 has a significant fourth moment even though all of its lower moments are equal to zero. Multipole interactions are beyond the scope of this book and we limit ourselves mainly to the dipole interactions. In general, the dipole-orientation forces may be of two types: interactions between ions and dipolar molecules, and interactions between dipolar molecules.

2.4.1   Interactions between ions and dipolar molecules (fixed and rotating)

When a polar molecule is near to an ion, it will tend to become oriented in the way suggested in Figure 2.2 *a* due to Coulombic ion–dipole interactions. The molecule may be

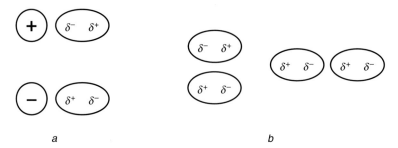

Figure 2.2   a. Schematic representation of the possible orientation of charges in an ion–dipole interaction, (+) and (–) are ionic charges and (δ^+) and (δ^-) are dipolar charges. **b.** Schematic representation of the possible orientation of charges in a dipole–dipole interaction.

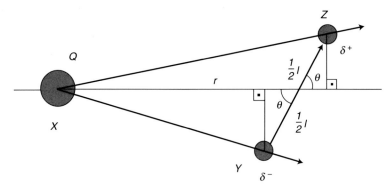

Figure 2.3   Geometric relations in a schematic ion–dipole interaction: (Q) is the charge of the ion X and (δ^+) at Z and (δ^-) at Y are the charges of dipolar molecule YZ having a dipole length of (l) and a dipole moment of ($\mu = \delta\, l$) subtending an angle θ to the line joining the two molecules. The X ion is at a distance of (r) away from the center of the dipolar molecule YZ.

revolving, but it will then spend more of its time in the direction shown than in any other. The potential energy will be minimized in this position, since the attraction between oppositely charged points would exceed the repulsion between the more distant like charges. An overall attraction of moderate strength will result. We can derive the interaction potential from the geometric considerations shown in Figure 2.3. In this figure, a charged ion (X) having a charge of $q = ze$ is at a distance of r away from the center of a polar molecule (YZ) which has a dipole length of l and a dipole moment of $\mu = \delta l$ subtending an angle θ to the line joining the two molecules. The total interaction energy will be the sum of the Coulomb energies of q with δ^- at Y and q with δ^+ at Z.

$$V(r) = -\frac{q\delta}{4\pi\varepsilon_o\varepsilon_r}\left[\frac{1}{XY} - \frac{1}{XZ}\right] \tag{30}$$

By applying plane geometry rules we can write

$$XY = \left[\left(r - \frac{l}{2}\cos\theta\right)^2 + \left(\frac{l}{2}\sin\theta\right)^2\right]^{1/2} \tag{31a}$$

$$XZ = \left[\left(r + \frac{l}{2}\cos\theta \right)^2 + \left(\frac{l}{2}\sin\theta \right)^2 \right]^{1/2} \tag{31b}$$

when $r \gg l$ then $XY \approx (r - l/2\cos\theta)$ and $XZ \approx (r + l/2\cos\theta)$ may be used. The interaction energy in this limit becomes

$$V(r,\theta) = -\frac{q\delta}{4\pi\varepsilon_o\varepsilon_r} \left[\frac{1}{\left(r - \frac{l}{2}\cos\theta \right)} - \frac{1}{\left(r + \frac{l}{2}\cos\theta \right)} \right] \tag{32}$$

When permanent dipole moments are considered by combining Equations (3) and (32), and applying the $r \gg l$ condition we obtain

$$V(r,\theta) = -\frac{q\delta}{4\pi\varepsilon_o\varepsilon_r} \left[\frac{l\cos\theta}{\left(r^2 - \frac{l^2}{4}\cos^2\theta \right)} \right] = -\frac{q\mu\cos\theta}{4\pi\varepsilon_o\varepsilon_r r^2} \tag{33}$$

This expression shows that ion–dipole interactions are much weaker than the ion–ion interactions. If we compare Equation (33) with Equation (22) we see that instead of q_2 in Equation (22) we have $(\mu\cos\theta)$ in Equation (33) and we have r^2 instead of r in the denominator. All of these parameters cause $V(r)$ to decrease. When a cation is near a dipolar molecule, maximum negative energy (attraction) will occur when the dipole points away from the ion ($\theta = 0°$ and $\cos\theta = 1$) as can be seen in Figure 2.2 *a*, and if the dipole points towards the ion ($\theta = 180°$ and $\cos\theta = -1$) the interaction energy is positive and the force is repulsive. At typical interatomic separations (0.2–0.4 nm) ion–dipole interactions are strong enough to bind ions to polar molecules and mutually align them. If the charge separation, *l*, is less than 0.1 nm then Equation (33) is valid for all intermolecular separations, however for greater *l* values in dipolar molecules, the deviations may be large and the interactions are always stronger than expected from Equation (33).

All the expressions given in Equations (30)–(33) are for fixed dipoles. In reality, all dipoles and ions are mobile and the dipole is rotating freely so that only *angle-averaged potentials* can be applied. In this case, the θ angle of the dipole in Figure 2.3 does not have a fixed value and is changing with time. This occurs at large intermolecular separations or in a medium of high ε_r, where the angle dependence of the interaction energy falls below the thermal energy, *kT*, and dipoles can now rotate more or less freely due to the presence of the weak interaction potentials. For this case, the *angle-averaged* potential energies are not zero, even though the value of $\cos\theta$ is zero when averaged over all of the space, because the Boltzmann weighting factor gives more weight to those orientations that have a lower (more negative) energy. Angle-averaged potential energies for the ion–dipole interactions can be calculated by applying the potential energy distribution theorem and it is found that

$$V(r) = -\frac{q^2\mu^2}{6(4\pi\varepsilon_o\varepsilon_r)^2 kTr^4} \tag{34}$$

for $kT > (q\mu/4\pi\varepsilon_o\varepsilon_r r^2)$ conditions. It should be noted that angle-averaged potential energies are only attractive and temperature dependent. Values calculated from Equation (34) may supersede the values calculated using Equation (33) at distances larger than

$r = \sqrt{q\mu/4\pi\varepsilon_o\varepsilon_r kT}$. In general, the distance for many monovalent ions dissolved in water is roughly 0.2 nm and thus the angle-averaged molecular interaction with the ion strongly restricts the movement of water molecules in the first shell around the ion and causes the difference in properties from those of bulk water values.

2.4.2 Interactions between dipolar molecules (fixed)

All neighboring polar molecules will tend to orientate themselves so as to minimize the interaction energy. This results in a weak dipole–dipole attraction, especially between organic liquids, as seen in Figure 2.2 b. A molecule (1) having a permanent dipole moment of μ_1 experiences an effect if there is a gradient of an electric field from another molecule (2) having a dipole moment of μ_2. Two such polar molecules are shown in Figure 2.4 in a fixed position (without rotating around the axis). In such a fixed orientation, if the charge separation distance, r_{12}, is much larger than the diameter of the molecules, the mutual potential energy of the two dipoles can be derived similarly to the procedure used for ion–dipole interactions,

$$V(r_{12}, \theta_1, \theta_2, \phi) = -\frac{\mu_1\mu_2}{4\pi\varepsilon_o\varepsilon_r r_{12}^3}[2\cos\theta_1\cos\theta_2 - \sin\theta_1\sin\theta_2\cos\phi] \tag{35}$$

where θ_1, θ_2 and ϕ are the angles, in polar coordinates, describing the orientations of the dipoles as shown in Figure 2.4. When $\theta_1 = \theta_2 = 0$, the two dipole molecules lie in line according to the above equation so that the maximum attraction occurs:

$$V(r_{12}, 0°, 0°, \phi) = -\frac{2\mu_1\mu_2}{4\pi\varepsilon_o\varepsilon_r r_{12}^3} \tag{36}$$

If two dipoles are free to rotate relative to each other, they will take up the head-to-tail (lowest energy) configuration as seen in Figure 2.2 b. For the same r_{12}, when the two dipoles align parallel to each other, that is $V(r_{12}, 90°, 90°, 180°)$ then the interaction potential energy is half of the maximum value. However, when r_{12} is not constant, contrary to the expectations that dipole–dipole attraction will prefer to orientate themselves in line, most dipolar molecules prefer to align in parallel so as to become significantly closer to each

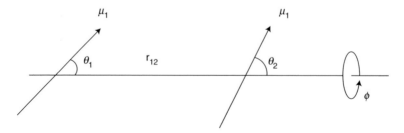

Figure 2.4 Geometric relations in a schematic dipole–dipole interaction: μ_1 is the permanent dipole moment of molecule (1) and μ_2 of molecule (2) and the charge separation distance, r_{12} is much larger than the diameter of the molecules. θ_1, θ_2 are the angles of each dipole in the polar coordinates, and ϕ is the rotation angle around the axis; all three angles describe the orientations of the two dipoles to each other.

other. Because they are anisotropic, that is they are longer along the direction of the dipole turning to cigar shaped molecules, this makes the parallel interaction more favorable. The smaller the r_{12} distance the stronger will be the interaction, due to the presence of r_{12}^3 in the denominator of Equation (36).

In general, independent dipole–dipole interactions are weaker than the ion–dipole interactions, and only very polar molecules are strong enough to bind to each other at normal temperatures in the liquid state. However, water is an exception to this due to its hydrogen-bonding properties. Hydrogen bonding may be considered to be a special type of directional dipole–dipole interaction (see also Section 2.8).

2.4.3 Keesom orientation interactions: interactions between dipolar molecules (rotating)

Equations (35) and (36) can only be used for fixed dipole interactions. However, in reality, all polar molecules are mobile and rotating freely so that only *angle-averaged potentials* can be applied similarly for ion–dipole interactions. The Boltzmann weighting factor gives more weight to the lower (more negative) energy orientations and angle-averaged potential energies are not zero. In a pure fluid, $\mu_1 = \mu_2$ and thermal agitation tends to make relative orientations random, while the interaction energy acts to favor alignment. We expect therefore, that as the temperature rises, the orientations become increasingly random until in the limit of very high temperature, the average potential energy due to polarity becomes vanishingly small. This expectation is confirmed by experimental evidence of polar gas behavior. In 1912, Keesom carried out statistical calculations averaging over all orientations with each orientation weighted according to its Boltzmann factor, in order to calculate the average potential energy between two dipoles at a fixed separation, r_{12}, and found that

$$V(r_{12}) = -\frac{\mu_1^2 \mu_2^2}{3(4\pi\varepsilon_o\varepsilon_r)^2 kTr_{12}^6} \tag{37}$$

for $kT > (\mu_1\mu_2/4\pi\varepsilon_o\varepsilon_r r_{12}^3)$ conditions. Thus, the Boltzmann-averaged interactions between two mobile and permanent dipoles are usually referred to as *Keesom orientation interactions*. Since they vary with the inverse sixth power of the separation distance, they contribute to the total *van der Waals interactions* between molecules (see also Section 2.6). The Keesom contribution becomes increasingly significant for small molecules having larger dipole moments.

Ion–quadrupole interactions vary with the inverse eighth power of the separation distance; and the quadrupole–quadrupole interactions vary with the inverse tenth power of the separation distance. The effect of quadruple moments on thermodynamic properties is already much less than that of dipole moments and the effect of higher multipoles is usually negligible. In general, neither ion–dipole nor dipole–dipole forces can produce long-range alignment effects in liquids.

In reality, the dipolar energies are not pair-wise additive in multi-molecular interactions. The head–to–tail configuration, which is the most stable one for two dipoles or quadrupoles, is not even very common in the lowest-energy configurations when three or more dipoles are present. Instead, a cluster of three dipoles has a lowest-energy configuration with the dipoles arranged in a triangle. Since the dipoles in a condensed phase usually form clusters containing three or more dipoles, an individual dipolar molecule cannot be positioned in the same lowest-energy orientation with respect to all of its nearest neighbors, resulting in diminished values of $V(r)$ for Keesom orientation interactions. Thus, every multimolecular interaction should be independently examined to find the lowest-energy configuration.

2.5   Induction Effects: Interactions Between Induced Non-polar and Polar Molecules

In deriving the above expressions, we have assumed that molecules are non-polarizable, that is, they interact without perturbing the charge distributions of each other. However in real systems, the electric fields emanating from nearby molecules induce dipole moments both in non-polar and polar molecules, resulting in *molecular polarizability*, α, as defined in Section 2.1.3. Actually, we have already seen polarization effects when considering the *relative permittivity*, ε_r (earlier called *dielectric constants*) measurements carried out in parallel-metal plate capacitors (or condensers) as detailed in both Section 2.1.3 relating to dipole moments and Section 2.3 relating to Coulomb interactions. The relative permittivity is a macroscopic property of the medium and it reflects the degree of polarization of the molecules in the medium by the applied local electric field, as given by Equation (8) in Section 2.1.3 and Equation (22) in Section 2.3. In this section, we will see the physical nature of the polarizability and will derive some important dipole moment expressions such as the Debye–Langevin equation (Equation (11)) from first principles.

2.5.1   Polarizability of non-polar molecules

All molecules are polarizable. When a non-polar molecule is subjected to an electric field, the electrons in the molecule are displaced from their ordinary positions so that the electron clouds and nuclei are attracted in opposite directions. The charges inside the molecules shift, a dipole is induced, and the molecule has a temporarily induced dipole moment, μ^i. One would expect that the molecular polarizability, α should increase with the molecular size because in larger molecules the electrons can be displaced over longer distances. In fact, α is proportional to the volume of the molecule. As a simplified example, a one-electron Bohr atom is shown in Figure 2.5 *a* in which an electron ($q = e^-$) moves around a nucleus ($q = e^+$) with a circularly symmetrical orbit having a radius, r, in the absence of an external field. When this non-polar atom is subjected to an external electric field, E, the electron orbit is shifted by a distance, l, from the nucleus as shown in Figure 2.5 *b*. The external force, F_{ext}, on the electron due to the field E can be calculated from Equation (5) so that

$$F_{ext} = Ee \qquad (38)$$

This F_{ext} must be balanced by the internal attractive Coulomb force between the displaced electron orbit and the nucleus as given by Equation (24) where $z_1 = z_2 = 1$, $\varepsilon_r = 1$ in a vacuum and by considering the shift angle, θ

$$F_{int} = \frac{e^2(\sin\theta)}{4\pi\varepsilon_o r^2} \tag{39}$$

At equilibrium, $F_{ext} = F_{int}$ and by plane geometry, $\sin\theta = l/r$, which gives

$$E = \frac{e(\sin\theta)}{4\pi\varepsilon_o r^2} \approx \frac{el}{4\pi\varepsilon_o r^3} \tag{40}$$

The induced dipole moment can be expressed by using the general dipole moment equation (Equation (3)), and can be related to molecular polarizability by using Equation (7) so that,

$$\mu^i = el = \alpha E \tag{41}$$

By combining Equations (40) and (41) we obtain

$$E \approx \frac{\mu^i}{4\pi\varepsilon_o r^3} \tag{42}$$

and for a non-polar gas molecule in free space, the polarizability can be expressed as

$$\alpha = 4\pi\varepsilon_o r^3 \tag{43}$$

The units of polarizability are $C^2 m^2 J^{-1}$ and it is related to the volume of the Bohr atom so that geometrically, $V_B = (4/3)\pi r^3$ and thus, $\alpha = 3 V_B \varepsilon_o$. This shows that there is a direct proportionality between the polarizability and the volume of a molecule. However, the calculation of the molecular radius from Equation (43) is not recommended because such calculations generally give values which are less than the real molecular radii. For example, a water molecule has a radius of 0.135 nm, but Equation (43) gives a value of 0.114 nm, which is 15% less than the real value.

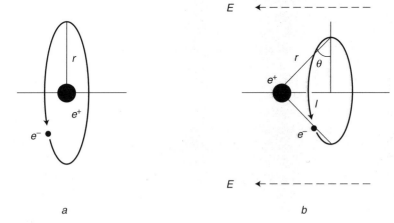

Figure 2.5 **a.** One-electron non-polar Bohr atom in which an electron (e^-) moves around the nucleus (e^+) with a circularly symmetrical orbit having a radius, r, in the absence of any external electric field. **b.** When this atom is subjected to an external electric field, E, the electron orbit is shifted by a distance, l, and angle, θ, from the nucleus.

2.5.2   Polarizability of polar molecules

In the above section, we considered the polarization of atoms and only non-polar gas molecules in free space. What happens to a dipolar molecule when it is subjected to an electrical field? We know that a freely rotating dipolar molecule has a time-averaged zero dipole moment. However, when an electrical field, E, is applied, the Boltzmann-averaged orientations of the rotating dipoles are affected so that an *orientation polarizability* is formed which will be weighted along the field. If the permanent dipole, μ, is at an angle θ to the field, E, at any instant, its time resolved dipole moment along the field is $(\mu \cos \theta)$ and the time-averaged induced dipole moment in free space is given (similarly to Equation (34) in Section 2.4.1) by

$$\mu^i = \frac{\mu^2 E}{3kT} \quad \text{for} \quad (kT \gg \mu E) \tag{44}$$

By combining Equations (41) and (44), the *orientation polarizability* of a dipolar molecule can be calculated as

$$\alpha_{\text{orient}} = \frac{\mu^2}{3kT} \tag{45}$$

and by combining Equations (43) and (45) we obtain the total polarizability of a dipolar molecule in an electric field that can be considered as the initial state of the *Debye–Langevin* equation

$$\alpha_{\text{T}} = \alpha + \frac{\mu^2}{3kT} = 4\pi\varepsilon_o r^3 + \frac{\mu^2}{3kT} \tag{46}$$

However there are exceptions: in a very high electrical field, such as when water molecules are placed near a small ion or when the temperature is very low where kT is very small, the orientation polarizability concept breaks down, and a dipolar molecule completely aligns along the field; however the induced polarizability still applies.

2.5.3   Solvent medium effects and excess polarizabilities

The above expressions were derived for the polarizabilities of molecules in free space or in a dilute gas (mostly air). However, we often encounter molecules interacting in a liquid solvent medium, which reduces the interaction pair potential by around ε_r or more; the extent of this reduction depends on several factors. First of all, the intrinsic polarizability and dipole moment of an isolated gas molecule may be different when it is itself in the liquid state, or alternatively when dissolved in a solvent medium. This is because of the difference in interaction strength and also the separation distance between molecules. Thus, the polarizability values are best determined by experiment. Second, a dissolved molecule can only move by displacing an equal volume of solvent from its path. If the molecule has the same polarizability as the solvent molecules, that is if no electric field is reflected by the molecule, it is invisible in the solvent medium and does not experience any induction force. Thus, the polarizability of the molecule, α, must represent the *excess* or *effective polarizability* of a molecule over that of the solvent. Landau and Lifshitz applied a continuum approach and modeled a molecule, i, as a dielectric sphere of radius, a_i, having a dielectric constant of ε_i, in a solvent medium. When such a dielectric sphere is polarized by an electric field of, E, in this solvent medium having a dielectric constant of ε_r, it will have an excess dipole moment of

$$\mu^i = 4\pi\varepsilon_o\varepsilon_r\left(\frac{\varepsilon_i - \varepsilon_r}{\varepsilon_i + 2\varepsilon_r}\right)a_i^3 E \tag{47}$$

and since $\mu^i = \alpha E$ from Equation (41), the excess total polarizability of the molecule in this solvent medium is

$$\alpha_{iT} = 4\pi\varepsilon_o\varepsilon_r\left(\frac{\varepsilon_i - \varepsilon_r}{\varepsilon_i + 2\varepsilon_r}\right)a_i^3 = 3\varepsilon_o\varepsilon_r\left(\frac{\varepsilon_i - \varepsilon_r}{\varepsilon_i + 2\varepsilon_r}\right)v_i \tag{48}$$

where $v_i = (4/3)\pi a_i^3$ is the volume of the dielectric sphere molecule. Since $\varepsilon_r = 1$ in free space, Equation (48) reduces to

$$\alpha_{iT} = 4\pi\varepsilon_o\left(\frac{\varepsilon_i - 1}{\varepsilon_i + 2}\right)a_i^3 = 3\varepsilon_o\left(\frac{\varepsilon_i - 1}{\varepsilon_i + 2}\right)v_i \tag{49}$$

and if the dielectric constant is high that is $\varepsilon_i \gg 1$, then its polarizability is roughly $\alpha \approx 4\pi\varepsilon_o a^3$ which is similar to Equation (43). Instead, if $\varepsilon_r > \varepsilon_i$ then the polarizability is negative, showing that the direction of the induced dipole in a solvent medium is opposite to that in free space.

On the other hand, Equation (49) can be used in the derivation of the classical Clausius–Mossotti equation (Equation (9)) so that, for an isolated gas molecule in free space, Equation (49) can be written as

$$\frac{\alpha}{3\varepsilon_o} = \left(\frac{\varepsilon_i - 1}{\varepsilon_i + 2}\right)v_i \tag{50}$$

where ε_i is now the dielectric constant of the gas molecules. Since $v_i = (V_m/N_A)$, then Equation (50) is the same as Equation (9). This derivation also shows the success of treating a molecule as a dielectric sphere in a medium.

2.5.4 Interactions between ions and induced non-polar molecules

When an ion and a non-polar molecule are separated by a distance, r, in the same medium, as shown in Figure 2.6 a, the electric field of the ion (see Equation (25), $E = q/4\pi\varepsilon_o\varepsilon_r r^2$) will form a shift of distance, l, in electron clouds of the non-polar molecule, and will tend to polarize it to form an induced dipole moment as long as the ion is sufficiently close

$$\mu^i = \alpha E = \frac{\alpha q}{4\pi\varepsilon_o\varepsilon_r r^2} \tag{51}$$

If, ΔE is the difference in E at either end of the dipole so that

$$\Delta E = \left(\frac{dE}{dr}\right)l \tag{52}$$

the resulting force on the neutral molecule is therefore $F = q\Delta E$ from Equation (38), and by combining with Equations (41) and (52) we obtain

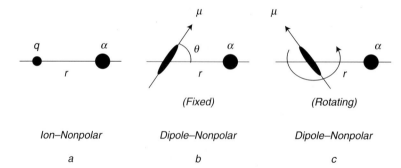

(Fixed)   (Rotating)

Ion–Nonpolar   Dipole–Nonpolar   Dipole–Nonpolar

a   b   c

Figure 2.6 **a.** Schematic representation of the interaction between an ion (q) and a nonpolar molecule (α) with a separation distance of r. The electric field of the ion will form a shift in electron clouds of the non-polar molecule and will tend to polarize it to form an induced dipole moment if the ion is near enough. **b.** Schematic representation of the interaction between a dipolar molecule having a fixed dipole, μ, and a non-polar molecule (α) with a separation distance of r. The fixed dipole, μ, is oriented at an angle, θ, to the line joining it to the polarizable molecule. **c.** Schematic representation of the angle-averaged interaction between a dipolar molecule having a rotating dipole, μ, and a non-polar molecule (α) with a separation distance of r.

$$F = q\left(\frac{dE}{dr}\right)l = \alpha E\left(\frac{dE}{dr}\right) \tag{53}$$

Then the pair potential energy can be calculated as

$$V(r) = -\int F dr = -\int \alpha E\left(\frac{dE}{dr}\right) dr = -\frac{\alpha E^2}{2} \tag{54}$$

By combining Equations (51) and (54), we obtain

$$V(r) = -\frac{q^2 \alpha}{2(4\pi\varepsilon_o\varepsilon_r)^2 r^4} \tag{55}$$

It is important to note that this is half of the pair potential that is expected for the inter-action of an ion with a molecule having a permanent dipole moment (rotating) similarly aligned [$V(r) = -\alpha E^2$, compare with Equation (34) in Section 2.4.1]. Since some energy is taken up in polarizing a molecule to induce a dipole moment rather than having a permanent dipole moment, it is reasonable that a weaker pair potential is obtained for the interactions between ions and induced non-polar molecules.

2.5.5 Interactions between ions and induced polar molecules

When an ion and an induced dipolar molecule interact, as shown in Figure 2.2 a, we should consider the total polarizability of this dipolar molecule (see Equation (46)), and then similar to the derivation of Equation (55), one obtains,

$$V(r) = -\frac{q^2 \alpha_T}{2(4\pi\varepsilon_o\varepsilon_r)^2 r^4} = -\frac{(ze)^2}{2(4\pi\varepsilon_o\varepsilon_r)^2 r^4}\left(\alpha + \frac{\mu^2}{3kT}\right) \tag{56}$$

2.5.6 *Interactions between dipolar molecules and induced non-polar molecules (fixed)*

When a dipolar and a non-polar molecule interact, as shown in Figure 2.6 b, the polarizing field arises from the permanent dipole and not from an ion. As a result, the applied electrical field is weaker than the interactions between ions and non-polar molecules. The magnitude of this electric field can be calculated for a fixed dipole, μ, oriented at an angle, θ, to the line joining it to the polarizable molecule,

$$E = \frac{\mu\sqrt{\left(1+3\cos^2\theta\right)}}{4\pi\varepsilon_o\varepsilon_r r^3} \tag{57}$$

and the pair potential for a fixed θ angle can be calculated by combining Equations (54) and (57) so that

$$V(r,\theta) = -\frac{\alpha E^2}{2} = -\frac{\mu^2\alpha\left(1+3\cos^2\theta\right)}{2\left(4\pi\varepsilon_o\varepsilon_r\right)^2 r^6} \tag{58}$$

For typical values of μ and α the strength of this interaction is so weak that it is not sufficient to mutually orient the molecules.

2.5.7 *Debye interactions: interactions between dipolar molecules and induced dipolar molecules (rotating)*

The above derivation for the fixed angle dipole–induced non-polar molecule interaction is not realistic and we need the angle-averaged interaction potential for the rotating dipole-induced non-polar molecule interactions as shown in Figure 2.6 c. Since the angle average of $\cos^2\theta = 1/3$ mathematically, Equation (58) becomes

$$V(r) = -\frac{\mu^2\alpha}{\left(4\pi\varepsilon_o\varepsilon_r\right)^2 r^6} \tag{59}$$

Debye derived a more general expression from Equation (59) for the interactions between dipolar molecules and induced dipolar molecules (rotating) in 1920. He found that when induction takes place, the pair potential energy between two different dipolar molecules each possessing permanent dipole moments of μ_1 and μ_2 and polarizabilities α_1 and α_1, can be expressed as,

$$V(r) = -\frac{\left[\mu_1^2\alpha_2 + \mu_2^2\alpha_1\right]}{\left(4\pi\varepsilon_o\varepsilon_r\right)^2 r^6} \tag{60}$$

These are often called *Debye induced dipole interactions*. It is interesting to note that the Keesom orientation interaction expression (Equation (37) in Section 2.4.3) may also be obtained from Equation (59) by replacing α with $\alpha_{\text{orien}} = \mu^2/3kT$. This fact also indicates the presence of induction in orientation interactions. Thus, both Keesom and Debye interactions vary with the inverse sixth power of the separation distance and they both contribute to the van der Waals interactions, which we will see in Section 2.6.

There are extensions to the Debye equation, for example, an electric field may also be caused by a permanent quadrupole moment. In that case, Equation (60) can be written as

$$V(r) = -\frac{3\left[\mu_1^2 \alpha_2 + \mu_2^2 \alpha_1\right]}{2r^8} \tag{61}$$

Quadrupole induction pair potential energy is less than dipole induction pair potential energy.

2.5.8 Unification of angle-averaged induction interactions

If we exclude the simple Coulomb interaction between two charges in a vacuum, all the other interactions we have considered so far have involved polarization effects, either explicitly for non-polar molecules, or implicitly for rotating dipolar molecules having permanent polar moment. It is possible to express all these angle-averaged interactions in a unified equation. If a charged dipolar molecule 1 having q_1 as its charge, α_1 polarizability and μ_1 permanent dipole moment, interacts with a second dipolar molecule 2 having α_2 as its polarizability and μ_2 permanent dipole moment, we may write an expression by combining all angle-averaged ion–dipole (Equation (34)), Keesom (Equation (37)), Debye (Equation (60)) and also ion–induced and induced–induced interactions,

$$V(r) = -\frac{\left[\frac{q_1^2}{2r^4} + \frac{3kT\left(\frac{\mu_1^2}{3kT} + \alpha_1\right)}{r^6}\right]\left(\frac{\mu_2^2}{3kT} + \alpha_2\right)}{\left(4\pi\varepsilon_o\varepsilon_r\right)^2} \tag{62}$$

If no ion is present in the system ($q_1 = 0$) then the above expression gives the sum of the Keesom-dipolar orientation and Debye-induction contributions to the total van der Waals forces between two molecules. A third contribution to van der Waals forces is also present, the *dispersion forces* (see Section 2.6.1).

Equation (62) can be used to compare the relative magnitudes of ionic, dipolar, and induction contributions to a molecular interaction, and a ratio of 800:3:1 was calculated, respectively, when $q_1 = 1\,e$, $\mu_1 = \mu_2 = 1\,D$, $\alpha_1 = \alpha_2 = (4\pi\varepsilon_o)3 \times 10^{-30}\,m^3$, and a molecular separation distance of $r = 0.5\,nm$ at $T = 300\,K$, were selected for an indicative example. This shows the importance of the presence of an ion in a medium. Equation (62) also shows the molecular volume effect such that the efficiency of a dipolar interaction depends on $(\mu^2/r^6 \approx (\mu/V)^2)$. rather than the absolute value of the dipole moment.

The unified expression can be used while commenting on very practical situations in solvents. Assume that two spherical solute molecules 1 and 2 having dielectric constants of ε_1 and ε_2 and radii of a_1 and a_2, respectively, are present in a solvent medium having a dielectric constant of ε_r. What will be the net resultant interaction? If we combine Equation (62) with the excess total polarizability expression (Equation (48)), given for molecules dissolved in a solvent medium, we obtain

$$V(r_{12}) = -\left[\frac{q_1^2}{8\pi\varepsilon_o\varepsilon_r r^4} + \frac{3kT\left(\dfrac{\varepsilon_1 - \varepsilon_r}{\varepsilon_1 + 2\varepsilon_r}\right)a_1^3}{r^6}\right]\left(\frac{\varepsilon_2 - \varepsilon_r}{\varepsilon_2 + 2\varepsilon_r}\right)a_2^3 \qquad (63)$$

for the resultant pair potential.

We can derive some conclusions from Equation (63):

1 The interaction type between solute molecules dissolved in a solvent medium may be attractive, repulsive or zero depending on the relative magnitudes of ε_1, ε_2 and ε_r.
2 The interaction between any two identical uncharged molecules ($\varepsilon_1 = \varepsilon_2$ and $a_1 = a_2$) is always attractive regardless of the nature of the solvent medium. (Note that we must use the square of the parentheses including ε_1 and ε_2 in Equation (63) for this case.)
3 When ions are present in the solvent medium, they are attracted to very polar molecules ($\varepsilon_2 > \varepsilon_r$) but repelled from non-polar or very low polar molecules ($\varepsilon_2 < \varepsilon_r$).

This approach depends on the general continuum model and can be used to predict the interaction behavior of solutes in solvents; however it cannot be used for quantitative calculations.

In summary, induced interactions are weak and more short-ranged. Induced interactions are not pair-wise additive, and this fact adds complications to the calculation of molecular interactions. For example a molecule midway between two similar ions does not experience any electrical field, and the interaction potential from the ion–induced dipole term is zero. If we use Equation (55) and assume pair-wise additivity in such a case, we will get a large and quite erroneous result for the interaction energy.

2.6   van der Waals Interactions

The classical insight concerns intermolecular forces, as clearly formulated by van der Waals in 1873 in his equation of state for gases, to explain why real gases do not obey the ideal gas law ($PV = nRT$),

$$\left(P + \frac{an^2}{V^2}\right)(V - nb) = nRT \qquad (64)$$

where P is the external pressure (Pa), V is the molar volume (m^3), R is the gas constant (R = 8.3145 J K^{-1} mol^{-1}), T is the absolute temperature (K), a is the attraction constant for molecules (Pa m^6 mol^{-2}), b is the actual volume of the molecules (m^3 mol^{-1}) and n is the number of moles. van der Waals subtracted the term (nb) from the total gas volume to account for the finite size of molecules, to find the free volume space that the gas molecules can actually occupy, and added the term (an^2/V^2) to the measured gas pressure to account for the attractive intermolecular forces, now known as *van der Waals forces*, which are long range in nature. van der Waals forces cause an attraction between the gas molecules and restrict them to hit the walls of the container with their full translation momentum thus decreasing the measured gas pressure from that of the ideal gas state.

The basis for the parameters a and b depends upon real facts: when a molecule is in close proximity to another, if there were no forces of attraction, gases would not condense to form liquids and solids; and in the absence of repulsive forces, condensed matter would not show resistance to compression. We shall now consider the attractive interaction part in some detail. It was realized that van der Waals forces consist of only long-range forces and the interaction pair potential, $V(r)$, decreases with the inverse sixth power of the distance between gas molecules, r^{-6}, for not too great a distance between interacting molecules (however, the interaction force, $F(r)$, decreases as r^{-7}). From the previous sections, we know that there are three main types of interaction to obey the rule of inverse sixth power of the distance between molecules, namely, Keesom (dipole–dipole angle-averaged orientation, see Section 2.4.3), Debye (dipole–induced dipolar (angle-averaged), see Section 2.5.6), and London (dispersion) interactions. As we have seen above, the Keesom and Debye interactions can be fully understood in terms of classical electrostatics, based on Coulomb and Boltzmann distribution laws; however, London dispersion forces are different and quantum mechanical in nature.

2.6.1   London dispersion interactions

All gases and liquids, including those whose molecules have zero dipole moment, and even the inert gases, show serious deviations from the ideal gas laws even at moderate pressures. Until about 1930, there was no adequate explanation for the forces acting between non-polar molecules. In 1930, London showed by using wave and quantum mechanics that some other types of intermolecular forces, the dispersion forces, are present. London stated that so-called non-polar molecules are, in fact, non-polar only when viewed over a period of time, and their time-averaged distribution of electrons is symmetrical; if an instantaneous photograph of such a molecule could be taken, it would show that, at a given instant, the oscillation of the electrons about the nucleus had resulted in distortion of the electron arrangement sufficient to cause a temporary dipole moment. The electrons circulate with extremely high frequency (of the order of $10^{16}\,s^{-1}$) and at every instant the molecule is therefore polar, but the direction of this polarity changes with the high frequency. This dipole moment is rapidly changing its magnitude and direction and therefore averages out to zero over a short period of time; however, these quickly varying dipoles produce an electric field which then induces dipoles in the surrounding molecules. The result of this induction is usually an attractive force between induced non-polar or dipolar molecules. The dispersion attraction force is then this instantaneous force averaged over all instantaneous configurations of the electrons in the molecule. Dispersion attraction force is proportional to $1/r^6$ and is always operative between all atoms, molecules and particles, even totally neutral ones such as helium, carbon dioxide and paraffinic hydrocarbons. Since London dispersion forces are always present, they play an important role in the properties of gases, liquids, thin films, adhesion, surface tension, physical adsorption, wetting, flocculation and structure formation of condensed macromolecules such as polymers and proteins. Dispersion interactions are also long-ranged (0.2–15.0 nm) and may be repulsive for very short spacing.

It is possible to give a qualitative summary of London's quantum mechanical treatment on how an instantaneous dipole moment would arise and on the magnitude of its interaction with neighboring molecules. First, the electron cloud of an atom or molecule can be considered to be an oscillator since whenever the electron cloud is unsymmetrically distributed, the attraction of the positive nucleus provides a restoring force, which increases with the displacement. Quantum mechanics shows that the energy of a simple harmonic oscillator is given by

$$V_n = h v_o \left(n + \frac{1}{2} \right) \tag{65}$$

where V_n is the potential energy of the oscillator in the nth quantum level, h is Planck's constant, n is an integer and v_o is the frequency of the oscillator corresponding to the wavelength of an electronic absorption band for the molecule in the ultraviolet region. The significance of the (1/2) term in Equation (65) is that the oscillator has a definite amount of energy even at absolute zero, and that the fluctuations in distribution of electrons persist even down at $0\,\mathrm{K}$. Hence the instantaneous dipole moment of the molecule exists even at $0\,\mathrm{K}$.

The magnitude of the instantaneous dipole is proportional to the polarizability, α, given by Drude:

$$\alpha = \left(\frac{q}{2\pi m v_o} \right)^2 \tag{66}$$

where q is the charge and m is the mass of the body (i.e. an electron) which oscillates with frequency v_o. The magnitude of the frequency, v_o, can be experimentally determined via polarizability measuring refractive index and also relative permittivity for the substance in question, and Equations (9), (11), (12) and (66) are used in the calculations. Indeed, the name *dispersion force* arises from the dispersion of the refractive index due to the frequency difference. Subject to certain simplifying assumptions, the dispersion potential energy, V_d, between two simple, spherically symmetrical molecules 1 and 2 at large distances, r, is given by

$$V_d = -\frac{3}{2} \frac{\alpha_1 \alpha_2}{(4\pi\varepsilon_o)^2 r^6} \left(\frac{h v_{o_1} h v_{o_2}}{h v_{o_1} + h v_{o_2}} \right) \tag{67}$$

For a molecule, the product, $h v_o$ is very nearly equal to its first ionization potential, I, which is the work required to be done to remove one electron from an uncharged molecule. Equation (67) is therefore usually written in the London dispersion form,

$$V_d = -\frac{3}{2} \frac{\alpha_1 \alpha_2}{(4\pi\varepsilon_o)^2 r^6} \left(\frac{I_1 I_2}{I_1 + I_2} \right) \tag{68}$$

If molecules 1 and 2 are of the same species, Equation (68) reduces to

$$V_d = -\frac{3}{4} \frac{\alpha^2 I}{(4\pi\varepsilon_o)^2 r^6} \tag{69}$$

A useful version of Equation (68) was given by Slater and Kirkwood in 1931,

$$V_d = -\frac{363 \alpha_1 \alpha_2}{r^6 \left[\left(\dfrac{\alpha_1}{n_1} \right)^{1/2} + \left(\dfrac{\alpha_2}{n_2} \right)^{1/2} \right]} \tag{70}$$

where n_1 and n_2 are the numbers of electron in the outer shells, r is in Å units, V_d is in kcal/mole and α_o is in Å^3.

Since, London dispersion energy is the potential energy between non-polar molecules, Equations (68) and (69) show that V_d is independent of temperature. On the other hand, the first ionization potentials of most substances do not differ very much from one another, so London's equation is more sensitive to the polarizability than it is to the ionization potential. We can rewrite Equations (68) and (69):

$$V_{d_{12}} = k\frac{\alpha_1\alpha_2}{r^6}, \quad V_{d_1} = k\frac{\alpha_1^2}{r^6}, \quad V_{d_2} = k\frac{\alpha_2^2}{r^6} \tag{71}$$

where k is a constant which is approximately the same for the three types of interactions, 1–1, 2–2, 1–2. It then follows that

$$V_{d_{12}} = \left(V_{d_1} V_{d_2}\right)^{1/2} \tag{72}$$

Equation (72) gives some theoretical basis for the frequently applied *geometric-mean* rule, which is so often used in equations of state for gas mixtures and theories for liquid solutions. Berthelot was the first to use the geometric mean rule:

$$\frac{a_{12}}{\left(a_{11}a_{22}\right)^{1/2}} \cong 1 \tag{73}$$

where the a terms are the attractive constants in the van der Waals equation, a_{12} being the constant for unlike molecules.

Although the London equation can only be derived using quantum mechanical perturbation theory, it is instructive to use a simple approach on how these interactions take place, using the one-electron Bohr atom where the shortest distance between the electron and proton is known as the *first Bohr radius*, r_B, at which the Coulomb energy $(e^2/4\pi\varepsilon_o r_B)$ is equal to $2h\upsilon_o$, so that we can calculate r_B as

$$r_B = \frac{e^2}{2(4\pi\varepsilon_o)h\upsilon_o} = 0.053\,\text{nm} \tag{74}$$

Since the product $h\upsilon_o$ is very nearly equal to its first ionization potential, I, we know that $\upsilon_o = 3.3 \times 10^{15}\,\text{s}^{-1}$ for a Bohr atom, so that $I = 2.2 \times 10^{-18}\,\text{J}$. The Bohr atom has no permanent dipole moment, however, at any instant there exists an instantaneous dipole moment of $(\mu_{instan} = r_B e)$ whose field will polarize a nearby neutral Bohr atom giving rise to an attractive interaction that is entirely analogous to the dipole–induced dipole interaction as given in Section 2.5.6. The energy of this interaction in a vacuum can thus be calculated using Equation (59) so that

$$V(r) = -\frac{\mu^2\alpha}{\left(4\pi\varepsilon_o\right)^2 r^6} = -\frac{\left(r_B e\right)^2\alpha}{\left(4\pi\varepsilon_o\right)^2 r^6} \tag{75}$$

where we can write $\alpha = 4\pi\varepsilon_o r_B^3$ by using Equation (43). By combining Equations (74), (75), and using the I parameter, the dispersion interaction energy can be written as

$$V(r) = -\frac{2\alpha^2 I}{\left(4\pi\varepsilon_o\right)^2 r^6} \tag{76}$$

Equation (76) is very similar to London's Equation (69), except for the (3/8) numerical factor which is necessary for atoms having more electrons than the Bohr atom. It can be understood from the above simple Bohr atom model that, although the dispersion forces are quantum mechanical in origin, the basis of interaction is still essentially electrostatic, as a kind of quantum polarization force.

The strength of London dispersion interactions is generally high. These interactions are more than $1\,kT$ at room temperature and cannot be neglected. Thus, many large molecules such as non-polar hydrocarbon liquids and solids can be held together solely by dispersion forces. When such molecules are solidified through a decrease in temperature they are called *van der Waals solids*, and they have weak, undirected physical bonds between them, low melting points and low heats of melting. The spherically symmetrical inert neon, argon and methane molecules form van der Waals solids with close-packed structures and having 12 nearest neighbors per atom at low temperatures (see Chapter 3, Figure 3.8). Their lattice energy may be calculated from pair potentials so that 12 molecules share 12 London bonds (or six full London bonds per molecule), but generally the attractions of more distant molecules are also included in the calculations and the factor of 6 rises to 7.22 to multiply the pair potential. Of course, other interactions are interfering, such as very short-range stabilizing repulsive interactions (see Section 2.7) and other attractive interactions such as fluctuating higher multipole interactions and from other adsorption frequencies; however these opposing effects partially cancel each other out so that the London dispersion interactions, which are calculated directly from Equations (68) and (69), fit well with molar experimental cohesive energy results for molecules with a diameter of less than about 0.5 nm. Nevertheless, Equations (68) and (69) cannot be applied to non-spherical and large molecules (>0.5 nm) such as alkanes, polymers, cyclic or planar molecules, because the dispersion force no longer acts between the centers of the molecules, but acts between the centers of electronic polarization within each molecule where the covalent bonds are located. In order to compute the attractions between such complex molecules, the molecular packing in the solid or liquid must be known and the contributions from the different parts of the molecule must also be considered separately. Thus London dispersion forces may be non-additive depending on the conditions. As an extension, the London dispersion interactions are also operative between spherical colloidal particles in a medium or spherical-flat surfaces, as will be shown in Chapter 7.

2.6.2 Correlation with van der Waals constants

In the classical van der Waals expression (Equation (64)) we considered the gas molecules to be hard spheres of diameter, σ. If we write this expression in terms of the molecular parameters of a gas then we have

$$\left(P + \frac{a}{v^2}\right)(v - b) = kT \tag{77}$$

where v is the gaseous volume occupied per gas molecule $v = (V_m/N_A)$ and k is the Boltzmann constant. The London dispersion pair interaction potential between these hard sphere molecules may be simplified from Equation (69) so that for $r > \sigma$

$$V_d(r) = -\frac{3}{4}\frac{\alpha^2 I}{(4\pi\varepsilon_o)^2 r^6} = -\frac{C_d}{r^6} \tag{78}$$

then, we can say for the London dispersion interaction coefficient C_d,

$$C_d = \frac{3\alpha^2 I}{4(4\pi\varepsilon_o)^2} \tag{79}$$

$V_d = \infty$ for $r < \sigma$. For two dissimilar molecules, the London dispersion interaction coefficient C_d, from Equation (68), is

$$C_d = -\frac{3\alpha_1\alpha_2}{2(4\pi\varepsilon_o)^2}\left(\frac{I_1 I_2}{I_1 + I_2}\right) \tag{80}$$

On the other hand, we may also use gas constants, a and b to calculate the dispersion pair potential. As we will see its derivation in Section 3.4.2, the van der Waals constant, a, can be expressed as

$$a = \frac{2\pi C_W}{(n-3)\sigma^{n-3}} = \frac{2\pi C_W}{3\sigma^3} \quad \text{for} \quad n = 6 \tag{81}$$

where C_W is the van der Waals interaction coefficient. If we want to write Equation (81) using the classical van der Waals terms, containing one molar volume of the gas, V, as in Equation (64) (by analogy with the a/v^2 term), we obtain

$$a = \frac{2\pi N_A^2 C_W}{3\sigma^3} \tag{82}$$

We know that the constant b is the volume of the rigid molecules ($dm^3 mol^{-1}$), which is the volume unavailable for the molecules to move in, and the *excluded volume* per mole = $4/3\pi\sigma^3$ because the spherical diameter, σ, is the closest that one molecule can approach another. Thus the constant, b, can be written in terms of molecular diameter so that (see also Section 3.4)

$$b = \frac{2\pi N_A \sigma^3}{3} \tag{83}$$

which is four times the spherical volume of the molecules which, as we know from geometry, is equal to $4/3\mu N_A (\sigma/2)^3$ By combining Equations (82) and (83) we obtain for the van der Waals interaction coefficient, C_W,

$$C_W = \frac{9ab}{4\pi^2 N_A^3} = 1.05 \times 10^{-76} ab \tag{84}$$

where, C_W is in units of $J m^6$, a is in $dm^6 atm\, mol^{-2}$, and b is in $dm^3 mol^{-1}$. Equations (79), (80) and (84) enable us to compare the C_d and C_W values obtained for the same molecule using different experimental parameters. For example, for the methane molecule, CH_4, $C_W = 101 \times 10^{-79} J m^6$ was found from a and b gas constants obtained experimentally, and $C_d = 102 \times 10^{-79} J m^6$ was found from α, ε_o and I measurements as seen in Table 2.3. However, for a large molecule such as CCl_4, having a diameter of $\sigma = 0.55\, nm$, the exper-

Table 2.3 Relative magnitudes of van der Waals interactions

Relative magnitudes of van der Waals interactions between two identical molecules at 0°C

Molecule	Permanent dipole moment (D)	Polarizability $(10^{-24}\,cm^3)$	Ionization potential (eV)	London dispersion coefficient $(10^{-79}\,Jm^6)$ (Equation 80)	Keesom polar orientation coefficient $(10^{-79}\,Jm^6)$ (Equation 86)	Debye induced coefficient $(10^{-79}\,Jm^6)$ (Equation 88)
He	0	0.20	24.7	1.2	0	0
Ne	0	0.39	21.6	4	4	0
Ar	0	1.63	15.8	50	0	0
CH_4	0	2.60	12.6	102	0	0
Cyclohexane	0	11.00	11.0	1560	0	0
CO	0.12	1.99	14.3	68	0.0034	0.057
HCl	1.08	2.63	12.7	106	11	6
HBr	0.78	3.61	11.6	182	3	4
HI	0.38	5.44	10.4	370	0.2	2
CCl_4	0	10.50	11.1	1520	0	0
CH_3Cl	1.87	4.56	11.3	282	101	32
$(CH_3)_2CO$	2.87	6.33	10.1	486	1200	104
NH_3	1.47	2.26	10.2	63	38	10
H_2O	1.85	1.48	12.6	33	96	10

Relative magnitudes of van der Waals interactions between two different molecules at 0°C

Molecule 1	Molecule 2			
CCl_4	Cyclohexane	1510	0	0
CCl_4	NH_3	320	0	23
$(CH_3)_2CO$	Cyclohexane	870	0	89
CO	HCl	83	0.21	2.3
H_2O	HCl	64	70	11
$(CH_3)_2CO$	NH_3	185	315	33
$(CH_3)_2CO$	H_2O	135	493	35
Ne	CH_4	19	0	0
HCl	HI	197	1	7
H_2O	Ne	11	0	1
H_2O	CH_4	58	0	9

imental value obtained from a and b gas properties is $C_W = 2969 \times 10^{-79}\,Jm^6$, which is much larger than $C_d = 1520 \times 10^{-79}\,Jm^6$ from α, ε_o and I measurements as seen in Table 2.3. This is because CCl_4 has a stronger interaction potential than can be accounted for by applying the London equation to the molecular centers.

2.6.3 Comparison of Keesom, Debye and London interactions in polar molecules

As we have seen, London dispersion interactions, Keesom dipole–dipole orientation interactions and Debye dipole–induced dipole interactions are collectively termed van der Waals interactions; their attractive potentials vary with the inverse sixth power of the intermolecular distance which is a common property. To show the relative magnitudes of dispersion, polar and induction forces in polar molecules, similarly to Equation (78) for London Dispersion forces, we may say for Keesom dipole-orientation interactions for two dissimilar molecules using Equation (37) that

$$V_P(r) = -\frac{\mu_1^2 \mu_2^2}{3(4\pi\varepsilon_o)^2 kT r_{12}^6} = -\frac{C_P}{r^6} \tag{85}$$

Thus, the Keesom dipolar orientation interaction coefficient, C_P, can be written as

$$C_P = \frac{\mu_1^2 \mu_2^2}{3(4\pi\varepsilon_o)^2 kT} \tag{86}$$

We may say for Debye dipole–induced dipole interactions for two dissimilar molecules using Equation (60) so that

$$V_1(r) = -\frac{\left[\mu_1^2 \alpha_2 + \mu_2^2 \alpha_1\right]}{(4\pi\varepsilon_o)^2 r^6} = -\frac{C_1}{r^6} \tag{87}$$

and the Debye induced interaction coefficient C_1, can be written as

$$C_1 = \frac{\left[\mu_1^2 \alpha_2 + \mu_2^2 \alpha_1\right]}{(4\pi\varepsilon_o)^2} \tag{88}$$

Then, the total van der Waals force is the sum of these three contributions

$$V_W(r) = -\frac{C_W}{r^6} = -\left[\frac{C_d + C_P + C_1}{r^6}\right] \tag{89}$$

and the van der Waals interaction coefficient C_W, for two dissimilar polar molecules, can be written as

$$C_W = -\frac{\left[\dfrac{3\alpha_1\alpha_2}{2}\left(\dfrac{I_1 I_2}{I_1 + I_2}\right) + \dfrac{\mu_1^2 \mu_2^2}{3kT} + \left(\mu_1^2 \alpha_2 + \mu_2^2 \alpha_1\right)\right]}{(4\pi\varepsilon_o)^2} \tag{90}$$

For the same molecules where 1 = 2, Equation (90) reduces to

$$C_W = -\frac{\left[\dfrac{3\alpha^2 I}{4} + \dfrac{\mu^4}{3kT} + 2\mu^2\alpha\right]}{(4\pi\varepsilon_o)^2} \tag{91}$$

The values in Table 2.3 indicate that the most important contribution to van der Waals interactions results from the London dispersion interactions. Keesom dipolar orientation interactions are only operative for strongly polar and hydrogen-bonding substances such as acetone, water and ammonia, because the contribution of Debye induction forces is small for these molcules. For two dissimilar molecules, we again notice that polar forces are not important when the dipole moment is less than 1 Debye, and the induction forces are always much smaller than the dispersion forces. Agreement between the C_W values obtained from gas a, b constants and Equations (90) and (91) are surprisingly good for some molecules, even for NH_3 and H_2O. For asymmetrical molecules it is not so good because of the error in the molecular diameter. However, when two dissimilar molecules are considered, the agreement between the C_W values obtained from gas a, b constants and Equations (90) and (91) is not good but is usually intermediate between the values of 1–1 and 2–2. The C_W coefficient for 1–2 is often close to the geometric mean of 1–1 and 1–2, but this rule breaks down many times especially for hydrogen-bonding molecules. This will also be discussed in Sections 2.8 and 2.9.

2.6.4   van der Waals interactions in a medium

As we already discussed in Section 2.5.3 for excess polarizabilities of molecules dissolved in a solvent, the London dispersion interactions between molecules in a solvent medium may be very different from those of isolated molecules in free space. The intrinsic permanent dipole moment, μ, and polarizability of an isolated gas molecule, α, may be different in the liquid state or when dissolved in a medium, and this can only be determined by experiment.

In all the above London dispersion interaction expressions, the gas molecules were assumed to have only single ionization potentials (one absorption frequency) in free space, and this I cannot be used for interactions in a solvent. The polarizability of a molecule varies with the change in frequency, so there is a definite need for an expression to compute the dispersion forces between molecules that have a number of different absorption frequencies or ionization potentials. In practice we need expressions to be applied to interactions of two molecules, 1 and 2, which are present in a condensed solvent medium, 3. In this case the polarizabilities of 1 and 2 are the *excess polarizabilities* as given by Equation (48) in Section 2.5.3. McLachlan, in 1963, developed a generalized theory on van der Waals interactions, including dispersion, orientation and induction interactions in a single expression. This theory is also suitable for application in solvents. If the dielectric constants are expressed by using the refractive index of the molecules, then McLachlan's equation may be given as

$$V(r) = -\frac{\sqrt{3}h v_e a_1^3 a_2^3}{2r^6} \frac{\left(n_1^2 - n_3^2\right)\left(n_2^2 - n_3^2\right)}{\sqrt{\left(n_1^2 + 2n_3^2\right)}\sqrt{\left(n_2^2 + 2n_3^2\right)}\left[\sqrt{\left(n_1^2 + 2n_3^2\right)} + \sqrt{\left(n_2^2 + 2n_3^2\right)}\right]} \tag{92}$$

where a_1 and a_2 are the radii of molecules (1) and (2), n_1 and n_1 are their refractive indices, n_3 is the refractive index of the solvent medium, h is Planck's constant and v_e is the common adsorption frequency, which is assumed for simplicity. If we consider the interaction of like molecules (1 = 2) in a solvent medium (3), Equation (92) reduces to

$$V(r) = -\frac{\sqrt{3}h v_e a_1^6}{2r^6} \frac{\left(n_1^2 - n_3^2\right)^2}{\left(n_1^2 + 2n_3^2\right)\left(2\sqrt{n_1^2 + 2n_3^2}\right)} \tag{93}$$

Equations (92) and (93) show that the presence of a solvent medium other than a free space much reduces the magnitude of van der Waals interactions. In addition, the interaction between two dissimilar molecules can be attractive or repulsive depending on refractive index values. Repulsive van der Waals interactions occur when n_3 is intermediate between n_1 and n_2, in Equation (92). However, the interaction between identical molecules in a solvent is always attractive due to the square factor in Equation (93). Another important result is that the smaller the $(n_1^2 - n_3^2)$ difference, the smaller the attraction will be between two molecules (1) in solvent (3); that is the solute molecules will prefer to separate out in the solvent phase which corresponds to the well-known *like dissolves like* rule. However there are some important exceptions to the above explanation, such as the immiscibility of alkane hydrocarbons in water. Alkanes have $n_1 \cong 1.30$–1.36 up to 5 carbon atoms, and water has a refractive index of $n_3 \cong 1.33$, and very high solubility may be expected from Equation (93) since the van der Waals attraction of two alkane molecules in water is very small. Nevertheless, when two alkane molecules approach each other in water, their entropy increases significantly because of the very high difference in their dielectric constants and also the zero-adsorption frequency contribution; consequently alkane molecules associate in water (or vice versa). This behavior is not adequately understood.

van der Waals interactions in a solvent medium are not generally pairwise additive. Interaction between any two solute molecules is affected by the other molecules nearby (many-body effects) and this prevents simple additivity of pair potentials. This effect on the total interaction is usually less than 20% but it can be positive or negative and may be important in some situations such as interactions between large particles and surfaces in a medium (see Chapter 7).

2.7   Repulsive Interactions, Total Interaction Pair Potentials

All the different kinds of interactions we have discussed so far have been attractive forces. There must also be some repulsive force, otherwise molecules would collapse. Two types of repulsive force have been considered in the preceding sections: the Coulomb repulsion between like-charged ions, and the repulsion between atoms and molecules brought too close together which are very short-range. When repulsion occurs between two ions it is generally called *Born repulsion*. For the second example, the repulsive forces increase very suddenly as two atoms or molecules approach each other very closely, this is due to the repulsion between electron clouds overlapping at very small separations. This repulsion, which increases very steeply with decreasing distance, is due to the Pauli principle, which forbids outer electrons of one molecule from entering occupied orbitals of the other. This repulsion is called *hard core or Born repulsion*. We will use the name *hard core repulsion* for the interactions between two uncharged molecules in order to discriminate them from ionic repulsions. These repulsion interactions are quantum mechanical in nature and there is no general expression for their distance dependence, but some empirical potential functions are derived. Hard core repulsions are responsible for the magnitude of the densities of solids and liquids.

2.7.1 *van der Waals radius*

The repulsive force between atoms and molecules is negligible beyond a certain distance, and it is a characteristic of such repulsive forces that they increase very suddenly at very small separations with the approach of two atoms or molecules towards each other. It is this repulsive force that gives the atom or molecule a well-defined boundary at the *van der Waals radius*, r_W. If we consider the atoms and molecules as incompressible *hard spheres* then the repulsive force becomes infinite at a certain interatomic or intermolecular separation. It is possible to determine the molecular *van der Waals packing radius* for gases from pressure–volume–temperature data, and also from viscosity, solubility and spectroscopic data; for liquids from self-diffusion and compressibility measurements; and for solids from X-ray and neutron diffraction data. The values obtained from these different methods can differ by as much as 30% because each method measures a slightly different property. Thus, when pressure–volume–temperature data are used the van der Waals radius may be calculated from the coefficient b of the van der Waals equation of state by using Equation (83) and taking $r_W = (\sigma/2)$; this gives the smallest value because gas molecules approach each other more closely than their equilibrium separations. However, if we determine r_W values from the mean molecular volume occupied in the liquid state by assuming close packing conditions (each molecule is surrounded by 12 nearest neighbors), so that we may use the expression

$$\frac{4}{3}\pi r_W^3 = 0.7405\left(\frac{M}{N_A\rho}\right) \tag{94}$$

then this results in the highest r_W values, because the mobile molecules in the liquid state are 5–10% farther apart than in their corresponding close packed crystalline solids. When non-spherical molecules are considered, the *effective van der Waals radius* can be calculated by considering the bond angles and covalent radii of individual atoms in the molecule. This method can only be used for small and nearly spherical molecules, it cannot be used for cigar shaped molecules such as long chain alkanes. When the van der Waals radii of ions are measured, they are known as the *bare ion radius*, this is different (smaller) than the *hydrated ion radius* in water. In general, the van der Waals radius of most small molecules will lie between 0.1 and 0.2 nm, and the bare radii of anions are larger than those of cations due to the fact that anions have additional electrons in their outer shell adding additional repulsion force to increase the repulsive radius of the anion.

2.7.2 *Repulsive pair potentials*

Conveniently, the repulsive pair potential of two hard sphere molecules is represented by an inverse-power law of the type:

$$V_r(r) = +\left(\frac{r_W}{r}\right)^m = +\frac{B}{r^m} \tag{95}$$

where B is a constant, m is an integer usually taken to be between 8 and 16 and r is the intermolecular separation distance. When $r > r_W$ the value of $V_r(r)$ is approximately zero, whereas when $r_W > r$ the value of $V_r(r)$ is approximately infinite.

Contrary to attractive interactions, repulsive interactions are directional and their orientation dependence has a large effect on the way that molecules can pack together when they solidify. This property determines the lattice structure, rigidity and especially the solid density and melting point. Melting points of solids depend on the geometry of how they pack together; that is if the repulsive interactions and their shapes allow them to comfortably pack together then they have high melting points, and if they are packed loosely, their melting points will be low.

2.7.3   Total intermolecular pair potentials, Mie and Lennard-Jones potentials

It is possible to obtain the total intermolecular pair potential by simply summing the attractive and repulsive potentials. In 1903, Mie proposed a semi-empirical interaction pair potential in the form

$$V_{total}(r) = V_{attractive}(r) + V_{repulsive}(r) = -\frac{A}{r^n} + \frac{B}{r^m} \tag{96}$$

where A, B, n and m are constants, and $m > n$ by taking both repulsive and London dispersion attractive potentials into account. Mie's potential applies to two non-polar, spherically symmetric molecules, which are completely isolated in free space. The Lennard-Jones potential is a special case of a Mie potential

$$V_{total}(r) = -\frac{A}{r^6} + \frac{B}{r^{12}} \tag{97}$$

where the magnitude of $n = 6$ is taken from the well-known inverse sixth power dependence of the separation distance in van der Waals interactions, and $m = 12$ is used for repulsive interactions. Lennard-Jones potentials are very useful and still in use today. As can be seen in Figure 2.1, $V = V_{max}$ when $r = r_e$ and $r = r_o$ (the intermolecular distance) when $V_{total} = 0$. Let us try an indicative calculation by using some normal values such as $A = 1 \times 10^{-77}\,J\,m^6$ and $B = 1 \times 10^{-134}\,J\,m^{12}$. We know that $V_{total} = 0$ at $r_o = (B/A)^{1/6} = 3.16 \times 10^{-10}\,m$ which can easily be calculated from Equation (96). The value of V_{max} can be obtained when $(dV/dr) = 0$. This occurs at $r = r_e = (2B/A)^{1/6} = 3.55 \times 10^{-10}\,m$. By rearrangement, we also obtain $V_{max} = -(A^2/4B) = -(A/2\,r_e^6) = -2.5 \times 10^{-21}\,J = 0.607\,kT$ at 25°C. It is also possible to show that $r_e/r_o = 2^{1/6} = 1.12$. As can be seen in Figure 2.1, the maximum attractive force, F_{max} occurs at $(d^2V/dr^2) = 0$, that is when $r = r_s = (26B/7A)^{1/6} = 3.94 \times 10^{-10}\,m$. Then we can find $(r_s/r_o) = (27/6)^{1/6} = 1.24$. Since $F_{max} = -(dV/dr) = -6A/r^7 + 12B/r^{13}$, then by inserting $r_s = (26B/7A)^{1/6}$ one obtains $F_{max} = -(126A^2/169B)(7A/26B)^{1/6} = -1.894 \times 10^{-11}\,N$ (maximum attractive force). However, the conventional laboratory balances can measure down to about $10^{-9}\,N$, and we need specialized equipment such as Surface Force Apparatus (SFA) or Atomic Force Microscopy (AFM) to measure F_{max} in this case.

The Lennard-Jones expression may have some weak points in it but it gives good results when compared with the experimental lattice energies of non-polar solids. This may be due to the cancellation of a variety of erroneous factors.

2.7.4 Application of total intermolecular pair potentials in a liquid medium

In condensed phases such as liquids, two molecules are not isolated but have many other molecules in their vicinity. If these two molecules are dissolved in another solvent medium then the situation may be very complicated, and only by introducing simplifying assumptions is it possible to construct a simple theory of dense media and a solution using Lennard-Jones or similar total intermolecular pair potentials. Some important facts, which should be taken into account before applying any simplifying assumptions, are given below:

1 As we have already discussed in Section 2.5.3 for excess polarizabilities of molecules dissolved in a solvent, and in Section 2.6.4 for van der Waals interactions in a medium, when two molecules 1 and 2 are dissolved in a medium 3, the van der Waals forces between them are reduced because of the dielectric screening of the medium. This reduction is particularly important for liquids with high dielectric constants. The attraction force is decreased by a factor of the medium's ε_r for Keesom and Debye interactions and by a factor of ε_r^2 for London dispersion interactions. This strong reduction in the attractive pair potential means that the contributions of molecules further apart tend to be relatively minor, and each interaction is dominated only by contributions from its nearest neighbors.

2 The dielectric constant (relative permittivity) is a macroscopic property. If molecules 1 and 2 approach each other so closely, there will be no room for a solvent molecule between them. Then we may question the use of the medium's ε_r in the denominator of the related expressions, because if any polarization takes place, this may be due to a much smaller effective value of ε_r. On the other hand, some associated solvents apply forces on solute molecules, which are determined mainly by the orientation of the molecules of 1 and 2 in the medium 3, so that the resultant distribution functions are not only functions of r but also dependent on the orientation angle.

3 If molecules 1 and 2 are in complete contact in liquid 3, the hard-core repulsion interaction must also be considered. This sometimes gives rise to positive values in potential expressions.

4 According to Archimedes' principle, a dissolved solute molecule can approach another only by displacing the solvent molecules from its path. The net force therefore depends on the attraction between the solute molecules and also the solvent molecules. Thus the pair energy cannot be determined just by V_{12} but also by V_{13}, V_{23}, and V_{33}.

5 The interactions may be pair-wise additive or may not be additive due to the presence of other identical molecules of 1 and 2 nearby.

6 Coulomb forces operate between ions in electrolyte solutions. Coulomb interactions are much longer range and stronger than van der Waals interactions. However this does not mean that van der Waals interactions can always be ignored in electrolyte solutions. As we saw in Section 2.4.1, there are interactions between ions and dipolar molecules; in Section 2.5.4 interactions between ions and induced non-polar molecules; and in Section 2.5.5 interactions between ions and induced polar molecules. All of these interactions affect the total interaction potential.

In summary, the application of pair potentials is not a straightforward process, but also not particularly complicated. One should be careful to think about all the intrinsic properties of the interacting molecules and the medium (molecular diameter, shape, dipole moment, polarizability, ionization potential, density, density distribution function, surface tension, viscosity, the distance between interacting molecules, the presence of association in the solvent, ionic strength of the solution etc.), and also the possible consequences after the interaction takes place, before starting to model a specific molecular interaction. Sometimes, the simplest form of pair potential may fit the experimental findings very well due to the cancellation of the errors.

2.8   Hydrogen-bonding Interactions

Hydrogen-bonding interaction is a donor–acceptor (or Lewis acid–base) interaction involving hydrogen atoms. They exist between electronegative atoms such as O, N, F, Cl, with H atoms covalently bound to similar electronegative atoms. Examples of groups in molecules possessing such H atoms are —OH (in water, alcohols, carboxylic acids), —NH (in primary and secondary amines, amides), HF and HCl. A hydrogen bond is formed if the covalently bound H atom comes into contact with a strongly electronegative group in another molecule, for example with an oxygen (in water, alcohols, carboxylic acids, ketones, esters, ethers), a nitrogen, fluorine or chlorine atom in other molecules. The presence of such interactions was first proposed by German chemists during the 1902–1914 period, and also by M.L. Huggins in 1919; later, in 1930, L. Pauling called the —O . . . H interaction a *hydrogen bond* for the first time.

Hydrogen bond formation is both quantum mechanical and electrostatic in nature and applies to a wide range of interactions. Formerly it was believed that a hydrogen bond was only quasi-covalent in nature and the H atom (or proton) was shared between two electronegative atoms. Now it is also believed that, especially for weak H-bonds, it is predominantly an electrostatic interaction.

In general, when an H-bond is formed between an AH molecule and a B molecule, this can be shown schematically as A—H . . . B. The A and B terms are used for hydrogen-bond donor and acceptor atoms by analogy with the Brönsted *acid* and *base* terms, respectively, because the donor–acceptor interactions between atoms and molecules are related to their acidity and basicity. An acid is a substance, in classical Brönsted terms, whose molecule can donate a proton, whilst a base is a substance whose molecule can accept a proton. This is the protonic concept of acid–base behavior and cannot be applied to hydrogen-bonding interactions directly because there is no complete proton transfer in this process. Some scientists even believe that there is no sharing of the H atom between two electronegative atoms, but it remains covalently bound to its parent atom. It was experimentally determined that the covalent bond between A—H is weakened after the A—H—B hydrogen bond forms.

As we know from Section 2.1, electronegativity is the power of an atom in a molecule to attract electrons to itself. When two atoms have different degrees of electronegativity, the bond between them will have partial ionic character. If electrons are available to occupy the resulting molecular orbitals, the bond will have some covalent character. Hydrogen bonds are formed when the electronegativity of A relative to H in an A—H covalent bond is such that it withdraws electrons from the H atom and leaves the proton partially unshielded. The result is a molecule with a localized positive charge, δ^+, which can link up with the concentration of negative charge, δ^-, elsewhere in another molecule of the same kind. The key factor is the small effective size of the poorly shielded proton, since electrostatic forces vary as the second power of separation distance, r^{-2}. The acceptor B is also an electronegative atom and must have a lone-pair of electrons or polarizable π electrons in order to interact with this donor A—H bond. In H-bonding, a delocalized molecular orbital formation takes place in which A, H, and B each supply one atomic orbital from which three molecular orbitals are constructed. Initially, 1s orbitals in A and H are used to form the A—H bond in the AH molecule, and the B orbital originally accommodates the lone pair on B. After the H-bonding takes place, there are four electrons to accommodate and they occupy the two lowest molecular orbitals of the AHB fragment. Since the most anti-bonding, uppermost orbital is vacant, the energy lowering for the formation of the H-bond is a feasible result. The distance between AH and B is very important, and H-bonds are formed when AH touches B since H-bonding depends on orbital overlap. When the molecular contact is broken, H-bonds are also broken.

Later, G.N. Lewis suggested the electronic concept for acid–base behavior. An acid is now defined as a substance whose molecule can accept a pair of electrons to form a bond, and a base, one whose molecule can donate a lone pair of electrons to form a bond. This approach has much wider applicability for all donor–acceptor interactions and was also used to define both former Brönsted protonic acids and bases, and also to define such molecules as BF_3 and $AlCl_3$ as acids, and aromatic hydrocarbons as bases. The Lewis acid–base definition can be successfully applied to H-bonding especially in non-aqueous mediums. Thus, H-bonds are formed when any hydrogen-bond donor group (N—H, O—H, F—H) is near a hydrogen-bond acceptor group (Lewis base solvents containing N, O, F or Cl atoms). While F^- ions are good H-bond acceptors, the F atom in a C—F bond is not, so there is no H-bonding in fluorocarbon solvents.

Hydrogen bonding is very important in aqueous media. Water is an unusual molecule because of the presence of H-bonds within it. Water has unexpectedly high melting and boiling points and latent heat of vaporization for such a small molecule. Without H-bonding all the oceans would evaporate at ambient temperature. Water also has association properties, very low compressibility, and an unusually high solubility capacity both as a solvent and as a solute. All of this strange behavior can only be explained by the presence of H-bonds in water (see Section 2.8.2).

Hydrogen bonding also takes place between alcohol molecules, greatly increasing their boiling points: ethanol, CH_3CH_2—OH, and methoxymethane, CH_3—O—CH_3, both have the same molecular formula, C_2H_6O. They have the same number of electrons, and similar molecular lengths. The van der Waals attractions (both dispersion forces and dipole–dipole attractions) in each will be much the same. However, ethanol has a hydrogen atom attached directly to an oxygen – and that oxygen still has exactly the same two lone pairs as in a water molecule. Hydrogen bonding can occur between ethanol molecules, although not as effectively as in water. The strength of this H-bonding is limited by the fact that there is only one hydrogen in each ethanol molecule with a sufficient δ^+ charge. In methoxymethane, the lone pairs on the oxygen are still there, but the hydrogens are not sufficiently charged for hydrogen bonds to form. The boiling points of ethanol and methoxymethane show the dramatic effect of the hydrogen bonding: ethanol (with hydrogen bonding) has a b.p. of 78.5°C and methoxymethane (without hydrogen bonding) has a b.p. of −24.8°C. It is important to realise that H-bonding exists *in addition* to van der Waals attractions. For example, if we compare the structure of pentane ($CH_3CH_2CH_2CH_2$—CH_3) with butan-1-ol ($CH_3CH_2CH_2CH_2$—OH), we see that they contain the same number of electrons, and are much the same length. However, the higher boiling point of the butan-1-ol (b.p. = 117°C) over pentane (b.p. = 36.3°C) is due to the additional H-bonding property.

Hydrogen bonding is also very important in biological systems with N and O atoms in their structure and which also contain water as their normal solvent. H-bonds are formed between biological macromolecules such as amino acids (the building blocks of proteins) and nucleic acids and bases, which are present in all living matter. Without H-bonds all living things would disintegrate into inanimate matter, and all wooden materials would collapse. Examples range from simple molecules such as $HOOC$—CH_2—NH_2 (glycine) to large molecules like proteins and DNA. The ability of main chain carbonyl oxygen bonds to form hydrogen bonds with the main chain amino groups leads to the possibility of forming different secondary structures, such as the alpha helix in DNA. When the proteins are crystallized, more than half the volume of the crystal is often still occupied by water, and the water molecules near the protein crystal surface are in an ordered structure due to the presence of the H-bonds; these are known as *bound water*. In biochemistry, it is determined that the strongest H-bonds are formed as salt bridges such as N—H . . . O$=$C in proteins and P—OH . . . O$=$P bonds in nucleic acids. We often encounter the weak H-bonds in biological molecules because they are not rigid and can easily be broken. Recently very weak H-bonds such as C—H . . . O and the minor components of three-center bonds have been examined to give a better understanding of interactions in living systems.

There is also an important property of H-bonds: H-bonding interactions are not dependent on intrinsic material properties and can be active between two surfaces having no H-bonds within them. Unlike van der Waals interactions, H-bonding interactions are essentially asymmetrical and can only be satisfactorily treated by taking that asymmetry into account.

2.8.1 Properties of hydrogen bonds

There are many types of hydrogen bonding varying in strength between 1 and 40 kJ mol^{-1} and with bond lengths from 0.12 to 0.32 nm. Very strong H-bonds are formed ($\approx$40 kJ mol^{-1}) which are weaker than but resemble covalent bonds (>150 kJ mol^{-1}). Very weak H-bonds can also be formed which are comparable with van der Waals interactions ($\approx$1 kJ mol^{-1}), and these weak H-bonds interact mostly with electrostatic forces. If H-bonding is present, it sometimes dominates the other intermolecular interactions. In general, most H-bonds are weak attractions with a binding strength about one tenth of that of a normal covalent bond. Some authors divide the H-bonding interactions into three categories, *strong* (14–40 kJ mol^{-1}), *moderate* (4–15 kJ mol^{-1}) and *weak* (<4 kJ mol^{-1}), and examine their properties within these categories in order to prevent any confusion.

Two hydrogen-bonded electronegative atoms can approach each other more closely than the sum of their molecular radii. In other words, the intermolecular distance between B . . . H is less than the value expected from summing the two van der Waals radii of the B and H atoms. Hydrogen bonds are directional and thus result in three- or two-dimensional structures, or one-dimensional chains.

Hydrogen bond strength decreases with the increase in temperature and thermal motion of the atoms involved, and after a certain temperature, the H-bonds break. Conversely, the formation of H-bonds liberates heat, and H-bond formation energies can be measured by determination of the enthalpy of mixing or dilution of donor–acceptor liquid mixtures by calorimetry, or by infrared shifts. Good correlation is obtained between H-bond enthalpies falling between 3 and 20 kcal mol^{-1} with discrepancies of 2–3 kcal mol^{-1} being found from different measurements. It was determined that H-bond energies are neither additive nor transferable.

The measurement of any physical property that is sensitive both to H-bonding and temperature provides access to thermodynamic properties of the H-bonds formed. The most important spectroscopic methods are infrared (IR), Raman, nuclear magnetic resonance (NMR) spectroscopy, X-ray diffraction, neutron diffraction and neutron inelastic scattering methods. In addition, some special techniques such as gas-phase microwave rotational spectroscopy and deuteron nuclear quadrupole coupling have also been applied successfully. More recently, computational chemistry has become an important tool for understanding H-bonding, and especially ab initio molecular orbital calculations are applied. This method seeks minimum energy intermolecular geometry by solving the wave equation using a linear combination of atomic orbitals known as the LCAO approximation. These atomic orbitals have to be expressed mostly by using Gaussian distribution functions to facilitate the computer calculations. The translational, rotational and vibrational energies have to be calculated in order to relate the computational chemistry results with experimentally measured enthalpies of gas-phase H-bonding interactions. H-bond energies in the gas phase can also be related to the difference in proton affinities for the donor and acceptor; one of these will be an ion and the other a molecule. Proton affinities can be calculated by applying the ab initio computational methods.

2.8.2 Hydrogen bonds in water

Water is the most important liquid on earth. As stated above, water has unexpectedly high melting and boiling points and latent heat of vaporization, for a small non-ionic molecule. This shows that there is one more interaction present than for normal polar molecules. Water molecules are exceptionally prone to form H-bonds because the four pairs of electrons around the O atom occupy sp^3 hybrid orbitals that project outwards as though towards the vertexes of a tetrahedron. Hydrogen atoms are at two of these vertexes, which accordingly exhibit localized positive charges, while the other two vertexes exhibit somewhat more diffuse negative charges. Thus, a water molecule has a tetrahedral coordination (four nearest neighbors per molecule) rather than a close-packed structure (12 nearest neighbors per molecule), and the bonds in water are directional. Each water molecule can therefore form H-bonds with four other water molecules; in two of these bonds the central water molecule provides the bridging protons, and in the other two the attached water molecules provide them, as seen in Figure 2.7. In bulk liquid water, each molecule can form a maximum of four hydrogen bonds with neighbors. It actually achieves an average of 3.4 bonds per molecule. However, owing to the thermal agitation, the H-bonds between

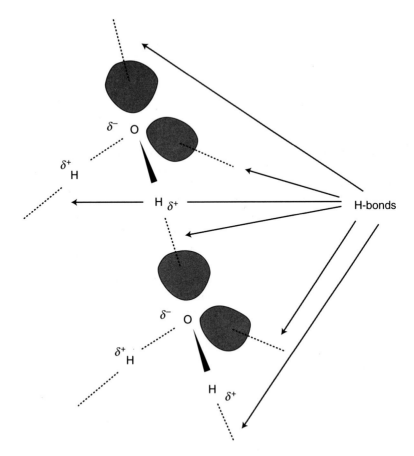

Figure 2.7 Hydrogen bonds between water molecules: (δ^+) and (δ^-) are charges on the hydrogen and oxygen atoms respectively. A water molecule has four nearest neighbors per molecule and can form hydrogen bonds with four other water molecules; in two of these bonds the central water molecule provides the bridging protons, and in the two other bonds, the attached water molecules provide the required protons.

adjacent water molecules are continually being broken and reformed (reorienting every $\sim 10^{-12}$ s) but even so, at any instant, the water molecules are combined in definite clusters which are made of four to seven water molecules that resemble short-lived ice-like lattices.

In the solid state (ice), these clusters are large and stable, and form the lattice crystals. The density of ice is less than the liquid water indicating that the molecules are farther apart than in the liquid. Thus, this indicates that the presence of H-bonds in ice results in a longer intermolecular distance between molecules than that of liquid water. In addition, the proton conductivity in ice is also higher than in liquid water showing that the ice lattice has some easy pathways for the movement of charges. Thus it is a good idea to measure the intermolecular distances in ice: the intramolecular distance between the covalently bound O—H atoms is about 0.100 nm, as expected, but the intermolecular distance between O . . . H atoms is found to be about 0.176 nm, which is much less than the 0.260 nm expected from summing the two van der Waals radii of the O and H atoms.

The characteristic hexagonal pattern of an ice crystal arises from the tetrahedral arrangement of the four H-bonds in which each water molecule can participate. Ice crystals have extremely open structures with only four nearest neighbors around each molecule. When many water molecules are involved in a large system, their equilibrium configuration can only be solved by using a computer and this process can also account for the highly open ice structure.

In liquid water, the tetrahedral ice structure is retained somewhat, but of course, the structure is more labile and disordered. Because the molecular clusters are smaller and less stable in the liquid state, water molecules on average are packed more closely together than are ice molecules (the average number of nearest neighbors increases approximately to five) and liquid water has the higher density; hence ice floats. This is known as *association* of liquid water, by bringing water molecules together to form three-dimensional *cluster* structures. Other H-bonding liquids such as formamide, alcohols, ammonia, HF etc. also show association properties. Formamide gives two-dimensional layered associated structures where two N—H . . . O═C bonds form these layers. Hydrogen fluoride and alcohols give one-dimensional chain structures.

As we know from carbon and silicone chemistry, only tetrahedral coordination allows the formation of a three-dimensional network, and thus the formation of an associated tetrahedral ice and liquid water structure is the main reason for all the unusual properties of water. Some examples are given:

(i) The density of liquid water shows a maximum at +4°C, which is an unusual property: the density of liquid water increases from 0°C to a maximum at +4°C as large clusters of water molecules are broken up into smaller ones which occupy less space in the water associate; only beyond +4°C does the normal thermal expansion of the liquid decrease its density with increasing temperature.

(ii) The high dielectric constant of liquid water increases as it solidifies by freezing into ice. This is contrary to most molecules whose liquid dielectric constant falls abruptly during freezing due to the decrease in molecular rotations with the decrease in thermal energy.

(iii) Water has a very low compressibility.

(iv) Water has unusual solubility properties both as a solvent and as a solute. Water dissolves ionic salts very readily because it is good at shielding charged atoms from one another by forming H-bonds with anions and also cations.

2.9 Hydrophobic and Hydrophilic Interactions

Some molecules such as hydrocarbons, oils, fats, inert atoms, fluorocarbons and some polymers that are incapable of forming H-bonds and do not like to interact with water, are called *hydrophobic* molecules. Hydrophobic means water-fearing or water-repelling. The more general name is *lyophobic* or *solvophobic*, which means solvent-repelling, and when the solvent is water, the solute is called *hydrophobic*. Similarly, if a solute is solvent-attracting, then it is called *lyophilic* or *solvophilic*, and if the solvent is water, the solute is called *hydrophilic*. Hydrophilic molecules are mostly polar and have a high H-bonding capacity. Hydrophobic and hydrophilic interactions are very important in surface and colloid chemistry and biochemistry, and they determine the three-dimensional shape of polymers and biologically important molecules such as proteins and cell membranes.

2.9.1 Hydrophobic effect

When water molecules come into contact with a non-polar hydrophobic molecule, one or more of the four charges of the water molecule will have to point towards the inert solute molecule, and thus be lost to H-bond formation. This is unfavorable thermodynamically and if the non-polar solute molecule is small, water molecules prefer to pack around it without giving up any of their H-bonding sites. Then, water molecules form cages around the hydrophobic molecule, known as *clathrate* and these are labile structures. Although the H-bonds in the cage structure are not stronger than in bulk water, the water molecules are more ordered. As we know from H-bonding interactions, in bulk water, each molecule can form an average of 3.4 H-bonds per molecule (between 3.1 and 3.5 bonds) with its neighbors. Thermodynamically, if a molecule is immobilized when it forms a bond, it must lose entropy, but in bulk water the molecules remain highly mobile. They are constantly rotating around the single bonds, and also breaking bonds and reforming them to new partners. The molecules in the bulk water therefore retain a high state of entropy, which is favorable from free energy considerations (see below). However, when a cage is formed around a small hydrophobic molecule, the water molecules actually have four coordination H-bonds thus rendering them even more ordered than in the bulk liquid, and their entropy decreases.

Water–hydrophobic molecule interactions are driven entropically rather than enthalpically and are affected greatly by temperature. The unfavorable entropy resulting from the ordering of water around a hydrophobic solute provides a driving force for solutes to aggregate and thereby reduces the hydrophobic material surface area available to the surrounding water molecules. In thermodynamic terms, the free energy of transfer of a hydrophobic compound from some reference state, such as its pure solvent state, into water, ΔG_{tr}, is made up of an enthalpy, ΔH_{tr}, and an entropy, ΔS_{tr}, term. At room temperature, the enthalpy of transfer from a hydrophobic solvent into water is negligible; the interaction enthalpies are nearly the same in both cases. The entropy term however is negative because water tends to form ordered cages around the hydrophobic molecule, and this leads to a decrease in total entropy, ΔS_{tr}. According to the free energy equation ($\Delta G_{tr} = \Delta H_{tr} - T\Delta S_{tr}$, see Chapter 3), if we assume that $\Delta H_{tr} \cong 0$ and ΔS_{tr} is negative, then ΔG_{tr} is positive, which means that a hydrophobic molecule does not favor being transferred into the water phase and thus will prefer to aggregate in a separate phase, and it is for this reason that hydrophobic compounds are so sparingly soluble in water. In practice, hydrophobic molecules, in contact with water, prefer to interact with each other instead of the water and try to minimize their total surface area and form a water-immiscible phase. For example, when we add some drops of a hydrocarbon to water, the drops combine to form a larger drop. The attractiveness of the water molecules for each other then has the effect of squeezing the oil drops together to form a larger drop. The ability of water to push together hydrophobic molecules is called the *hydrophobic effect* and this figures prominently in biochemistry.

At high temperatures (~110°C), because of the contribution of thermal motions, the clathrate cages are broken, and the entropy contribution tends towards zero. Thus, when the temperature increases, the water molecules around the hydrophobic groups become disoriented and the hydrophobic groups attract each other more strongly. The enthalpy of transfer, however, is now positive (and ΔG_{tr} may also be unfavorable according to the temperature, as we see from the free energy equation). In summary, because the temperature dependences of entropy and enthalpy are not the same, there is some temperature at which the hydrophobic effect is strongest, and it decreases at temperatures above and below this temperature. The decrease in the strength of the hydrophobic effect with decreasing temperatures is probably the major cause of cold denaturation in proteins.

In biology, the hydrophobic effect is considered to be the major driving force for the folding of globular proteins and forming protein–protein interactions. At the hydrophobic protein surface there is an interesting constraint: the water molecules next to this surface cannot make hydrogen bonds to the protein (except very few donor or acceptor sites), but they can still make bonds to neighboring water molecules. However, here they either lose a couple of hydrogen bonds, or orient themselves in a position that will place their hydrogen bonding sites facing the solvent. In thermodynamic terms their entropy is decreased. Then, the water molecules next to the surface actually achieve about three hydrogen bonds each, but in order to do this they are much more constrained than those in the bulk solution. These water molecules are partially *frozen*; they have less mobility than molecules in the bulk water. When two proteins come together to make the interface, this layer of *frozen* water is released from each surface. The water molecules can now make H-bonds in all directions, and they have more freedom, i.e. increased entropy and a lower (more favorable) free energy.

2.9.2   Hydrophobic interactions

Hydrophobic interaction describes the strong attraction between hydrophobic molecules or surfaces in water. This is often stronger than their attraction in free space. This is contradicted by the prediction of a decrease in the van der Waals attraction in water medium. If we compare hydrophobic interactions with other molecular interactions already present, the interaction between a hydrophobic molecule and water is actually attractive, due to dispersion interactions, and these dispersion interactions between water and hydrophobic molecules are not very different from those between water–water or hydrophobic–hydrophobic. However, the interaction of water with itself is much more attractive due to the H-bonding interactions, and water then has the effect of squeezing the hydrophobic molecules to attract each other to form a larger aggregate. When two hydrophobic molecules come together, the water is also ejected into the bulk thereby reducing the total free energy of the system.

The attraction between hydrocarbons or fluorocarbons in air (mainly van der Waals) increases very much when we place these molecules in water. For example, for two contacting methane molecules in free space the interaction pair potential energy is -2.5×10^{-21} J, whereas in water it is -14×10^{-21} J. On the other hand, experimental evidence suggests that the hydrophobic attraction between macroscopic surfaces is operative at a surprisingly long range, and it decays exponentially with a characteristic decay length of 1–2 nm in the range 0–10 nm, and sometimes more gradually farther out. At small separations it has been determined that the hydrophobic attraction is much stronger than the van der Waals attraction. The magnitude of the hydrophobic attraction decreases with reducing hydrophobicity in the solute molecule. It is also found experimentally that, for different hydrocarbons, the surface areas of the hydrocarbons determine the magnitude of the free energy transfer. An increase in the temperature can increase the hydrophobic attraction. It has been assumed that long-range hydrophobic attraction forces are responsible for the coalescence of gas bubbles in water, since there are strong similarities between water at a hydrophobic surface and at the liquid/vapor interface. This would suggest that the rate of coalescence of bubbles is related to the hydrophobicity of the gas within the bubbles. Hydrophobic particle attachment to rising air bubbles is the basic mechanism in *froth flotation* which is used to separate the hydrophobic and hydrophilic particles in the mineral industry.

Hydrophobic attraction is also very important in understanding molecular self-assembly, micelle-formation, biological membrane structure and protein conformations, which will be discussed in Chapters 5, 7, 9 and 10.

2.9.3 Hydrophilic interactions

A *hydrophilic* molecule or a hydrophilic group of a molecule is generally polar and capable of H-bonding, enabling it to dissolve more readily in water than in oil or other hydrophobic solvents. It has a strong affinity with water; tending to dissolve in, mix with, or be wetted by water. When two solute molecules repel each other in water it is called *hydrophilic interaction*. However there is no phenomenon known as the *hydrophilic effect*.

Strongly hydrated ions, and electronegative atom (such as N and O) containing molecules are hydrophilic. These groups should associate with the H-bonded water network. However, contrary to some claims, a polar group is not necessarily always hydrophilic. The most important hydrophilic molecules are alcohols, glycols, glycerol, glucose and other sugars, urea, soluble proteins, polyethylene oxide and acrylic acid and methacrylic acid containing polymers. The most important hydrophilic groups are carboxylate ($-COO^-$), sulfonate ($-SO_3^-$), sulfate ($-SO_4^{-2}$), phosphate ester ($-OPO_2^-O-$) anions; dimethyl ammonium $[-N^+(CH_3)_2]$, trimethyl ammonium $[-N^+(CH_3)_3]$ cations; and hydroxyl ($-OH$), amine ($-NH_2$), amine oxide $[-NO(CH_3)_2]$, suphoxide ($-SOCH_3$), and phosphine oxide $[-PO(CH_3)_2]$ polar compounds. Hydrophilic groups prefer to be in contact with water rather than with each other. Hydrophilic group containing molecules are often *hygroscopic* in that they absorb water from water vapor present in the air or in other gases. When dissolved, water molecules surround the hydrophilic molecules and this layer is called a *hydration shell*. In contrast to the effect of hydrophobic molecules, hydrophilic molecules have a disordering effect on water molecules, surrounding them, thus causing drastic effects on other solute molecules dissolved in the water. For example, when urea is dissolved in water, previously dissolved proteins unfold.

Certain polar groups, which are good hydrophilic candidates, do not show any hydrophilic properties when attached to a long hydrocarbon (hydrophobic) chain. These are alcohol (—OH), ether (—OCH$_3$), mercaptan (—SH), amine [—NH(CH$_3$)], amide (—CONH$_2$), nitro (—NO, —NO$_2$), aldehyde (—CHO) and ketone (—COCH$_3$) groups. The hydrophobic portion of the molecule prevents their hydrophilic activity. Most surfactants and biological lipids are made of a combination of hydrophilic and hydrophobic groups. These molecules are called *amphiphilic* molecules, one end containing a hydrophilic group while the rest of the molecule is hydrophobic, usually a long hydrocarbon chain (see Section 5.4). Such molecules form spherical micelle structures in water due to the fact that their hydrophobic parts associate with one another whilst the hydrophilic parts expand to interact with the water.

References

1. Israelachvili, J. (1991). *Intermolecular & Surface Forces* (2nd edn). Academic Press, London.
2. Hirschfelder, J.O., Curtiss, C.F. and Bird, R.B. (1954). *Molecular Theory of Gases and Liquids.* Wiley, New York.
3. Pauling, L. (1960). *The Nature of the Chemical Bond* (3rd edn). Cornell University Press, Ithaca.
4. Verwey, E.J.W. and Overbeek, J.Th.G. (1948). *Theory of Stability of Lyophobic Colloids.* Elsevier, Amsterdam.
5. Adamson, A.W. and Gast, A.P. (1997). *Physical Chemistry of Surfaces* (6th edn). Wiley, New York.
6. Lyklema, L. (1991). *Fundamentals of Interface and Colloid Science* (vols. I and II). Academic Press, London.
7. Erbil, H.Y. (1997). Interfacial Interactions of Liquids. In Birdi, K.S. (ed). *Handbook of Surface and Colloid Chemistry.* CRC Press, Boca Raton.
8. Stokes, R.J. and Evans, D.F. (1997). *Fundamentals of Interfacial Engineering.* Wiley-VCH, New York.
9. Jeffrey, G.A. (1997). *An Introduction to Hydrogen Bonding.* Oxford University Press, Oxford.
10. Pimental, G.C. and McClellan, A.L. (1960). *The Hydrogen Bond.* Freeman, San Francisco.
11. Smyth, C.P. (1955). *Dielectric Behavior and Structure.* McGraw-Hill, New York.

Chapter 3
Thermodynamics of Interfaces

3.1 Introduction of Thermodynamical Concepts

Thermodynamics is the branch of science that is concerned with the principles of energy transformation in *macroscopic* systems. Macroscopic properties of matter arise from the behavior of a very large number of molecules. Thermodynamics is based upon experiment and observation, summarized and generalized in the *Laws of Thermodynamics*. These laws are not derivable from any other principle: they are in fact improvable and therefore can be regarded as assumptions only; nevertheless their validity is accepted because exceptions have never been reported. These laws do not involve any postulates about atomic and molecular structure but are founded upon observation about the universe as it is, in terms of instrumental measurements. In order to represent the state of a gas or a liquid or a solid system, input data of average quantities such as temperature (T), pressure (P), volume (V), and concentration (c) are used. These averages reduce the enormous number of variables that one needs to start a discussion on the positions and momentums of billions of molecules. We use the thermodynamic variables to describe the state of a system, by forming a *state function*:

$$P = f(V, T, n) \tag{98}$$

This simply shows that there is a physical relationship between different quantities that one can measure in a gas system, so that gas pressure can be expressed as a function of gas volume, temperature and number of moles, n. In general, some relationships come from the specific properties of a material and some follow from physical laws that are independent of the material (such as the laws of thermodynamics). There are two different kinds of thermodynamic variables: *intensive variables* (those that do not depend on the size and amount of the system, like temperature, pressure, density, electrostatic potential, electric field, magnetic field and molar properties) and *extensive variables* (those that scale linearly with the size and amount of the system, like mass, volume, number of molecules, internal energy, enthalpy and entropy). Extensive variables are additive whereas intensive variables are not.

In thermodynamic terms, the object of a study is called the *system*, and the remainder of the universe, the *surroundings*. Amounts of the order of a mole of matter are typical in a system under consideration, although thermodynamics may remain applicable for considerably smaller quantities. The imaginary envelope, which encloses the system and separates it from its surroundings is called the *boundary* of the system. This boundary may serve either to isolate the system from its surroundings, or to provide for interaction in specific ways between the system and surroundings. In practice, if a reactor is used to carry out a chemical reaction, the walls of the reactor that are in contact with the thermostated liquid medium around the reactor may be assumed to be the surroundings of the experimental system. For particles such as colloids, the medium in which they are immersed may act as the surroundings, provided nothing beyond this medium influences the particle.

An *isolated system* is defined as a system to or from which there is no transport of matter and energy. When a system is isolated, it cannot be affected by its surroundings. The universe is assumed to be an isolated system. Nevertheless, changes may occur within the system that are detectable using measuring instruments such as thermometers, pressure gauges etc. However, such changes cannot continue indefinitely, and the system must eventually reach a final static condition of *internal equilibrium*.

If a system is not isolated, its boundaries may permit exchange of matter or energy or both with its surroundings. A *closed system* is one for which only energy transfer is permitted, but no transfer of mass takes place across the boundaries, and the total mass of the system is constant. As an example, a gas confined in an impermeable cylinder under an impermeable piston is a closed system. For a closed system, which interacts with its surroundings, a final static condition may be reached such that the system is not only internally at equilibrium but also in *external equilibrium* with its surroundings. A system is in *equilibrium* if no further spontaneous changes take place at constant surroundings. Out of equilibrium, a system is under a certain stress, it is not relaxed, and it tends to equilibrate. However, in equilibrium, the system is fully relaxed. If a system is in equilibrium with its surroundings, its macroscopic properties are fixed, and the system can be defined as a given *thermodynamic state*. It should be noted that a *thermodynamic state* is completely different from a *molecular state* because only after the precise spatial distributions and velocities of all molecules present in a system are known can we define a *molecular state* of this system. An extremely large number of molecular states correspond to one thermodynamic state, and the application of *statistical thermodynamics* can form the link between them.

An *open system* is a system in which both matter and energy transport are permitted through the boundaries. When a closed or an open system is displaced from an equilibrium state, to another state, it undergoes a *process*, a change of state, which continues until its properties attain new equilibrium values. During such a process the system may be caused to interact with its surroundings so as to interchange energy in the forms of heat and work. Except for isolated systems, all systems can be transformed from one thermodynamic state to another by changing the properties of the surroundings.

3.1.1 Thermodynamical expressions for closed systems

The *First Law of Thermodynamics* is the law of conservation of energy; it simply requires that the total quantity of energy be the same both before and after the conversion. In other words, the total energy of any system and its surroundings is conserved. It does not place any restriction on the conversion of energy from one form to another. The interchange of heat and work is also considered in this first law.

The total energy of a system is called its *internal energy*, U. The internal energies is the sum of the potential and kinetic energies of the molecules and sub-molecular particles composing the system, and it is a state function so that its value depends only on the properties that determine the current system and is independent of how that state has been prepared. If the system is not moving, the changes in internal energy include only the changes in the potential energy of the system and also the energy transferred as heat. (In chemistry, the potential energy of the system is the sum of the potential energy of the molecules in the system and also the energy changes caused by the rearrangements of molecular configurations.) If the system is moving, the kinetic energy of this translation is also added to U.

In principle, the internal energy of any system can be changed, by heating or doing work on the system. The First Law of Thermodynamics requires that for a closed (but not isolated) system, the energy changes of the system be exactly compensated by energy changes in the surroundings. Energy can be exchanged between such a system and its surroundings in two forms: *heat* and *work*. Heat and work have the same units (joule, J) and they are ways of transferring energy from one entity to another. A quantity of heat, Q, represents an amount of energy in transit between a system and its surroundings, and is not a property of the system. Heat flows from higher to lower temperature systems. Work, W, is the energy in transit between a system and its surroundings, resulting from the displacement of external force acting on the system. Like heat, a quantity of work represents an amount of energy and is not a property of the system. Temperature is a property of a system while heat and work refer to a process. It is important to realize the difference between temperature, heat capacity and heat: *temperature*, T, is a property which is equal when heat is no longer conducted between bodies in thermal contact and can be determined with suitable instruments (thermometers) having a reference system depending on a material property (for example, mercury thermometers show the density differences of liquid mercury metal with temperature in a capillary column in order to visualize and measure the change of temperature). The *heat capacity* at constant pressure, $C_P = \left(\dfrac{dU}{dT} \right)_P$ corresponds to how much the temperature changes as heat is added to a body that remains in the same state.

Suppose any closed system (thus having a constant mass) undergoes a process by which it passes from an initial state to a final state. If the only interaction with its surroundings is in the form of transfers of heat, Q, and work, W, then only the internal energy, U, can be changed, and the First Law of Thermodynamics is expressed mathematically as

$$\Delta U = U_{final} - U_{initial} = Q + W \tag{99}$$

where Q and W are quantities inclusive of sign so that when the heat transfers from the system or work is done by the system, we use negative values in Equation (99). Processes where heat should be given to the system (or absorbed by the system) $(Q > 0)$ are called *endothermic* and processes where heat is taken from the system (or released from the system) $(Q < 0)$ are called *exothermic*. The total work performed on the system is W. There are many different ways that energy can be stored in a body by doing work on it: volumetrically by compressing it; elastically by straining it; electrostatically by charging it; by polarizing it in an electric field $\bar{E}$; by magnetizing it in a magnetic field $\bar{H}$; and chemically, by changing its composition to increase its chemical potential. In surface science, the formation of a new surface area is also another form of doing work. Each example is a different type of work – they all have the form that the (differential) work performed is the change in some extensive variable of the system multiplied by an intensive variable. In thermodynamics, the most studied work type is pressure–volume work, W_{PV}, on gases performed by compressing or expanding the gas confined in a cylinder under a piston. All other work types can be categorized by a single term, *non-pressure–volume work*, W_{non-PV}. Then, W is expressed as the sum of the pressure–volume work, W_{PV}, and the non-pressure–volume work, W_{non-PV}, when many types of work are operative in a process, which we will see in Section 3.2.1.

Equation (99) states that the internal energy, ΔU depends only on the initial and final states and in no way on the path followed between them. In this form, heat can be defined as *the work-free transfer of internal energy from one system to another*. Equation (99) applies both to *reversible* and *irreversible* processes. A *reversible process* is an infinitely slow process during which departure from equilibrium is always infinitesimally small. In addition, such processes can be reversed at any moment by infinitesimal changes in the surroundings (in external conditions) causing it to retrace the initial path in the opposite direction. A reversible process proceeds so that the system is never displaced more than differentially from an equilibrium state. An *irreversible process* is a process where the departure from equilibrium cannot be reversed by changes in the surroundings. For a differential change, Equation (99) is often used in the differential form

$$dU = dQ + dW \tag{100}$$

for reversible processes involving infinitesimal changes only. The internal energy, U is a function of the measurable quantities of the system such as temperature, volume, and pressure, which are all state functions like internal energy itself. The differential dU *is* an exact differential similar to dT, dV, and dP; so we can always integrate $[\int_i^2 f(U)dU]$ expression. The Euler reciprocity relation is often used in a number of derivations of thermodynamic equations from exact differentials. If we apply Euler reciprocity, i.e. for the case, $U = f(T, V)$, then the total differential, dU can be defined in terms of the partial derivatives and the differentials of independent variables:

$$dU = \left(\frac{\delta U}{\delta T}\right)_V dT + \left(\frac{\delta U}{\delta V}\right)_T dV \tag{101}$$

On the other hand, dQ and dW are inexact differentials because they cannot be obtained by differentiation of a function of the system.

All the state variables and the changes in thermodynamic quantities during a process are measurable in principle. The value of ΔU is measurable, but the absolute values of U cannot be obtained. Thus, the thermodynamic data are reported with respect to certain internationally agreed *standard or reference state* values. Normally, a temperature of 25°C and a pressure of 1 bar $= 10^5$ pascal (Pa $=$ N/m^2) are taken as standard conditions, and for solutions, a *molar concentration, c,* of 1 mol/dm^3 is used as a reference state.

Fluids (liquid or gas) are known as *PVT* systems, where the macroscopic properties at internal equilibrium can be expressed as functions of temperature, pressure and composition only. In accepting this model, one assumes that the effects of fields (e.g., electric, magnetic or gravitational) are negligible and that surface and viscous-shear effects are unimportant. The *PVT* system serves as a satisfactory model in an enormous number of practical applications. As we know, for a gas confined in an impermeable cylinder under an impermeable piston, the contribution to dW is the pressure–volume work, $dW_{PV} = -PdV$, where P is the external pressure and V is the volume of the system. There is no non-pressure–volume work, W_{non-PV} in this system, so that $W = W_{PV}$. If the system is expanded, $dV > 0$ then the work is done by the system $dW < 0$; hence the minus sign appears in the $dW = -PdV$ expression. In order to maintain a mechanical equilibrium, the inner and outer pressures must be equal. If a system is expanding, the pressure of the surroundings (external pressure) must be slightly lower than the pressure of the system, by an amount dP. As $P \gg dP$, we may assume that $(P + dP)dV \cong PdV$, and in a reversible process the pressure of the system may be assumed nearly equal to the external pressure $(P = P_{system})$. However, for an irreversible process the pressure of the system is not defined and only the external pressure can be used in the calculations.

If no work is done during a process ($dW = 0$) then the exchanged heat, Q, would be equal to ΔU, according to Equations (99) and (100), and would only show the increase or decrease in the internal energy of the system by the exchange of heat. If only the volume expansion ($P\Delta V$) work is done by the system when the external pressure is held constant then, Equation (99) can be written as

$$\Delta U = U_2 - U_1 = Q_P - P\Delta V = Q_P - P(V_2 - V_1) \tag{102}$$

where Q_P is the heat absorbed at constant pressure, P. Then, by rearrangement,

$$Q_P = (U_2 + PV_2) - (U_1 + PV_1) \tag{103}$$

During most experimental calorimetric measurements, the external pressure is kept constant (generally under approximately 1 bar) rather than the volume. For such constant pressure conditions, a new term, *enthalpy*, H, is defined as a new thermodynamic function,

$$H \equiv U + PV \tag{104}$$

H represents the available thermal energy at constant pressure. By combining Equations (103) and (104) we obtain

$$Q_P = H_2 - H_1 = \Delta H \tag{105}$$

Since, U, P and V are all state functions, H is also a state function. In differential form, Equation (104) may be written as

$$dH = dU + d(PV) = dU + PdV + VdP \tag{106}$$

and since we know $dU = dQ - PdV$ from the First Law, it follows that

$$dH = dQ + VdP \tag{107}$$

and hence $dP = 0$ at constant pressure, which gives $dH = (dQ)_P$ as the enthalpy definition implies.

For many systems, the non-pressure–volume work term, W_{non-PV} is generally used (see Section 3.2 for surface area expansion) and we need to consider other ways that work can be done on a body. The rate at which work is done is always of the form $dW = F\,dx$ where dx is an extensive property and represents the change in the *extent* of some quantity, and F is a force that resists this change. This should be familiar from, for example, springs and mechanical objects. We might also think of dx as the *change in the charge held in a capacitor*, or the *change in the number of sodium ions in an anode*, or the change in the number of solvent molecules in a polymer. Each one of these examples represents a different way for a material to store internal energy – and each one has an intensive variable (generalized) force associated with it.

The *Second Law of Thermodynamics* is concerned with the direction of natural processes. In nature, rivers run from the mountains to the sea, never in the opposite way. In general, when non-relaxed systems are left to themselves, they always spontaneously change towards equilibrium. The directions of these processes are neither given by the First Law nor by the direction in which the total energy changes. From the First Law alone, it would be possible for a river to flow upstream, obtaining the required energy by extracting heat from the surroundings, but of course this never happens. Therefore there must be another independent law. The *Second Law of Thermodynamics* was derived to fill this gap, and this dictates the direction of processes by introducing a new state function the *entropy, S*. In combination with the First Law, the Second Law enables us to predict the natural direction of any process; in other words, it gives information as to whether or not the specified change can occur spontaneously.

There are many statements of the Second Law, two of the most common are:

1 No process is possible that consists solely of the transfer of heat from one temperature to a higher one.
2 It is impossible for a system operating in a cycle and connected to a single heat reservoir to produce positive work in the surroundings.

The Second Law of Thermodynamics is concerned mainly with reversible processes. It was realized that some work is always wasted as heat during laboratory experimentation and in industrial processes. In other words, not all of the work that is done on a body can be stored as internal energy and some of it leaks out as heat. The amount of wasted work is minimized when a process is carried out *reversibly*. A reversible process happens very slowly and the system is always in equilibrium (i.e., the intensive variables are uniform). As we know, there is a limiting process in mathematics when minimizing something, and we can define a new state variable, *entropy, S*, based on this limiting idealization of wasted heat. In order to formulate the *entropy*, we need to define the cyclic processes and their relation to the state properties. A cyclic process is one where the property returns to its initial state, so that the initial state and the final state are the same. A state property is such that the sum of the changes of that property in a cyclic process is zero. For example, the sum of changes in internal energy of a system in a cycle is given by $\oint dU = 0$. However, this is not true for non-state functions such as work and heat transfer, so that $\oint dW \neq 0$ and $\oint dQ \neq 0$, so we need even more information to integrate these, namely the path, and consequently dW and dQ are not perfect differentials but can be solved by *path-dependent integrals*. Mathematically, for a cyclic process, these path-dependent integrals can be related to the internal energy as follows:

$$\oint dU = 0 = \oint dW + \oint dQ = \oint (dW + dQ) \tag{108}$$

Now, we need to define the relation between entropy and reversible cyclic processes. In 1824, Sadi Carnot proved that in any cyclic reversible process which takes place between two heat reservoirs one hot, Q_H, and one cold, Q_C:

$$\frac{Q_H}{T_H} + \frac{Q_C}{T_C} = 0 \tag{109}$$

It is clear that the left-hand side of Equation (109) is simply the sum over the cycle of the quantity (Q/T). Thus, it can be written as the cyclic integral of the differential quantity for reversible systems:

$$\oint \frac{dQ_{rev}}{T} = 0 \tag{110}$$

This quantity is the exact differential of some state property and it is later defined as the *entropy* of the system:

$$dS = \frac{dQ_{rev}}{T} \tag{111}$$

where S is the entropy of the system and T is the absolute temperature of the system. The entropy is also an intrinsic property and necessarily an extensive quantity for systems at internal equilibrium. Since entropy is a state property, similar to Equation (110), $\oint dS = 0$ is also valid for reversible systems from fundamental principles.

A system plus its surroundings constitutes an isolated system and the entropy concept should be applied to both the system and its surroundings for a better understanding of the rules of spontaneous process formation. Depending on the conditions, the entropy of the surroundings, dS_{sur} may be equal to or different from dS_{system}. If we consider a system in thermal and mechanical contact with its surroundings at the same temperature, and if the system is not in equilibrium with its surroundings, then an infinitesimal transfer of heat, $dQ_{system} = -dQ_{sur}$ from the surroundings to the system may take place (for example heat transfer occurs in a gas confined in an impermeable cylinder with the movement of the impermeable piston until the in and out pressure equilibrium is reached). This heat transfer is accompanied by a change in the entropy of the system and also of the surroundings. This process may be reversible or irreversible and we need to derive expressions for both processes. In a reversible process, any heat flow between system and surroundings must occur with no finite temperature difference; otherwise the heat flow would be irreversible towards the cooler side. In reversible isothermal processes, $dQ_{system} = -dQ_{sur}$. Now, if we assume that the surroundings consist of reservoir of constant volume ($dW = 0$ from $dV = 0$, as usual in a metal cylinder containing a confined gas), from the First Law of Thermodynamics we have $dU_{sur} = dQ_{sur}$. For the surroundings, as dU_{sur} is a state function, it is independent of whether the process is reversible or irreversible. Since the expression $dU_{sur} = dQ_{sur,irrev}$ is also valid, then we have $dQ_{sur,rev} = dQ_{sur,irrev}$, and we may write

$$dS_{sur} = \frac{dQ_{sur,rev}}{T} = \frac{dQ_{sur,irrev}}{T} \tag{112}$$

(We must realize that Equation (112) is not valid for dS_{system}.) Now, we may connect dS_{system} with dS_{sur} by considering the total entropy. The total entropy is the sum of the entropy of the system and its surroundings ($\Delta S_{tot} = \Delta S_{system} + \Delta S_{sur}$). In a reversible process, since $dQ_{system} = -dQ_{sur}$, and the temperature is constant, then $dS_{tot} = 0$ and integration gives $\Delta S_{tot} = 0$. However, any irreversible change of state is accompanied by a change in both dS_{system} and dS_{sur}. The special case of an *adiabatic* (no heat transfer) irreversible process leads to a general result: in the Carnot Cycle, it was proved that for any irreversible adiabatic closed process, $dS_{sys,irrev} > 0$, and we may then state that *the total entropy of an isolated system increases during the course of a spontaneous change*, as a new form of the Second Law of Thermodynamics. As we know, all real processes are irreversible, and their entropies are increasing as these processes are occurring in an isolated system. The entropy of an isolated system is maximized at equilibrium when the process ceases. For example, if two parts of an isolated system are at different temperatures, heat will flow from the hot part to the cold part until the temperatures of both parts are equalized; and at this stage, the entropy of the isolated system is maximized. Since $dS_{tot} \geq 0$, we may write, for any irreversible process, $(dS_{system} + dS_{sur} \geq 0$, and this may be rearranged as, $dS_{system} \geq -dS_{sur}$. Since the heat supplied to the system comes from the surroundings, that is $dQ_{system} = -dQ_{sur}$, or for irreversible processes, $dQ_{system,irrev} = -dQ_{sur,irrev}$, and by combining these with Equation (112) we obtain

$$dS_{system} \geq \left(-\frac{dQ_{sur,irrev}}{T} = \frac{dQ_{system,irrev}}{T} \right) \qquad (113)$$

This is known as the *Clausius inequality* and has important applications in irreversible processes. For example, $dS > (dQ/T)$ for an irreversible chemical reaction or material exchange in a closed heterogeneous system, because of the extra disorder created in the system. In summary, when we consider a closed system and its surroundings together, if the process is reversible and if any entropy decrease takes place in either the system or in its surroundings, this decrease in entropy should be compensated by an entropy increase in the other part, and the total entropy change is thus zero. However, if the process is irreversible and thus spontaneous, we should apply Clausius inequality and can state that there is a net increase in total entropy. Total entropy change approaches zero when the process approaches reversibility.

After realizing that the entropy of an isolated system is maximized at equilibrium, the next question arises: what is maximized at equilibrium in physical terms? If we consider the mixing of two ideal gases, A and B, why do we naturally obtain mixed gases from unmixed gases, but we can never obtain unmixed gases from mixed gases? The answer lies in the probability of these processes. The probability of mixing all the A and B molecules in the whole container is very high but the probability that all the A molecules will be in the left half and all the B molecules will be in the right half of the container is extremely small. Thus, the increase in the entropy of an isolated system proceeding toward equilibrium is related to the system's going from a state of low probability to one of high probability. At equilibrium, the thermodynamic state of an isolated system is the most probable state.

In addition, there is a relation between entropy and *disorder*: disordered states have higher probabilities than ordered states. In general, the changes that are accompanied by an increase in entropy result in increased molecular disorder. Thus, entropy is also a measure of the molecular disorder of the state. Although disorder may be related to entropy qualitatively, the amount of disorder is a subjective concept and it is much better to relate entropy to probability rather than to disorder. Such concepts can be described in terms of *thermodynamic probabilities* (Ω) in statistical mechanics. The entropy of a system is a function of the probability of the thermodynamic state of this system, $S = f(\Omega)$. We know from statistical mathematics that only logarithmic functions satisfy probabilistic equations, so that we may use

$$S = k \ln \Omega + a \tag{114}$$

for this purpose, where a is an additive constant. L. Boltzmann defined the Ω parameter as the number of ways that microscopic particles can be distributed among the *states* accessible to them,

$$\Omega = \frac{n!}{(n_1!)(n_2!)(n_3!) \cdots} \tag{115}$$

where, n is the total number of particles and n_1, n_2, n_3 etc. represent the numbers of particles in *states* 1, 2, 3 etc. (these *states* are generally considered as the energy states of the molecules or particles). Boltzmann's definition relates the entropy of a system to the number of different ways that the system can store a fixed amount of internal energy, so that Ω is the number of distinguishable ways that the system can be arranged with a fixed amount of internal energy.

From Equation (114), the entropy difference between initial (1) and final (2) states may be given as

$$\Delta S = S_2 - S_1 = (k \ln \Omega_2 + a) - (k \ln \Omega_1 + a) = k \ln \frac{\Omega_2}{\Omega_1} \tag{116}$$

By using Stirling's statistical approximation, Botzmann reached the equation for the isothermal, isobaric, irreversible mixing of two perfect gases having equal volumes:

$$\Delta S = k N_A \ln 2 = R \ln 2 \tag{117}$$

The proportionality constant, k, is the *Boltzmann constant* which relates kinetic energy to temperature, and it is the ideal gas constant R divided by Avogadro's number ($k = R/N_A$). The Boltzmann constant governs the distribution of molecules among energy levels and also the thermodynamic systems among quantum states.

The application of the $[S = k \ln \Omega + a]$ equation to situations other than mixing of perfect gases requires knowledge of quantum and statistical mechanics via partition functions in other ways; these can be found in statistical thermodynamics textbooks. As an example, we may interpret the heat flow in terms of probability. The heat flow occurs via collisions between molecules of the hot part with molecules of the cold part. It is more probable that the high-energy molecules of the hot part lose some of their energy to the low-energy molecules of the cold part in such collisions. Thus, internal energy is transferred from the hot to the cold until the thermal equilibrium is attained. This indicates that entropy is related to the distribution or spread of energy among the available molecular energy levels. If an isolated system (i.e., fixed U) is able to undergo some change such that the number of states increases, it will do so. The interpretation of this statement is that if all observations are equally probable, the state of the system with the most numerable observations is the one that is most likely to be observed. Thus, the distribution of energy, which is directly related to the entropy, determines the direction of spontaneity in irreversible processes. When the process approaches equilibrium then the system has the most probable distribution of energy.

Temperature is a way of measuring the average molecular energy of materials. The increase of temperature in a system increases the disorganization either in terms of location or in terms of the occupation of their available translational, rotational, and vibrational energy states. The gain of heat stimulates disorderly motion in the surroundings, whereas work transfer stimulates uniform motion of atoms in the surroundings; it does not change the degree of disorder, so does not change the entropy. As we stated above, macroscopic properties such as temperature, pressure and volume are manifestations of behavior of countless microscopic particles, such as molecules, that make up a finite system. Because of the enormous number of particles contained in any system of interest, such a description must necessarily be statistical in nature. In pure substances, entropy may be divided into three parts:

1 Translational degrees of freedom (e.g., monatomic ideal gas molecules moving in space).
2 Rotational degrees of freedom (e.g., non-spherical molecules in fluids).
3 Vibrational degrees of freedom (e.g., non-spherical fluid molecules and solids).

For pure materials, it is easy to monitor the entropy–molecular disorder relationship during phase changes. For example, in fusion, the system changes from the highly ordered arrangement of a crystal lattice to the irregular molecular arrangement of the liquid state. In vaporization, the molecules are released from the confined motion of the liquid state. In the gas phase, as the temperature of a substance is increased an increasingly chaotic and disordered motion of the molecules occurs. In non-pure substances (e.g., in a solution) another degree of freedom arises which relates to the ways that the system can be mixed up. The entropy change associated with mixing liquids, gases and also expanding gases can be related to the increased freedom of position in space of the individual molecules.

From the scientific definition point of view, there is a slight difference between our continuum thermodynamics definition of the Second Law and its statistical mechanical version so that the continuum thermodynamics definition of the Second Law states that an observation of decreased universal entropy is *impossible* in isolated systems; however the statistical mechanical definition says that an observation of universal increased entropy is *not probable*.

The *Third Law of Thermodynamics* relates the change of entropy to temperature, stating that the limiting value of the entropy of a system can be taken as zero as the absolute value of temperature approaches zero. Thus, the absolute entropy is zero for all perfect crystalline substances at absolute zero temperature, and from this definition it is clear that entropy has a universal reference state, while enthalpy and free energy quantities do not. The Third Law allows us to calculate the absolute value of entropy by integrating Equation (111) so that

$$\Delta S = \int_{T=0}^{T} \frac{dQ_{rev}}{T} \tag{118}$$

The entropy concept allows us to define two new extensive thermodynamic variables: the *Helmholtz Free Energy*, F, which is the maximum amount of work a system can do at constant temperature (isothermal changes); and the *Gibbs Free Energy*, G, which is the maximum amount of work a system can do at constant pressure (isobaric changes) and is a minimum for closed systems at equilibrium with a fixed temperature and pressure.

The Helmholtz free energy parameter, F, is defined as

$$F \equiv U - TS \tag{119}$$

and for differential changes, it becomes

$$dF = dU - TdS - SdT \tag{120}$$

Since, for reversible cycles in the isothermal pressure–volume change of pure substances we know that $dQ_{rev} = TdS$ and $dW_{rev} = -PdV$, from the First and Second Laws of Thermodynamics, we may then obtain by rearrangement

$$dU_{rev} = TdS - PdV \tag{121}$$

and by combining Equations (120) and (121), for reversible processes

$$dF_{rev} = -PdV - SdT \tag{122}$$

since T is constant in an isothermal change, that is $dT = 0$, where F is defined, then

$$dF_{rev} = -PdV = dW_{rev} \tag{123}$$

is obtained. Since a system does maximum work during a reversible process, this is in conjunction with the definition of the Helmholtz free energy parameter, F, as the maximum amount of work a system can do at a constant temperature. Thus, F_{rev} is also defined as the *maximum work* function. As Equation (123) implies, when both T and V are constant ($dV = 0$), $dF_{rev} = 0$, so the Helmholtz free energy parameter minimizes at equilibrium in a process that takes place in a closed system capable of doing only P–V work.

On the other hand, for transformations under constant pressure (isobaric changes), the Gibbs free energy parameter, G, is defined

$$G \equiv H - TS = U - TS + PV \tag{124}$$

For both Helmholtz and Gibbs free energies the term *free* indicates that only the portion $(U - TS)$ of the internal energy is free to perform work. For differential changes in Gibbs free energy and enthalpy, it becomes

$$dG = dH - TdS - SdT \tag{125}$$

Since, for reversible cycles, in an isobaric temperature–volume change of pure substances, $dH_{rev} = TdS + VdP$, then by rearrangement, one obtains

$$dG_{rev} = VdP - SdT \tag{126}$$

since P is constant at isobaric change, that is $dP = 0$, where G is defined, then

$$dG_{rev} = -SdT \tag{127}$$

is obtained. As Equation (127) implies, when both P and T are constant ($dT = 0$), $dG_{rev} = 0$, so that the Gibbs free energy parameter minimizes at equilibrium in a process that takes place in a closed system capable of doing only P–V work. If a process takes place at constant pressure and temperature, then both Helmholtz and Gibbs free energy functions may be used together. By combining Equations (104), (119) and (124), one obtains

$$G \equiv F + PV \tag{128}$$

For differential changes, this becomes

$$dG = dF + PdV + VdP \tag{129}$$

and for isothermal and isobaric changes, it gives

$$dG = dF + PdV \tag{130}$$

For reversible changes, the above derived $dU = TdS - PdV$, $dH = TdS + VdP$, $dF = -SdT - PdV$ and $dG = -SdT + VdP$ equations are the fundamental thermodynamic property expressions for a homogeneous fluid of constant composition. Implicit in these are the following:

$$T = \left(\frac{\partial U}{\partial S}\right)_V = \left(\frac{\partial H}{\partial S}\right)_P \tag{131}$$

$$P = -\left(\frac{\partial U}{\partial V}\right)_S = -\left(\frac{\partial F}{\partial V}\right)_T \tag{132}$$

$$V = \left(\frac{\partial H}{\partial P}\right)_S = \left(\frac{\partial G}{\partial P}\right)_T \tag{133}$$

$$S = -\left(\frac{\partial F}{\partial T}\right)_V = -\left(\frac{\partial G}{\partial T}\right)_P \tag{134}$$

In addition, four Maxwell equations result from application of the Euler reciprocity relation (Equation (101) was derived for internal energy, U) and the exact differentials rule $\frac{\partial}{\partial y}\left(\frac{\partial z}{\partial x}\right) = \frac{\partial}{\partial x}\left(\frac{\partial z}{\partial y}\right)$. Maxwell equations are used occasionally in thermodynamics for the processes that take place in closed systems:

$$\left(\frac{\partial T}{\partial V}\right)_S = -\left(\frac{\partial P}{\partial S}\right)_V \tag{135}$$

$$\left(\frac{\partial T}{\partial P}\right)_S = \left(\frac{\partial V}{\partial S}\right)_P \tag{136}$$

$$\left(\frac{\partial P}{\partial T}\right)_V = \left(\frac{\partial S}{\partial V}\right)_T \tag{137}$$

$$\left(\frac{\partial V}{\partial T}\right)_P = -\left(\frac{\partial S}{\partial P}\right)_T \tag{138}$$

3.1.2 *Thermodynamical expressions for open systems*

All the above thermodynamic relations can be used for homogeneous closed systems consisting of a single phase of a pure substance which does not exchange matter with its surroundings, although it may exchange energy. As we know, matter and energy transport are only permitted in an *open system* where the composition of the system varies. Thermodynamic expressions were developed to apply to varying compositions in open systems regardless of the cause of the composition changes. When we alter the composition, we change the internal energy of an open system. For instance, we know that NaCl salt has a maximum solubility in water (it is about 300 g/l). If we place a salt crystal in an open system, a fixed volume of water, we perform chemical work by dissolution of the salt crystal in water to form a single phase solution by changing the internal energy of the system.

However, if the solution is saturated previously, and we add the same amount of salt to this solution, then we form a salt precipitate in the solution forming two phases, and thus the change in internal energy is different for this case from the first process.

The internal energy for a single homogeneous open system, which can exchange matter as well as energy with its surroundings, is a function of S, V and also composition,

$$U = f(S, V, n_1, n_2, \ldots, n_m) \tag{139}$$

where, n is the number of moles of various species and m is the number of species present in the system. Then, the total differential of the internal energy for a single homogeneous open system (similar to Equation (101)) can be written as

$$dU = \left(\frac{\delta U}{\delta S}\right)_{V,n_i} dS + \left(\frac{\delta U}{\delta V}\right)_{S,n_i} dV + \sum_i \left(\frac{\partial U}{\partial n_i}\right)_{V,S,n_j} dn_i \tag{140}$$

Now, a new parameter called the *chemical potential*, μ_i, of a substance, i, present in the system, in terms of constant volume and entropy is defined as

$$\mu_i = \left(\frac{\partial U}{\partial n_i}\right)_{V,S,n_j} \tag{141}$$

where the subscript, n_j, signifies that the amounts (mole numbers) of all other substances present in the system (other than i) are constant. By combining Equations (131), (132) and (141), we may rewrite Equation (140) in the form

$$dU = TdS - PdV + \sum_i \mu_i dn_i \tag{142}$$

which is the fundamental thermodynamic equation for an open system. Similarly we may derive three other fundamental equations for an open system, by using Equations (106), (121), (122) and (126)

$$dH = TdS + VdP + \sum_i \mu_i dn_i \tag{143}$$

$$dF = -SdT - PdV + \sum_i \mu_i dn_i \tag{144}$$

$$dG = -SdT + VdP + \sum_i \mu_i dn_i \tag{145}$$

Then, similar to Equation (141), it is possible to define the chemical potential in terms of other thermodynamic parameters from Equations (143)–(145), so that

$$\mu_i = \left(\frac{\partial U}{\partial n_i}\right)_{V,S,n_j} = \left(\frac{\partial H}{\partial n_i}\right)_{P,S,n_j} = \left(\frac{\partial F}{\partial n_i}\right)_{V,T,n_j} = \left(\frac{\partial G}{\partial n_i}\right)_{T,P,n_j} \tag{146}$$

It should be pointed out that there are many one-phase open systems in which the composition is changing due to irreversible chemical reactions. Two or more-phase open systems are also present if the irreversible interphase transport of matter takes place in the closed system. In chemistry, the independent variables of T and P are mostly used. Consequently, the chemical potential is generally expressed as the partial molar

Gibbs free energy as given in the far right-hand side of Equation (146). The chemical potential, μ_i, is also an intensive property and its value depends on T, P and the composition of the system.

When mixtures and solutions are considered, an equation is required to describe the changes in P, T and n_i. If we integrate Equation (142) between zero and finite mass at constant P, T and n_i, it gives

$$U = TS - PV + \sum_i \mu_i n_i \tag{147}$$

Differentiation of this equation results in a more general expression of dU,

$$dU = TdS + SdT - PdV - VdP + \sum_i \mu_i dn_i + \sum_i n_i d\mu_i \tag{148}$$

By combining Equations (142) and (148), we obtain the Gibbs–Duhem equation:

$$SdT - VdP + \sum_i n_i d\mu_i = 0 \tag{149}$$

For constant T and P, this can be simplified to

$$\sum_i n_i d\mu_i = 0 \tag{150}$$

which is a fundamental equation of solutions. Equation (150) shows that if the composition varies, the chemical potentials do not change independently, but in a related way. For example, in a system of two constituents, Equation (150) can be written as

$$n_1 \, d\mu_1 + n_2 \, d\mu_2 = 0 \tag{151}$$

Rearranging we have

$$d\mu_2 = -\frac{n_1}{n_2} d\mu_1 \tag{152}$$

Equation (152) shows that a simultaneous change in dμ_2 occurs if dμ_1 changes by the variation in composition.

In order to understand the equilibrium properties of gas and liquid solutions we should explain how the chemical potential of a solution varies with its composition. For gases in a closed system, $dG = VdP - SdT$, as we know from Equation (126). When the temperature is constant, $dT = 0$, we can calculate the Gibbs free energy at one pressure in terms of its value at another pressure

$$G_2 - G_1 = \int_{P_1}^{P_2} VdP \tag{153}$$

If the gas obeys ideal gas law, then it is easy to integrate Equation (153), since ($V = nRT/P$) so that

$$G_2 - G_1 = nRT \int_{P_1}^{P_2} \frac{dP}{P} = nRT \ln\left(\frac{P_2}{P_1}\right) \tag{154}$$

In order to have a standard Gibbs free energy value, G^o, we set $G_1 = G^o$ when $P_1 = P^o = 1$ bar, then we have

$$G(P) = G^o + nRT \ln\left(\frac{P}{P^o}\right) \tag{155}$$

Now since chemical potential in a closed system is defined as $\mu = (\partial G/\partial n)_{T,P}$, we can write chemical potential terms instead of Gibbs free energy terms,

$$\mu = \mu^o + RT \ln\left(\frac{P}{P^o}\right) \tag{156}$$

where μ^o is defined as the *standard chemical potential*, the molar Gibbs free energy of the pure gas at 1 bar.

For real pure gases, we cannot apply the $(V = nRT/P)$ expression and we cannot integrate Equation (153). However, we need to preserve the form of expressions that have been derived for the ideal thermodynamic system. In order to adapt Equation (156) for real gases, the replacement of the true measurable pressure, P, with another effective pressure term called *fugacity, f*, was carried out in classical thermodynamics:

$$\mu = \mu^o + RT \ln\left(\frac{f}{P^o}\right) \tag{157}$$

Fugacity is known as the *escaping tendency* of molecules and has the same units as pressure. We know that for an ideal gas, the pressure arises solely from the kinetic energy of the molecules and there are no intermolecular interactions. However, for real gases, intermolecular interactions are present and we need to express them in thermodynamic terms. Now, if we can define a *hypothetical standard state* in which all the intermolecular interactions have been extinguished, this may serve as a basis for the expression of other states. Then, it is assumed that at $P^o = 1$ bar, any real gas behaves ideally with only the kinetic energy of molecules accounting for its pressure, without any intermolecular interactions taking place. This is the hypothetical standard state of a real gas. (Alternatively, we may select $P^o = 0$ bar (at which it certainly behaves ideally), as the standard state of a real gas instead of $P^o = 1$ bar, however we refrain from this choice because, for this case, when $P \to 0$, then $\mu \to -\infty$, mathematically this is not scientific.) Now we need to relate the fugacity to the measurable pressure:

$$f = \phi P \tag{158}$$

where ϕ is the dimensionless *fugacity coefficient* which depends on the entire effect of intermolecular interactions of the real gas. It is clear that $\phi \to 1$ as $P \to 0$, and $f \to P$ as $P \to 0$. The fugacity coefficient, ϕ, can be calculated from the measurable *compressibility factor* of gases, Z ($Z = PV/nRT$) by using the expression

$$\ln \phi = \int_0^P \left(\frac{Z-1}{P}\right) dP \tag{159}$$

and by applying numerical (or graphical) integration from the experimental data of a real gas, or if the $Z = f(P)$ function is known, the analytical integration of Equation (159) is also possible.

For liquid solutions, we need to express how the chemical potential of a solution varies with its composition. In order to derive a useful expression, we should remember that, in equilibrium, the chemical potential of a substance in the liquid phase must be equal to the chemical potential of this substance in its vapor phase, from the Gibbs Rule (see Section 3.1.3 for the proof). If we denote A as the solvent in the solution, the chemical potential of the vapor of pure solvent A in the gas phase can be obtained from Equation (156)

$$\mu_A^*(\text{vapor}) = \mu_A^o + RT \ln\left(\frac{P_A^*}{P^o}\right) \tag{160}$$

where the superscript * denotes the quantities related to pure substances and P_A^* is the vapor pressure of pure A. From the Gibbs rule, we know that $\mu_A^*(\text{liquid}) = \mu_A^*(\text{vapor})$, and for $P^o = 1$ bar, then we may write

$$\mu_A^*(\text{liquid}) = \mu_A^o + RT \ln P_A^* \tag{161}$$

Now, if we dissolve a solute B in pure solvent A, we obtain an A–B solution. The dissolution of B in A reduces the vapor pressure from P_A^* value to P_A. For this solution, similar to Equation (161), we may write

$$\mu_A(\text{solution}) = \mu_A^o + RT \ln P_A \tag{162}$$

If we subtract Equation (161) from Equation (162) to eliminate the standard chemical potential term, μ_A^o, we have

$$\mu_A = \mu_A^* + RT \ln\left(\frac{P_A}{P_A^*}\right) \tag{163}$$

In a series of experiments, F. Raoult measured the vapor pressures of solvents and solutions with changing concentrations of solutes in them, and found that some ideal liquid solutions obey the following equation, named Raoult's law:

$$X_A = \left(\frac{P_A}{P_A^*}\right)_{\text{ideal}} \tag{164}$$

where X_A is the mole fraction of the solvent where $X_B = 1 - X_A$. Then Equation (163) rearranges to

$$\mu_A = \mu_A^* + RT \ln X_A \tag{165}$$

Some liquid mixtures obey Raoult's law, but most of the solutions deviate from this. Thus, the name *ideal liquid solutions* is defined for solutions that obey Raoult's law. Similarly to real gases, real solutions that do not obey Raoult's law should also be expressed by thermodynamical equations. Since we can still measure the vapor pressure of real solutions, instead of the mole fraction X_A term in Equation (164), we may write

$$a_A = \frac{P_A}{P_A^*} \tag{166}$$

where a_A is the *activity* of the solvent. The standard state of any solvent is defined as the pure liquid at 1 bar when $X_A = 1$. Activity is a kind of *effective mol fraction* as the fugacity is an *effective pressure*. Then, for real solutions, Equation (165) turns into

$$\mu_A = \mu_A^* + RT \ln a_A \tag{167}$$

Now, we need to relate the activity with the measurable concentration of the solution:

$$a_A = \varphi_A^x X_A = \varphi_A^c c_A \tag{168}$$

where φ is the dimensionless *activity coefficient* which depends on the entire effect of inter-molecular interactions of the real solution, and c is the molar concentration term. Activities may be expressed in many concentration terms such as mol fraction, X; molarity, c; and molality etc. It is clear that $\varphi_A^x \to 1$ as $X_A \to 1$, and $a_A \to X_A$ as $X_A \to 1$, and the same applies for other concentrations, too.

Equations (160)–(168) apply to solvent activities. For solute activities, a_B, a similar thermodynamic derivation gives a final expression, similar to Equation (167)

$$\mu_B = \mu_B^* + RT \ln a_B \tag{169}$$

In summary, fugacity and activity coefficients were introduced in thermodynamics in order to explain deviations from ideal behavior and thus serving as correction factors for non-ideal behavior.

3.1.3   Equilibrium between phases in heterogeneous closed systems

As stated above, a system is in *equilibrium* if no further spontaneous changes take place at constant surroundings and if the same state can be reached from different directions. In general, there are two kinds of material equilibrium; phase equilibrium and chemical reaction equilibrium. In surface and colloid science, thermodynamic systems generally contain more than one *phase*. As stated in Section 1.1, a phase is defined as a homogeneous form of matter that can be physically distinguished from any other such phase by an identifiable interface. A phase equilibrium consists of the same chemical species present in different phases. We will only consider phase equilibrium in this section. Thus, chemical potentials deserve special attention in surface thermodynamics because material transport from one phase to another is a common phenomenon. A *heterogeneous closed system* is made up of two or more phases, with each phase considered as an open system within the overall closed system. The heterogeneous system is in internal equilibrium with respect to the three processes of heat transfer, boundary displacement and mass transfer. In order to have thermal and mechanical equilibrium in the system, the temperature and pressure must be uniform throughout the whole heterogeneous mass. If a closed multiphase system, having uniform T and P, is not initially at internal equilibrium with respect to mass transfer (or with respect to chemical reaction) then the changes occurring in the system are irreversible and must necessarily bring the system closer to an equilibrium state. We may write from the First Law (Equation (100)) for any irreversible process

$$dQ_{sys,irrev} = (dU - dW)_{sys,irrev} \tag{170}$$

If we combine the Clausius inequality given in Equation (113) with Equation (170), we obtain for both irreversible and reversible systems

$$TdS_{system} \geq dQ_{system} = (dU - dW)_{system} \tag{171}$$

and for the mechanical equilibrium where $dW = -PdV$, for the system parameters, we may write

$$0 \geq dU + PdV - TdS \tag{172}$$

This inequality applies to all incremental changes towards the equilibrium state, and the equality holds at the equilibrium state where any change is reversible. It follows immediately that when S and V are constant,

$$0 \geq (dU)_{S,V} \tag{173}$$

Since $G = U + PV - TS$, from its definition, that is $dG = dU + PdV + VdP - TdS - SdT$. If T and P are constant in the entire heterogeneous system, we may write $dG_{T,P} = dU + PdV - TdS$, and by combining with Equation (172), we obtain

$$0 \geq dG_{T,P} \quad \text{or} \quad 0 \geq d(U + PV - TS)_{T,P} \tag{174}$$

This expression shows that all irreversible processes occurring at constant T and P proceed in a direction such that the total Gibbs energy of the system decreases. Thus the equilibrium state of a heterogeneous closed system is the state with the minimum total Gibbs energy attainable at the given T and P. At the equilibrium state, differential variations may occur without producing a change in G. This is the meaning of the equilibrium criterion, $dG_{T,P} = 0$. These equations may be applied to a closed, non-reactive, many-phase system.

Gibbs was the first to prove that in a heterogeneous (multiphase) closed system, the chemical potential of every phase is equal to the chemical potential of the other phase, which is in equilibrium with it:

$$\mu_i^\alpha = \mu_i^\beta = \ldots = \mu_i^\pi \tag{175}$$

where superscripts α, β and π identify the phases. It was proved that a chemical species could be spontaneously transported from a phase of larger chemical potential to one of lower chemical potential. The criteria for internal thermal and mechanical equilibrium simply require uniformity of temperature and pressure throughout the system. In these equilibrium conditions, the Gibbs free energy of the system is minimized, and the chemical potential of any transportable species is uniform throughout the system.

We may derive Equation (175) for a two-phase (α and β phases) closed system having an interfacial flat boundary area A between them. The system as a whole is isolated and the internal energy, U, volume, V and all the number of moles, n_i s, are fixed. If infinitesimal changes occur in U, V and n_i of phases α and β, such changes are possible only provided that $dU^\alpha = -dU^\beta$, $dV^\alpha = -dV^\beta$ and $dn_i^\alpha = -dn_i^\beta$, in the given boundary conditions. In a heterogeneous (multiphase) closed system, there is no heat exchange with the surroundings and the entropy can only be changed by spontaneous changes in the system. Before the system reaches the equilibrium, $dS > 0$ and this entropy increase can be statistically shown as a change towards a more probable state, that is a state of higher Ω, as we see in Equation (115). When the system reaches equilibrium, S_{U,V,A,n_i} becomes maximum and hence $(\partial S)_{U,V,A,n_i} = 0$. If we consider only the equilibrium between two bulk phases (this means that we do not consider any special changes at the dividing interface which we will see in Section 3.2), we may write for equilibrium conditions

$$(\partial S)_{U,V,A,n_i} = 0 = (\partial S^\alpha + \partial S^\beta)_{U,V,A,n_i} \tag{176}$$

When we apply Euler's reciprocity relation to obtain the total differential of entropy by using Equation (176) for a constant interfacial flat boundary area, A, between them, we have

$$0 = \left(\frac{\partial S^\alpha}{\partial U^\alpha}\right)_{V^\alpha, n_i^\alpha} dU^\alpha + \left(\frac{\partial S^\alpha}{\partial V^\alpha}\right)_{U^\alpha, n_i^\alpha} dV^\alpha + \sum_i \left(\frac{\partial S^\alpha}{\partial n_i^\alpha}\right)_{U^\alpha, V^\alpha, n_{j\neq1}^\alpha} dn_i^\alpha$$
$$+ \left(\frac{\partial S^\beta}{\partial U^\beta}\right)_{V^\beta, n_i^\beta} dU^\beta + \left(\frac{\partial S^\beta}{\partial V^\beta}\right)_{U^\beta, n_i^\beta} dV^\beta + \sum_i \left(\frac{\partial S^\beta}{\partial n_i^\beta}\right)_{U^\beta, V^\beta, n_{j\neq1}^\beta} dn_i^\beta \tag{177}$$

We may write from Equation (131),

$$\frac{1}{T} = \left(\frac{\partial S}{\partial U}\right)_{V, n_i} \tag{178}$$

and from Equation (121),

$$\frac{P}{T} = \left(\frac{\partial S}{\partial V}\right)_{U, n_i} \tag{179}$$

and from Equation (142), for constant U, V and n_j,

$$\frac{\mu_i}{T} = -\left(\frac{\partial S}{\partial n_i}\right)_{U, V, n_{j\neq1}} \tag{180}$$

Then, by combining Equations (177)–(180) for two phases followed by the introduction of the boundary conditions we have

$$0 = \left(\frac{dU^\alpha}{T^\alpha}\right) + \left(\frac{PdV^\alpha}{T^\alpha}\right) - \sum_i \left(\frac{\mu_i^\alpha dn_i^\alpha}{T^\alpha}\right)$$
$$+ \left(\frac{dU^\beta}{T^\beta}\right) + \left(\frac{PdV^\beta}{T^\beta}\right) - \sum_i \left(\frac{\mu_i^\beta dn_i^\beta}{T^\beta}\right) \tag{181}$$

By rearrangement we obtain,

$$\left(\frac{dU^\alpha}{T^\alpha} + \frac{dU^\beta}{T^\beta}\right) + \left(\frac{P^\alpha dV^\alpha}{T^\alpha} + \frac{P^\beta dV^\beta}{T^\beta}\right) - \sum_i \left(\frac{\mu_i^\alpha dn_i^\alpha}{T^\alpha} + \frac{\mu_i^\beta dn_i^\beta}{T^\beta}\right) = 0 \tag{182}$$

We can also derive Equation (182) by rearranging the fundamental equation of open systems (Equation (142)) so that we may write for the homogeneous α bulk phase

$$dS^\alpha = \frac{dU^\alpha}{T^\alpha} + \frac{P^\alpha dV^\alpha}{T^\alpha} - \sum_i \left(\frac{\mu_i^\alpha}{T^\alpha}\right) dn_i^\alpha \tag{183}$$

and if we write Equation (183) for the β bulk phase, and combine with Equation (176), we can obtain Equation (182) again. Now, since $dU^\alpha = -dU^\beta$, $dV^\alpha = -dV^\beta$ and $dn_i^\alpha = -dn_i^\beta$ in the given boundary conditions, by rearranging Equation (182) we have

$$\left(\frac{1}{T^\alpha} - \frac{1}{T^\beta}\right) dU^\alpha + \left(\frac{P^\alpha}{T^\alpha} - \frac{P^\beta}{T^\beta}\right) dV^\alpha - \sum_i \left(\frac{\mu_i^\alpha}{T^\alpha} - \frac{\mu_i^\beta}{T^\beta}\right) dn_i^\alpha = 0 \tag{184}$$

If the two-phase system is in thermal equilibrium, so that $T = T^\alpha = T^\beta$, then Equation (184) reduces to

$$\left(\frac{P^\alpha - P^\beta}{T}\right) dV^\alpha - \sum_i \left(\frac{\mu_i^\alpha - \mu_i^\beta}{T}\right) dn_i^\alpha = 0 \tag{185}$$

If the pressures of all the phases are constant, as $P = P^\alpha = P^\beta$, the left side of Equation (185) vanishes and then it is clear that Equation (175) ($\mu_i^\alpha = \mu_i^\beta$) is valid. This means that at equilibrium there is no driving force for the transport of matter between phases. It is important to realize that all the above equilibrium conditions can only apply to flat interfaces, and if the interfaces are curved, then P^α and P^β differ by an amount of capillary pressure, ΔP, which depends on the radius of curvature (see Section 4.3).

There is another derivation of Equation (175). As we know, Equation (145) indicates that at equilibrium, for a chemical species in a single phase, at constant T and P, $\sum_i \mu_i^\alpha dn_i^\alpha = 0$, because $dG^\alpha = 0$. If there is more than one phase, then we may naturally write

$$\sum_\alpha \left(\sum_i \mu_i^\alpha dn_i^\alpha \right) = 0 \text{ for equilibrium, at constant } T \text{ and } P. \text{ Thus, for two materials 1 and 2, in}$$

two phases of α and β, at equilibrium, we may write, $\mu_1^\alpha dn_1^\alpha + \mu_2^\alpha dn_2^\alpha + \mu_1^\beta dn_1^\beta + \mu_2^\beta dn_2^\beta = 0$. If we consider the equilibrium of only one of the chemicals, we may write, $\mu_1^\alpha dn_1^\alpha + \mu_1^\beta dn_1^\beta = 0$. When n_1 moles flow from the α phase to the β phase, we may write, $-dn_1^\alpha = dn_1^\beta$ from the material balance in a closed system, giving $\mu_1^\alpha dn_1^\alpha - \mu_1^\beta dn_1^\alpha = (\mu_1^\alpha - \mu_1^\beta)dn_1^\alpha = 0$. Since, $dn_1^\alpha \neq 0$, then we must have $(\mu_1^\alpha - \mu_1^\beta) = 2$, resulting in $(\mu_1^\alpha = \mu_1^\beta)$. If the system is irreversible, since $dG < 0$, then we may write, $\mu_1^\alpha dn_1^\alpha + \mu_1^\beta dn_1^\beta < 0$, and by rearrangement we have $\mu_1^\beta dn_1^\beta < -\mu_1^\alpha dn_1^\alpha$ or $(\mu_1^\beta < \mu_1^\alpha)$ if $(-dn_1^\alpha = dn_1^\beta)$ is applied. This shows that any substance flows spontaneously from a phase with higher chemical potential (μ_1^α) to a phase with lower chemical potential (μ_1^β), until the chemical potential has been equalized. Similarly to temperature governing the flow of heat, chemical potential governs the flow of matter from one phase to another.

3.2 Gibbs Dividing Interface

3.2.1 Thermodynamical definition of an interface

While we were deriving Equations (173)–(185), we considered only the thermodynamical equilibrium between two bulk phases, we did not consider any special changes at the dividing interface (boundary), or the effect of the variation of the interfacial area. We assumed strictly homogeneous thermodynamic systems with their intensive properties constant throughout each phase. The former approach is a departure from reality because when phases α and β are in contact with each other, it is clear that both phases are not strictly homogenous throughout real systems. The atoms and molecules present in the interfacial regions undergo some changes, resulting in a concomitant change in the internal energy. So these atoms or molecules at the surface should be treated as being somehow different from those of the bulk, in thermodynamical terms. Thus in real systems, there is a finite distance across an interface over which the properties gradually change from those of one adjacent bulk phase to those of the other.

The three-dimensional region of contact between phases α and β is called the *interphase region* or *interfacial layer*. It is generally assumed that if the ions are not present, this region is a few molecules in thickness (approximately 1–2 nm) and only an extremely small fraction of the molecules in the system are present in the interfacial region due to geometrical constraints. (If ions are present the interfacial thickness is much larger because the concentration of ions might vary over a large distance in the solvent.) Consequently, in most non-electrolyte solutions, the influence of surface effects on the thermodynamic properties of the bulk system is essentially negligible. However, in some cases where the surface area-to-volume ratio is high, such as in nanotechnology, colloidal systems or gas adsorption or catalysis on porous material processes, the surface effects are significant if not substantial.

As we know, intermolecular interactions lower the internal energy, and the molecules at the surface of a liquid in contact with its own vapor have a higher average internal energy than molecules in the bulk phase, because they experience fewer attractions at the surface compared with molecules in the bulk liquid phase. One must do work to increase the surface area of a liquid, because such an increase in surface area means an increase in the more high-energy molecules at the surface, at the expense of fewer and low-energy molecules in the bulk liquid. Since the interfacial internal energy cannot be measured by any direct experimental process, we need some indirect methods to determine the internal energy in the interphase region. One option is to apply the idea of treating the interface as a very thin layer and defining the properties of this thin layer. However, this is not an easy task because we face a major problem in that we do not know which part of the total energy of the system with the two contacting phases must be attributed to the interface, U^S, and which part to the bulk phases, U^α and U^β.

This is a rather more complex problem because the interfacial layer is not infinitesimally thin and some free energy is stored in this layer. There is a gradient of molecular density, composition, enthalpy, entropy, electrical potential (for charged molecules) and many other properties in this interfacial transition layer, as shown in Figure 3.1 *a* as a property–distance plot. Actually, U^S is located in a layer of certain thickness, Δx, and some assumption must be made if we want to define U^S in thermodynamical terms, because it is impossible to decide physically where phase α ends and phase β begins. The thickness of this transition layer, Δx_γ, for any two immiscible phases is shown in Figure 3.2 *a* and is dependent on the molecular nature of phases α and β, and also on external factors such as temperature and pressure. It has been found experimentally that this interfacial layer is usually a few molecules in thickness for most non-electrolytes.

J. W. Gibbs was the first to propose a method to idealize the interface as a mathematical dividing surface. He assumed that phases α and β are separated by an infinitesimally thin boundary layer, known as the *Gibbs dividing plane*, as seen in Figures 3.1 *b*. and 3.2 *b*. The α and β phases are considered to have their bulk properties up to this plane, and any excess internal energy is supposed to be entirely located in this plane. Gibbs treated this thin layer as a *quasi-two-dimensional phase* having no volume, but excess extensive interfacial quantities (see Section 3.2.5). The Gibbs dividing plane concept is a departure from the physical reality but it is consistent and allows us to apply thermodynamics to surface processes. However, the next questions then arise: where do we locate the Gibbs dividing plane, $x_{\gamma 0}$, in the transition layer between phases, and what is the surface free energy of this layer?

3.2.2 *Physical description of a real liquid interface*

We have already seen that the boundary region between two bulk phases has a thickness, Δx. Usually in the liquid–vapor and some liquid$_1$–liquid$_2$ interfaces the boundary layers are

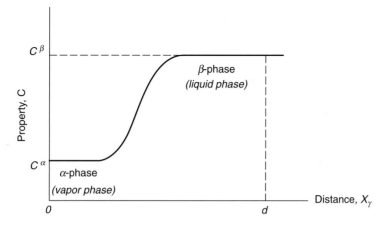

a. Property change in the vicinity of a real interface between two phases

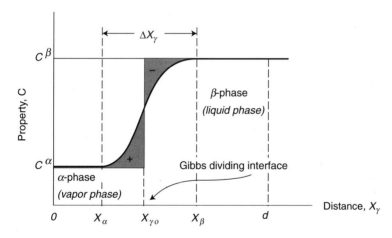

b. Idealization as a mathematical *Gibbs dividing interface*

Figure 3.1 **a.** The variation of a molecular property, C with the distance, x in the interfacial transition layer between α and β phases. **b.** The same plot after the mathematical *Gibbs dividing plane* is located at point $x_{\gamma o}$. Δx_{γ} is the thickness of the transition layer.

only of the order of the dimension of one molecule – a *monolayer*. Rarely, several molecular layers can be involved – a *multilayer*. This boundary region is necessarily static but the dynamics of the situation should not be forgotten. The interfacial boundary region in real liquids is in a very turbulent state. For a liquid–vapor interface, the liquid is in equilibrium with its vapor and there is a two-way and balanced traffic of molecules hitting and condensing on the surface from the vapor phase, and of molecules evaporating from the bulk phase. Kinetic theory of gases and liquids shows that, under zero external pressure (full vacuum), the mass of molecules that strikes one m² per second is

$$m_s = P\sqrt{\frac{M_w}{2\pi RT}} \qquad (186)$$

where P is the vapor pressure (Pa), M_w molecular mass (kgmol^{-1}), T is the absolute temperature in Kelvin, and R is the gas constant, 8.3144 (m^3 Pa/mol K). For liquid water at equilibrium, when the evaporation rate is equal to the condensation rate, at 20°C the vapor pressure is 2333 Pa and $m_s = 0.253\,\mathrm{g\,cm}^{-2}$ so that each cm^2 of liquid experiences (0.253/18) $N_A = 8.47 \times 10^{21}$ arrivals and departures of molecules per second. A water molecule has an average area of $10\,\text{Å}^2 = 10^{-15}\,\mathrm{cm}^2$, and the size of the closely packed water molecules permits only 10^{15} molecules to be present in a 1 cm^2 layer at any moment. Thus the average life of each water molecule in the surface is only $10^{15}/8.47 \times 10^{21} \cong 1.18 \times 10^{-7}$ sec. There is also a violent traffic between the surface region and the adjacent layers of the liquid. It is evident that although there is extremely violent agitation of most of the liquid surfaces, the attractive forces between molecules are able to maintain the surface intact within a few molecular thicknesses. Because of the fluidity of the liquid molecules, the boundary region of liquids is assumed to be *homogeneous*; that is, its properties do not vary from place to place across the surface. (However, this is not true for solid surfaces, their boundary regions are *heterogeneous*, and the chemical nature and adsorption characteristics of solids vary from place to place across the surface (see Chapter 8).)

The differences between interfacial and bulk molecular interaction energies are due mainly to the two-dimensional geometry of the surface and also to differences in interfacial structure and differences in magnitude of the molecular interactions at the interface, from those of the bulk. In principle, it would be possible to calculate the energy of cohesion between molecules within a single phase if the potential energy functions and the spatial distributions of all the atoms and molecules were known. Moreover, if the complete

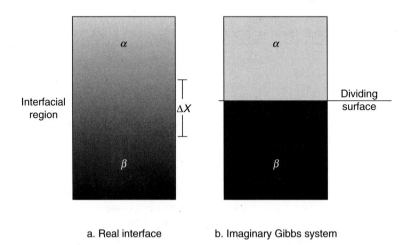

a. Real interface   b. Imaginary Gibbs system

Figure 3.2   *a.* Schematic representation of the interfacial transition layer between the α and β phases: Δx_γ is the thickness of the transition layer. ***b.*** Schematic representation of the interfacial transition layer between the α and β phases after the imaginary *Gibbs dividing plane* is located.

geometrical deformation at an interface is known, it would also be possible to calculate the energy of adhesion for this interface. However, in practice we are far away from such success. Firstly, the potential energy functions are not well enough known, especially for unsymmetrical polyatomic molecules and electron donor–acceptor interacting molecules. Secondly, the structure of liquids is not completely understood (see Chapter 4). The interfacial structure and geometry between two liquids is even less understood. Consequently, at present, it is necessary to apply semi-empirical approaches to interfacial free energy problems. In order to do this, we have to define the surface free energy and surface tension concepts in terms of thermodynamics.

3.2.3 Definitions of surface and interfacial free energy, and surface and interfacial tension

Surface free energy, γ_1, of material (1) is the work that should be supplied to bring the molecules from the interior bulk phase to its surface to create a new surface having a unit area ($1\,m^2$). Its dimension is energy per unit area, J/m^2 in the SI system (mJ/m^2 is generally used in surface science to keep the numerical values the same as for the previously used $ergs/cm^2$ values).

When we consider two immiscible phases, having a flat interface between them, we can define the *interfacial free energy*, γ_{12}, as the work that should be supplied to bring the molecules from both of the interior bulk phases (1) and (2) to the contact boundary to create a new (1–2) interface having an area of $1\,m^2$. For this case, the intermolecular forces acting on the surface molecules are similar to those encountered in the liquid–vapor system, however as a result of the replacement of the vapor by a condensed phase, the mutual attraction of unlike molecules across the interface becomes much more effective. Interfacial free energy may also be called more correctly, the *excess interfacial free energy*, which the molecules possess by virtue of their being in the interface.

The *surface tension*, γ_1, of a material (1) is the force that operates inwards from the boundaries of its surface perpendicularly, tending to contract and minimize the area of the surface. Its unit is force per unit length (N/m, or usually mN/m). The surface tension and the surface free energy of substances are dimensionally equivalent (N/m = J/m^2 = kg/s^2) and for pure liquids in equilibrium with their vapor, the two quantities are numerically equal. However these two terms are different conceptually as we will see in Section 3.2.5, and surface free energy is regarded as the fundamental property in thermodynamical terms, and surface tension would be taken simply as its equivalent if there is no adsorption on a surface. On the other hand, the unit of surface tension (N/m), is the two-dimensional analogue of the bulk pressure unit (N/m^2), and thus surface tension may be regarded as a two-dimensional negative pressure (see Sections 3.3 and 5.5).

When we consider two immiscible phases and an interface between them, we should define the *interfacial tension*, γ_{12}, as the force that operates inwards from the boundaries of a surface perpendicularly to each phase, tending to minimize the area of the interface. The *interfacial free energy* between liquids is dimensionally equivalent and numerically equal to their *interfacial tension*.

3.2.4 Surface free energy and surface tension of liquids

In liquids, the cohesion forces keep the molecules close to each other, and translational and rotational motion of molecules takes place within the liquid with considerable freedom. As we see in Figure 3.3, each molecule is surrounded by others on every side in the interior of the bulk liquid and is attracted in all directions resulting in a zero sum of force vectors. However, the molecules at the surface are attracted inwards, and also to each side by their neighbors; but there is no outward attraction to balance the inward pull, because there are very few molecules outside. Hence every surface molecule is subject to a strong inward attraction perpendicular to the surface. The surface molecules are continuously moving inwards more rapidly than interior molecules, which move upwards to take their places. This process decreases the number of molecules in the surface, and this diminishes the liquid surface area; this surface contraction continues until the interior accommodates the maximum possible number of molecules. The inward attraction normal to the surface causes the surface to be under a state of lateral tension as shown schematically in Figure 3.4. In this figure, if we assume that the surface molecules act similarly to simple pulleys, then the direction of inward pull may be changed to a lateral pull at right angles and thus cause a tension at the surface molecules, leading to the concept of *surface tension*. Thus, for a plane surface, the surface tension can be defined as the force acting parallel to the surface and at right angles to a line of unit length anywhere in the surface. This attraction makes the liquid behave as though surrounded by an invisible membrane skin, although there is actually no such skin in real systems (see below). At equilibrium conditions,

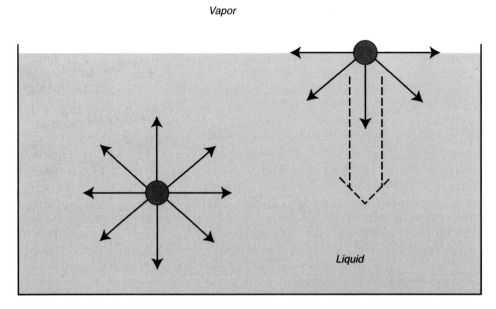

Figure 3.3 Schematic representation of a liquid molecule in the bulk liquid and at the surface. A downward attraction force is operative on the surface molecule due to the lack of liquid molecules above it.

Vapor

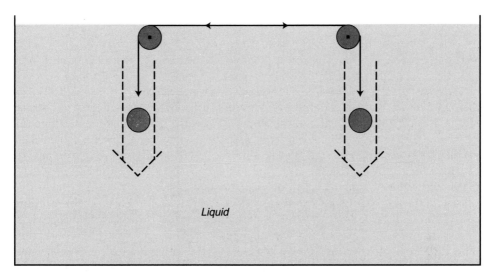

Liquid

Figure 3.4 Schematic representation of the surface tension between two liquid molecules at the surface: these molecules act similarly to simple pulleys, and the downward attraction forces cause a tension between the surface molecules.

without any external forces acting on the drop, the surface area is the smallest possible for a given liquid volume, which is geometrically a sphere, as we saw in Section 1.3. However if gravity, magnetic, electrical or other forces are also present, then shape distortions may happen according to the extent of the applied force and the size of the drop (for example gravity may flatten the large drops).

Since a liquid contracts spontaneously, this fact shows that there is free energy associated with it. We can measure the surface free energy of a liquid by performing work against the surface tension of its molecules by bringing the molecules from the interior to the surface. For example, if we insert a metal frame in a liquid, and pull it out slowly, a liquid film (or, if we use a soap or detergent solution, a soap film) is formed in the metal frame, as can be seen in Figure 3.5. We can expand the surface area of this liquid film by applying a perpendicular pulling force infinitesimally as dF_x, and the system moves from equilibrium, which leads to an infinitesimal increase in interfacial area, dA. If dF_x is slow enough, the newly formed area will have time to relax completely and has the equilibrium value of γ^{fil} everywhere. If the equilibrium conditions can be reached in a short time, and if the liquid film is sufficiently thick so that the two liquid–air interfaces are sufficiently far apart from each other to be considered as independent interfaces, then

$$\gamma^{\text{film}} = 2\gamma \tag{187}$$

According to Newton mechanics, the infinitesimal work to pull the frame in Figure 3.5 is $dW = F_x\, dx$. Since, the acting force, F_x, must be balanced by the force of surface tension along the length of the wire, l, of the two sides contacting the two film surfaces

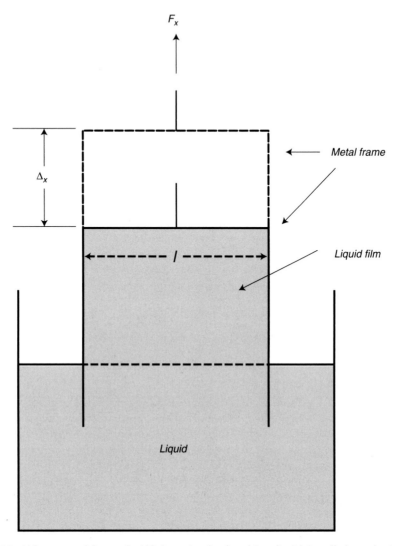

Figure 3.5 When a metal frame of width *l*, previously placed in a liquid, is pulled out slowly, with a perpendicular force of F_x, to a distance Δx, a liquid film is formed in the frame having an interfacial area of $\Delta A = l\Delta x$. The opposite surface tension balances the F_x force.

$$F_x = \gamma^{\text{film}} l = 2\gamma l \tag{188}$$

and we know from simple geometry that $dA = (2l)dx$, for the total area of two independent film surfaces. Then the work performed against the surface tension of liquid molecules can be given as

$$dW = (2\gamma l)dx = \gamma dA \tag{189}$$

and by combining Equations (188) and (189), we have

$$\gamma = \frac{dW}{dA} = \frac{F_x}{2l} \tag{190}$$

expressing both surface free energy (as work) and surface tension definitions. We may also prove that Equation (190) is valid with simple reasoning: the number of molecules in the interphase region is proportional to the surface area, A, and by increasing A by dA we increase the number of molecules in the interphase too. Since we must do positive work to increase the number of molecules in this region at the expense of their attractions to the bulk liquid, this work should be directly proportional to dA with a proportionality constant, say γ, thus giving Equation (190). The stronger the intermolecular attractions in a liquid the greater is the value of γ. (We will define surface free energy as a thermodynamic term more properly by applying the Gibbs and Helmholtz free energy functions in Section 3.2.5.)

In practice, we can only measure the liquid–air interfacial tension instead of the real surface tensions of liquids, in room conditions (see Chapter 6). This is not theoretically applicable to real liquid surface tension, which must be strictly measured in liquid–vacuum conditions. However, since liquids will continually evaporate in a high (or complete) vacuum condition, it is physically impossible to measure their real surface tension. Nevertheless, the air molecules above the liquid are very dilute at low or moderate pressures, and air is generally assumed to be inert to any liquid (the interaction between the air molecules and the liquid molecules is neglected). On the other hand, some scientists have proposed that most liquids have a saturated film of their own vapor at the liquid surface at room temperature, which forms very rapidly and instantaneously, and thus we always measure the *liquid–saturated vapor* surface tension instead of *liquid–air* surface tension. This may be a feasible scientific explanation, but it needs experimental proof and is worth investigating further.

Several authors have in the past stretched the term surface tension to imply that liquids have a saturated film in their surfaces, some mechanism such as a stretched membrane or *contractile skin*. However we should be careful since this view can lead to great problems when the structure of the so-called *skin* is considered in terms of molecules. Some writers propose that the surface molecules have their force-fields so deflected that they form a kind of linked skin in the surface, and the attractions between the surface molecules are directed along the surface instead of equally in all directions. This is wrong on two points:

1 Such a deflection of the force-field would be highly improbable unless the molecules are capable of a very special orientation in the surface layer.
2 If such a closely knitted skin were to be formed spontaneously, it would prevent rather than help the contraction of the surface, because the particularly strong linkage between the closely knitted surface molecules would tend to keep them at the surface rather than expel them from it as expected from a real contracted surface.

In addition, we cannot consider any concept of a skin at a liquid surface or at interfaces between immiscible liquids due to the following facts. The interfaces between liquid–air or immiscible liquid$_1$–liquid$_2$ normally possess a positive free energy. The interfaces between miscible liquids have a negative free energy and the molecules mix by a diffusion mechanism across the interface. If we suppose that a positive surface tension is due to a contractile skin formation at the interface, then any interface with a negative surface tension ought to have an expanding skin at the interface and should spontaneously extend itself during mixing. This would cause folding and puckering of the interfacial skin as it expands. Experimentally we know that miscible liquids do not mix by folding and puckering the interface between them; they only diffuse through the interface. Thus, in reality, the driving force of positive surface tension (or surface free energy) is the presence of an inward attractive force, exerted on the surface molecules by the underlying molecules. If the surface tension became negative, it would mean that the inward attractive force would be replaced by some other force tending to push the molecules outwards, away from the liquid. This is why the interfacial tension between two miscible liquids is negative and the vanishing of the positive interfacial tension is the condition of complete miscibility.

The surface area expansion process in Figure 3.5 must obey the basic thermodynamic reversibility rules so that the movement from equilibrium to both directions should be so slow that the system can be continually relaxed. For most low-viscosity liquids, their surfaces relax very rapidly, and this reversibility criterion is usually met. However, if the viscosity of the liquid is too high, the equilibrium cannot take place and the thermodynamical equilibrium equations cannot be used in these conditions. For solids, it is impossible to expand a solid surface reversibly under normal experimental conditions because it will break or crack rather than flow under pressure. However, this fact should not confuse us: surface tension of solids exists but we cannot apply a reversible area expansion method to solids because it cannot happen. Thus, solid surface tension determination can only be made by indirect methods such as liquid drop contact angle determination, or by applying various assumptions to some mechanical tests (see Chapters 8 and 9).

3.2.5 *Thermodynamics of Gibbs dividing interface and surface excess functions*

As already mentioned in Section 3.2.1, the Gibbs dividing interface is an imaginary mathematical interface enabling us to analyze the interfacial region between two phases. In this interfacial region between a liquid with its vapor, or between two immiscible liquids, the density of matter and of energy and entropy undergoes a gradual transition, as shown in Figures 3.1 *a* and 3.2 *a*. For example, for a single-component, two-phase system such as a liquid in equilibrium with its vapor, the density increases continuously from the low values in the bulk vapor phase (α-phase) to the high values in the bulk liquid phase (β-phase). (For two-component systems containing a solute and a solvent, these figures are redrawn in Section 5.6 as Figures 5.3 *a* and 5.3 *b*.) In order to apply thermodynamics, Gibbs compared a real system with a reference system formed by an imaginary Gibbs dividing surface, in which it is assumed that all the extensive properties of two bulk phases are unchanged up to the dividing surface, as shown in Figures 3.1 *b* and 3.2 *b*. There will be an *excess* number of moles of each component, energy and entropy (may be positive or negative) in the real system, as compared with the Gibbs reference system; all of these *excess* quantities are known as *surface excesses* of the various extensive properties of the system and are denoted by a superscript *S*. The surface excess of any component is defined as the amount by which the total quantity of that component in the actual system, given in Figures 3.1 *a* and 3.2 *a*, exceeds that in the idealized system, given in Figures 3.1 *b* and 3.2 *b*. As we know from Section 3.2.1, the *Gibbs dividing interface* is defined as an infinitesimally thin boundary layer with no volume ($V^S = 0$) and zero thickness. The various properties of the real system are then given by

$$V = V^\alpha + V^\beta \tag{191}$$

$$U = U^\alpha + U^\beta + U^S \tag{192}$$

$$S = S^\alpha + S^\beta + S^S \tag{193}$$

$$n_i = n_i^\alpha + n_i^\beta + n_i^S \tag{194}$$

where V, U, S and n_i are the total volume, internal energy, entropy and number of moles of component i, respectively; α and β refer to the two bulk phases of the reference system and U^S, S^S and n_i^S are all surface excess quantities.

When α and β phases are in contact, it is assumed throughout that α, β and the interface are in equilibrium with each other. However, the interphase region undergoes some changes, resulting in a concomitant change in the internal energy. This cannot be directly measured by experiment. However, the difference in internal energy between the two phases in contact and the same phases apart is accessible. But we do not know which part of the total energy of the system with the two contacting phases is attributed to the interface, U^S and which part to the bulk phases, U^α and U^β.

The concentration of the component i may be different in two phases so that $n_i^\alpha \neq n_i^\beta$ (but dictated by the condition of $\mu_i^\alpha = \mu_i^\beta$ at equilibrium, of course). Also, there is no requirement that the concentration should be uniform in the vicinity of the interface. Now, if we consider a binary immiscible solution, $i = 1, 2$, where the α phase mainly consists of 1 and the β phase mainly consists of 2; we should be careful that $\alpha \neq 1$ and $\beta \neq 2$ because of the presence of the number of moles of 1 and 2 in the interphase region. It is clear that $n_1 = n_1^\alpha + n_1^\beta + n_1^S$ and $n_2 = n_2^\alpha + n_2^\beta + n_2^S$, so that we can find the number of moles of each component in the interphase region by applying the $n_i^S = n_i - n_i^\alpha - n_i^\beta$ equation. We have many analytical techniques to determine the concentrations, as given by, $c_i = n_i/V$ but unfortunately we cannot use the $c_i^\alpha = n_i^\alpha/V^\alpha$ and $c_i^\beta = n_i^\beta/V^\beta$ equations because we do not know which part of the total volume, V can be used as V^α and which part V^β without knowing the exact location of the Gibbs dividing plane. The conclusion is that the value of the interfacial excess of a given compound, n_i^S, depends on the location of the Gibbs dividing plane. In differential form, Equation (192) can be written as

$$dU = dU^\alpha + dU^\beta + dU^S \tag{195}$$

By combining with the fundamental thermodynamic equation for open systems (Equation (142)), we can rewrite Equation (195)

$$dU = T^\alpha dS^\alpha - P^\alpha dV^\alpha + \sum_i \mu_i^\alpha dn_i^\alpha + T^\beta dS^\beta - P^\beta dV^\beta + \sum_i \mu_i^\beta dn_i^\beta + dU^S \tag{196}$$

Now we need a thermodynamic expression for the internal energy in the interfacial region dU^S. As we have already explained in Section 3.1.1, there are mainly two types of work function; one is pressure–volume work, W_{PV}, and the other is the *non-pressure–volume work*, W_{non-PV} comprising all other work types. The total work can be written as

$$dW = dW_{PV} + dW_{non-PV} = -PdV + dW_{non-PV} \tag{197}$$

By combining Equation (197) with the First Law of Thermodynamics (Equation (100)) we obtain

$$dU = dQ - PdV + dW_{non-PV} \tag{198}$$

For a reversible system at equilibrium, $dQ_{rev} = T\,dS$, and then we may modify the fundamental equation for an open system as,

$$dU_{rev} = TdS - PdV + dW_{non-PV} + \sum_i \mu_i dn_i \tag{199}$$

In every interfacial region, there is a tendency for mobile surfaces to decrease spontaneously in area. The system has a free energy and is doing work to reduce its interfacial area, and since $dA < 0$ for the decrease of area, it is obvious that dW_{non-PV} will be negative, as expected from a system doing work. Thus, the work spent by the system to decrease the surface area for an amount of unit area ($1\,m^2$) can be given as

$$dW_{non-PV} = \gamma\,dA \tag{200}$$

By combining Equations (199) and (200), we can write for a general expression of the *surface excess* of the internal energy

$$dU^S = T^S dS^S - P^S dV^S + \gamma dA^S + \sum_i \mu_i^S dn_i^S \tag{201}$$

Then, the surface tension (or interfacial tension) can be expressed thermodynamically from Equation (201) as

$$\gamma = \left(\frac{\partial U^S}{\partial A^S} \right)_{S^S, V^S, n_i^S} \tag{202}$$

and recalling that in the Gibbs dividing interface analysis $V^S = 0$ for the interfacial region, then the volume terms can be discarded in Equation (201). Unfortunately, Equation (202) is impractical in experimental respects, because it is a very difficult task to conduct an experiment in which we can expand the surface area at constant volume and entropy conditions.

As we have already stated, the value of n_i^S depends on the location of the dividing surface, and this can be seen in Figure 3.1 *b*. The volume of the infinitesimally thin slice taken in parallel in the interfacial region may be calculated as $dV = A\,dx_\gamma$ in this figure, where A is the cross-sectional area = real interfacial area between two planar, flat phases. The total number of moles in the system, n_i, can be obtained from $c_i = n_i/V$ by integrating the equation of the curve, $c = f(x_\gamma)$, from 0 to d. If the homogeneity of the bulk phases α and β persists up to the Gibbs dividing surface at $x_{\gamma0}$, then using the $c_i^\alpha = n_i^\alpha/V^\alpha$ expression, n_i^α can be calculated by A times the rectangular area under the lower horizontal line between 0 and $x_{\gamma0}$, from simple geometry. Analogously, n_i^β can be calculated by A times the rectangular area under the higher horizontal line between $x_{\gamma0}$ and d (without considering the $c = f(x_\gamma)$ curve). However this is not true in reality, because the amounts n_i^α and n_i^β are equal to A times the rectangular areas under the lower and upper horizontal lines between 0 and x_α, and between x_β and d, respectively, plus their amounts in the interphase region between x_α and x_β. Consequently, we have to calculate n_i^S by using the $c = f(x_\gamma)$ curve present. To the left of the $x_{\gamma0}$, under the $c = f(x_\gamma)$ curve, above the c^α horizontal line, we see that there is additional n_i which we should add; and at the right of x_{γ_0} we see that some n_i is absent over the $c = f(x_\gamma)$ curve, under the c^β horizontal line, which we should subtract. Since $n_i^S = n_i - (n_i^\alpha + n_i^\beta)$, the surface excess amount, n_i^S is equal to A times the difference between the area under the $c = f(x_\gamma)$ curve and the rectangular areas under the horizontal lines. Therefore this area difference is equal to the shaded areas in Figure 3.1 *b* so that the shaded area at the left of x_{γ_0} will be plus and the shaded area to the right of x_{γ_0} will be minus, to find n_i^S. Arbitrarily, if we locate the dividing plane so that the positive and negative shaded areas are equal to each other, then we have $n_i^S = 0$. We may check this on the plot, if the dividing plane in Figure 3.1 *b* moves to the right, the positive area will increase, and the negative area will decrease giving a positive n_i^S value; and if the dividing plane moves to the left then n_i^S becomes negative. Similar arguments show that S^S and U^S also depend on the location of the dividing plane. We will see the results of the choice of the location of the Gibbs dividing plane in adsorption processes given in Sections 3.3, 5.6 and 8.3.

The Helmholtz free energy parameter, $dF = dU - T\,dS - S\,dT$ was given in Equation (120), and if we combine Equations (120) and (201) we obtain

$$dF^S = \gamma\,dA^S - P^S dV^S - S^S dT^S + \sum_i \mu_i^S dn_i^S \tag{203}$$

On the other hand, the total Helmholtz free energy of the whole system, is the sum of the Helmholtz free energy of the α, β and the interfacial phases, given as $dF = dF^\alpha + dF^\beta + dF^S$, and we may write by combining Equations (120) and (142)

$$dF = -S^\alpha dT^\alpha - P^\alpha dV^\alpha + \sum_i \mu_i^\alpha dn_i^\alpha - S^\beta dT^\beta - P^\beta dV^\beta + \sum_i \mu_i^\beta dn_i^\beta + dF^S \tag{204}$$

If the volume, temperature and the number of moles are taken as constant, then the explicit thermodynamic definition of surface tension may be written from Equations (203) and (204),

$$\gamma = \left(\frac{\partial F^S}{\partial A^S}\right)_{V^S,T^S,n_i^S} = \left(\frac{\partial F}{\partial A}\right)_{V,T,n_i} \tag{205}$$

for a plane interface. This means that γ is the isothermal reversible work done to extend the interface by unit area at constant V and n_i. For a pure, one-component material, we can write, $\gamma = (\partial F/\partial A)_{V,T}$ from Equation (205), where the mass of the material is held constant during the extension of the surface.

On the other hand, the Gibbs free energy function is defined as $G \equiv U + PV - TS$ from Equations (104) and (124), which may be expressed as, $dG = dU + PdV + VdP - TdS - SdT$, and the *excess Gibbs free energy*, G^S, of the interfacial region in a reversible process, for a completely plane interface can be expressed as

$$dG^S = V^S dP^S - S^S dT^S + \gamma\,dA^S + \sum_i \mu_i^S dn_i^S \tag{206}$$

The total Gibbs free energy of the whole system is the sum of the Gibbs free energy of the α, β and the interfacial phases, $dG = dG^\alpha + dG^\beta + dG^S$ and is analogous to the Helmholtz free energy derivation given in Equation (205), and if the pressure, temperature and the number of moles are taken as constant in the plane interphase region, we may write for the surface tension

$$\gamma = \left(\frac{\partial G^S}{\partial A^S}\right)_{P^S,T^S,n_i^S} = \left(\frac{\partial G}{\partial A}\right)_{P,T,n_i} \tag{207}$$

We can also derive the *surface excess entropy* from Equations (204) and (206) so that

$$S^S = -\left(\frac{\partial G^S}{\partial T^S}\right)_{P^S,A^S,n_i^S} = -\left(\frac{\partial F^S}{\partial T^S}\right)_{V^S,A^S,n_i^S} \tag{208}$$

and the surface excess pressure can be found similarly from Equations (205)–(207)

$$P^S = -\left(\frac{\partial F^S}{\partial V^S}\right)_{A^S,T^S,n_i^S} \tag{209}$$

In summary, by the use of Equations (202), (205) and (207), we are able to realize the effect of variation of interfacial area that separates the various phases on the quantity of total interaction, as shown schematically in Figures 3.6 *a* and 3.6 *b*. The total surface area in Figure 3.6 *b* is much larger than the same material in Figure 3.6 *a* and, of course, the extent of molecular interaction in Figure 3.6 *b* is much larger. We should recall that previously no distinction was made between the systems in equilibrium that have an abundance of surface area and those that do not.

There is a dispute about the equality of thermodynamic (surface free energy) and mechanical (surface tension) γ terms in some books. Some authors extend the surface tension notion so that if the interfacial area A is expanded at constant P and T, then γ can be interpreted as $(\partial U^S/\partial A^S)_{P^S,T^S,n_i^S}$, but this is incorrect due to the fact that $dG \neq dU$. By combining Equations (124) and (207), we may write in strict thermodynamical terms

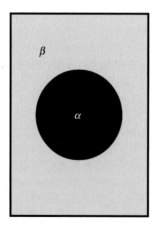

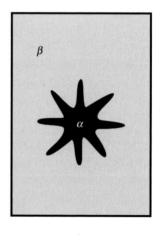

a   b

Figure 3.6 Schematic representation of the interfacial area that separates the same α and β phases for varying total interfacial areas: **a.** The interfacial area is minimal if phase α is a sphere forming a minimum total interaction energy with phase β. **b.** When the same volume of phase α has such a shape with a large interfacial area with phase β, then the total interaction energy is much larger than the case in **a.**

$$\gamma = \left(\frac{\partial U^S}{\partial A^S}\right)_{P^S,T^S,n_i^S} + P\left(\frac{\partial V^S}{\partial A^S}\right)_{P^S,T^S,n_i^S} - T\left(\frac{\partial S^S}{\partial A^S}\right)_{P^S,T^S,n_i^S} \tag{210}$$

and if we assume that mechanical surface tension, γ is equal to $(\partial U^S/\partial A^S)_{P^S,T^S,n_i^S}$ this means that we neglect the terms $P(\partial V^S/\partial A^S)_{P^S,T^S,n_i^S}$ and $T(\partial S^S/\partial A^S)_{P^S,T^S,n_i^S}$. The first one can be neglected because $dV^S = 0$ in the Gibbs convention, and in addition, it is also negligible in real systems where the volume expansion may take place, because this term has only a very minor effect in magnitude due to the fact that V^S is very small compared with the other terms. However the second (entropic) term is not negligible because there is a considerable entropy increase from the increase of the interfacial area in normal conditions and consequently the mechanical surface tension interpretation is thermodynamically wrong. A constant entropy process may only perform in adiabatic conditions and the entropic term may only be neglected for this case, when the mechanical interpretation would have been correct.

If we integrate the Helmholtz free energy equation for the interface (Equation (203)) holding constant the intensive properties T, P, μ_i and γ, we have

$$F^S = \gamma A^S - P^S V^S + \sum_i \mu_i^S n_i^S \tag{211}$$

Since $V^S = 0$ in the Gibbs convention, we will have, $F^S = \gamma A^S + \sum_i \mu_i^S n_i^S$ so that, unless $\sum_i \mu_i^S n_i^S$ is zero, the Helmholtz free energy is not equal to surface tension in Equation (211). Since $G \equiv F + PV$ from Equation (128), then we have $[G^S = F^S]$ according to the Gibbs convention, indicating that unless $\sum_i \mu_i^S n_i^S$ is zero, the Gibbs free energy is not equal to surface tension either. The $\sum_i \mu_i^S n_i^S$ term is zero when there is no adsorption on the interface, and consequently, the Gibbs and Helmholtz free energies may be equal to the surface tension for pure liquids according to the Gibbs convention but are never equal for solutions.

The effects of the variation of temperature on the surface tension and surface excess internal energy can be predicted from surface thermodynamics. At constant pressure ($dP = 0$) with varying temperature, Equation (206) can be written as

$$(dG^S)_P = [\gamma dA^S - S^S dT^S + \sum_i \mu_i^S dn_i^S]_P \tag{212}$$

Similar to the derivation of the Maxwell equations (Equations (135)–(138)) by applying Euler's reciprocity theorem, for $G^S = f(T^S, A^S, n_i^S)$ one obtains

$$\left(\frac{\partial \gamma}{\partial T}\right)_{A^S,n_i^S} = -\left(\frac{\partial S^S}{\partial A}\right)_{T^S,n_i^S} \tag{213}$$

When the temperature increases, the molecular interactions weaken, and the surface tension generally decreases. Since $\left(\dfrac{\partial \gamma}{\partial T}\right)_{A^S,n_i^S}$ is almost always negative, then the right-hand side of Equation (213) is positive, so that the entropy increases when the surface area expands at a constant temperature. We may write from Equation (213)

$$dS^S = -\left(\frac{\partial \gamma}{\partial T}\right)_{A^S,n_i^S} dA^S \tag{214}$$

and recalling $V^S = 0$ in the Gibbs convention, by combining Equations (201) and (214) for the constant n_i^S condition, one obtains

$$dU^S = \left[\gamma - T^S\left(\frac{\partial \gamma}{\partial T^S}\right)_{A^S,n_i^S}\right] dA^S \tag{215}$$

By integrating Equation (215) for a surface area expansion from 0 to $1\,m^2$, at a constant temperature, we obtain

$$U^S = \gamma - T^S\left(\frac{\partial \gamma}{\partial T^S}\right)_{A,n_i} \tag{216}$$

showing that the excess surface internal energy, U^S, is almost always larger than the surface free energy, γ. (U^S is more easily related to molecular models.) However, U^S is less affected by temperature changes than is γ, and U^S is generally assumed to be almost temperature-independent.

Similar to the derivation of the Gibbs–Duhem equation, it is also possible to show the dependence of surface tension on the chemical potentials of the components in the interfacial region. If we integrate Equation (201) between zero and a finite value at constant A, T and n_i, to allow the internal energy, entropy and mole number to almost from zero to some finite value, this gives

$$U^S = T^S S^S + \gamma A^S + \sum_i \mu_i^S n_i^S \tag{217}$$

Differentiation of this equation results in a more general expression of dU^S,

$$dU^S = T^S dS^S + S^S dT^S + \gamma dA^S + A^S d\gamma + \sum_i \mu_i^S dn_i^S + \sum_i n_i^S d\mu_i^S \tag{218}$$

By combining Equations (201) and (218), we obtain

$$S^S dT^S + A^S d\gamma + \sum_i n_i^S d\mu_i^S = 0 \tag{219}$$

which gives the dependence of surface tension on the chemical potentials of the existing components as an analogy to the Gibbs–Duhem equation for the hypothetical surface phase in the Gibbs model system.

When the temperature is constant, it simplifies into

$$d\gamma = -\frac{\sum_i n_i^S d\mu_i^S}{A^S} \tag{220}$$

Equation (220) is very important for the adsorption of matter on surfaces (see Section 3.3).

If we want to define the surface excess chemical potential, μ_i^S, in terms of other surface excess parameters, we may write from Equations (201), (203) and (206)

$$\mu_i^S = \left(\frac{\partial U^S}{\partial n_i^S}\right)_{S^S,V^S,A^S,n_j^S} = \left(\frac{\partial G^S}{\partial n_i^S}\right)_{T^S,P^S A^S,n_j^S} = \left(\frac{\partial F^S}{\partial n_i^S}\right)_{T^S,V^S,A^S,n_j^S} \tag{221}$$

In equilibrium, the surface chemical potential is equal to the bulk ($\mu_i^S = \mu_i$), and the chemical potential change on addition of any material is independent of the α, β, interface phases.

In the above section, we outlined the treatment given by Gibbs in 1878 where V^S is assumed to be 0. Later, in 1936, Guggenheim derived an alternative formulation, in which the interfacial region is thought of as a separate phase, of small but finite thickness and hence of volume, V^S, which leads to similar equations to Gibbs'. In Guggenheim's treatment the quantities V^S, G^S, F^S, n_i^S, etc. are then the variables of the interphase of finite thickness, and cannot be called excess functions. The two treatments are equivalent, both models having advantages and disadvantages, which is used as a matter of convenience. (When the Guggenheim convention for surfaces was applied, the Gibbs and Helmholtz free energies were found to be unequal to surface tension even for pure liquids, but the proof of this is beyond the scope of this book.)

Unfortunately, there are several limitations to the above arguments for the Gibbs convention. First of all, we should keep in mind that Gibbs or Guggenheim's treatments apply only to plane fluid interfaces. If the interface has a finite curvature, the problem is considerably more complex as we will see in the capillarity sections (Sections 4.3 and 4.4). Second, if the surface is a solid, it can withstand an anisotropic stress, and it is a very complex matter to apply thermodynamics to this case. Third, in the derivation of all the above equations we have assumed that the system is in thermodynamic equilibrium. However, this is not always true. The minimization of the excess surface free energy, either in Gibbs or Helmholtz functions as given in Equations (204) and (207), is not an equilibrium condition, unless the surface itself can be considered to be a completely closed system; this is not generally the case, especially for liquid systems containing insoluble monolayers at the surface. Thus, these monolayers are generally not stable, but they may either be metastable or not in any equilibrium at all (see Section 5.5). Lastly, for a pure liquid, which is in equilibrium with its saturated vapor at a plane interface, where n_i is constant, the surface tension at constant pressure and temperature can be found by integrating Equation (207) to give

$$\gamma_o = \left(\frac{\partial G_o^S}{\partial A^S} \right)_{p^S, T^S} = \frac{\Delta G_o^S}{A^S} \tag{222}$$

where subscript o shows the limitation to a pure single-component system. For multi-component systems especially for solute–solvent solutions this relation is not applicable, a point which has been overlooked by several researchers in the field (see Chapters 5 and 6). For pure liquids, the *specific surface excess entropy*, S_o^S, is given as $[S_o^S = -(\mathrm{d}\gamma/\mathrm{d}T)_{P,A}]$, where S_o^S is the entropy of a unit area of surface liquid less the entropy of the same amount of bulk liquid (this equation is derived in Section 4.5).

3.3 Thermodynamics of Adsorption

As we have already seen in Section 3.2.4, a pure liquid decreases its surface free energy by diminishing its surface area to the minimum possible, with molecules leaving the surface for the interior under the action of the inward attractive force exerted on the surface molecules. Surface molecules orient themselves so that the functional groups with the largest field of force point inwards. However, in the case of solutions of two or more substances, the surface minimization process is different and more complex, because different molecules present in a solution have different intensities of attractive force-fields and also have different molecular volumes and shapes. As a result, the molecules that have the greater fields of force tend to pass into the interior, and those with the smaller force-fields remain at the surface, thus causing a concentration difference for any solute between the bulk of the solution and the surface region. At the same time the surface area will decrease to the minimum permitted by the external constraints on the system. The surface layer of a solution, in comparison with the interior, will therefore be more concentrated in the constituents that have smaller attractive force-fields, and thus whose intrinsic surface free energies are the smallest. This concentration difference of one constituent of a solution at a surface is known as *adsorption*. In other words, *adsorption* is the partitioning of a chemical species between a bulk phase and an interface. The adsorption term is used both for the amount of molecules (moles or number of molecules, weights, volumes) accumulated per unit area $(1\,m^2)$ and for the *process of molecular movement* to the interface. *Desorption* is the reverse of the adsorption process, showing that the molecules are leaving the interface towards the solution. The material in the adsorbed state is called the *adsorbate*, and the material which has the potential to be adsorbed is called *adsorptive*. The substance on which adsorption takes place is defined as the *adsorbent*. There are various forms of adsorption: gas adsorption at a liquid surface from a pure gas or gas mixture (see Section 5.5), solute adsorption at a liquid surface from the interior of a solution (see Section 5.5), and gas and liquid adsorption at a solid surface (see Sections 8.3 and 9.2). In the case of gas adsorption, when a solid is in equilibrium with a gas, the gas is usually more concentrated in the surface region and is said to be *positively adsorbed*. However, if a species is less concentrated in the interfacial region than in the bulk, it is said to be *negatively adsorbed*. For example the surface concentration of some ions is less than in the bulk for some simple inorganic electrolytes. Adsorption is different from *absorption* where a species penetrates and is dissolved throughout a liquid or a solid bulk phase. However, the two processes may take place simultaneously; for example, ethanol vapor may dissolve in bulk liquid water and also be positively adsorbed at the water surface.

In molecular energy terms, adsorption occurs when a molecule loses sufficient energy to the atoms in a surface by exciting them vibrationally or electronically to become effectively bound to the surface. An ensemble of adsorbed molecules is called an *adlayer* (or *monolayer* if only a single molecular layer forms), and the average time of stay of a molecule upon the surface is called the *mean stay time*.

If a system consists of more than one component, then the surface tension may vary with the solution composition. In practice, if the solvent has a larger surface tension than the solution, then the solute is adsorbed at the surface. This is valid for most aqueous solutions because water has a high surface tension of $72.8\,mNm^{-1}$. Thus, the water molecules have stronger attractive force-fields than most of the solutes, and thus water molecules move inwards more rapidly than solute molecules, leaving the surface more concentrated in the solute molecules causing the *positive* adsorption. It is possible to obtain very considerable decreases in surface tension by adsorption of a low surface tension solute in any high surface tension solvent. In such cases, the adsorption may proceed so far that the surface layer consists almost entirely of the solute molecules (a monolayer) with the smaller field of force. However, the reverse is not true, if we add a very high surface tension solute to a low-surface-tension solvent; the solute molecules will move to the interior forming large aggregates (or droplets), and the surface layer will tend to consist mainly of pure solvent, containing few solute molecules; then the surface tension is raised but barely above that of the pure solvent. Positive adsorption always leads to a lowering of the solvent surface tension.

Surfactant molecules in aqueous solutions have a great tendency to adsorb from the solution, and their concentration in the water–air interface is generally much higher than in the bulk aqueous solution, considerably decreasing the surface tension of the solution (see Sections 5.5 and 5.6). In extreme cases, the surface tension of the solution may be reduced to about $25\,\mathrm{mNm^{-1}}$, which is nearly the surface tension of paraffin hydrocarbons, the surface layer then consisting almost entirely of long hydrocarbon chains. On the other hand, when inorganic salts are dissolved in water, negative adsorption usually takes place, and the surface tension of the solution may be raised by only a few $\mathrm{mNm^{-1}}$.

3.3.1 Gibbs adsorption isotherm

One of the main objectives in surface science is the prediction of the amount of substance that is adsorbed at an interface. Adsorption and interfacial free energy are related through the *Gibbs Adsorption Law*. If we define Γ_i as the *excess moles of the component, i,* adsorbed at the interface per unit area of an interphase,

$$\Gamma_i \equiv \frac{n_i^S}{A^S} \tag{223}$$

then by combining Equations (220) and (223) we obtain the *Gibbs adsorption isotherm* expression at a constant temperature,

$$d\gamma = -\sum_i \Gamma_i d\mu_i^S \tag{224}$$

The excess moles term, Γ_i of any component is defined as the amount by which the total quantity of that component in the actual system given in Figures 3.1 *a* and 3.2 *a* exceeds that in the idealized system given in Figures 3.1 *b* and 3.2 *b*. For a two-component system, Equation (224) can be reduced to

$$-d\gamma = \Gamma_1 \, d\mu_1^S + \Gamma_2 \, d\mu_2^S \tag{225}$$

where the subscript 1 refers to the solvent and 2 to the solute. We must recall that $\mu_1^S = \mu_1$ and $\mu_2^S = \mu_2$, from the Gibbs phase rule. The terms Γ_1 and Γ_2 in Equation (225) are both unknown and, as noted earlier, can be defined relative to an arbitrarily chosen Gibbs dividing surface, as a plane of infinitesimal thickness. However, in thermodynamic terms Equation (225) holds regardless of the location of the Gibbs dividing surface. In theory, any experimentally measurable property must be independent of the location of the dividing plane, but we have no better tool to apply the thermodynamic argument for choosing the location of the Gibbs dividing plane; we may then look for some practical considerations for the preference. In order to obtain physically meaningful quantities, the most widely used convention in dealing with solutions is to place the Gibbs dividing plane so that the surface excess of the solvent is zero, $\Gamma_1 = 0$, and then Equation (225) reduces to

$$\Gamma_2 = -\left(\frac{d\gamma}{d\mu_2}\right)_T \tag{226}$$

The activity of a solute in a solution, a_2, was given in Equation (169) so that $\mu_2 = \mu_2^* + RT^S$ ln a_2, and its total differentiation gives $d\mu_2 = d\mu_2^* + RT^S$ d ln a_2. Since the standard chemical potential of component 2 in the solution, μ_2^*, is a constant quantity, then $d\mu_2^* = 0$ and we have $[d\mu_2 = RT^S$ d ln $a_2]$, so that by applying the [d ln a = da/a] rule from mathematics, Equation (226) becomes

$$\Gamma_2 = -\left[\frac{d\gamma}{(RT^S d \ln a_2)}\right]_{TS} = -\frac{a_2}{RT^S}\left(\frac{d\gamma}{da_2}\right)_{TS} \tag{227}$$

where Γ_2 is the *excess moles of the solute* adsorbed per unit area at the interphase. Equation (227) is the *Gibbs adsorption isotherm* which is one of the most important equations in surface science. Using this, the calculation of the number of surface excess moles is possible by measuring the variation of the surface tension of a solution or vice versa. Equation (227) shows that if the surface tension of a solution decreases as the solute activity increases (so that the $(\Delta\gamma/\Delta a_2)$ term becomes negative giving a positive Γ_2), then there is a *positive adsorption* (or *positive surface molar excess*) of the solute at the interphase region. If the solution is ideal, as $\varphi_2^x = 1$ in Equation (168) $[a_2 = \varphi_2^x X_2]$, then Equation (227) becomes

$$\Gamma_2 = -\frac{X_2}{RT^S}\left(\frac{d\gamma}{dX_2}\right)_T \tag{228}$$

where X_2 is the mole fraction of the solute. In addition, since $a_2 = \varphi_2^c c_2$, if the solution is dilute enough, then we assume $\varphi_2^c = 1$ and $a_2 = c_2$ giving

$$\Gamma_2 = -\frac{c_2}{RT^S}\left(\frac{d\gamma}{dc_2}\right)_T \tag{229}$$

where c_2 is the molar concentration (mol l^{-1}) of the solute. Equations (227)–(229) are widely used to evaluate the extent of adsorption in dilute solutions from surface tension measurements. The determination of the slope of the plot of γ versus the logarithm of concentration (or activity) is the first step to calculate the surface excess of the solute. (The choice of the concentration units is immaterial for this calculation due to the presence of a concentration term in both numerator and denominator of the fraction.) In systems such as liquid–air and liquid$_1$–liquid$_2$, where γ is directly measurable, the Gibbs adsorption equation may be used to determine the surface concentration. In solid–gas systems where the surface concentration can be measured directly but γ cannot, the Gibbs adsorption equation may be used to calculate the lowering of γ which would not otherwise be possible.

The thickness of adsorbed layers depends on the extent of molecular interactions, but usually films of one molecule thickness, which are called *monolayers*, form at the liquid–vapor and liquid–liquid interfaces upon adsorption. In gas adsorption on solids, several molecular layers, which are called *multilayers*, form at high pressures, and *monolayers* can only be formed if the gas pressure is sufficiently low. If van der Waals forces are operative during the adsorption process, it is called *physical adsorption* or *physiosorption* (see Section 8.3.1), whereas if chemical bonds are formed during the adsorption process, then it is called *chemical adsorption* or more preferably *chemisorption* (see Section 8.3.2).

The adsorption at liquid–vapor and liquid–liquid interfaces is generally physical in nature and the adsorbed molecules may easily be desorbed from the surface by lowering the bulk concentration of the adsorbate. For example, when we dilute the aqueous solution of ethanol with water, the adsorbed ethanol molecules will desorb, and the surface tension of the solution rises. On the other hand, chemisorbed molecules are much more difficult to desorb.

If we assume that the surface layer is monomolecular, then Γ_2 can be converted into the area per molecule of the solute. Alternatively, if we know the molecular areas of the solvent and the solute from other sources, the relative number of solute and solvent molecules in the surface monolayer can be calculated. Adsorbed amounts are easily measured, either directly or from depletion of the liquid (or gas) phase, to determine the unadsorbed material concentration, but the total interfacial area, A, must also be known in order to obtain the Γ value. Since nearly every surface has a roughness, the evaluation of A is generally difficult and requires special techniques.

We should note that the evaluation of Γ_2 depends on assigning a particular position to the dividing surface. This is because the amount of each component in the α or β phases in the idealized Gibbs system given in Fig. 3.2 *b* clearly depends on the precise position of the Gibbs dividing plane in the interphase region. When we assume that the shaded areas are equal for the solvent, $\Gamma_1 = 0$, this means that $n_1^S = 0$, and we are taking identical numbers of moles of solvent in two portions of the solution, one from the surface region and the other from the bulk, so that there would be no surface excesses, because of [n_1(surface) = n_1(bulk)]. This can be shown in Figure 3.1 *b* such that $x_{\gamma o}$ is located in a place where the shaded area to the left of $x_{\gamma o}$ is equal to the shaded area to the right of $x_{\gamma o}$. In these conditions, the surface excess of the adsorbed molecules, Γ_2, is defined as $\Gamma_2 = [n_2$ (surface) $- n_2$ (bulk)]$/A^S$, and if there is a positive surface excess, this shows that n_2 (surface) $> n_2$ (bulk). We may apply this argument to any thermodynamical parameter such as G or U. In this case we must locate the Gibbs dividing plane at a place where G^S or U^S equals zero. As a numerical illustration we may check the surface of a 50–50% mole methanol–water solution. If we take a slice of the surface region deep enough to contain some of the bulk solution and find, let us say, 600 moles of methanol and 200 moles of water for A m^2 of the flat interfacial area, then we can find Γ_2 (methanol) just by comparing with the bulk solution. Since there are 200 moles of methanol per 200 moles of water in the bulk solution, then Γ_2 (methanol) can be calculated as $(600 - 200)/A = 400/A$, as a positive surface excess. If we try the reverse process, that is if we check the negative surface excess (surface deficiency) of water, then we have to compare the sample concentration with 600 moles of water per 600 moles of methanol in the bulk, so that Γ_1 (water) can be calculated as $(200 - 600)/A = -400/A$ as a negative surface excess. Surface excess values Γ_1 or Γ_2 are algebraic quantities depending on the moles, volumes, weights of the materials, and may be positive or negative depending on the convention chosen for Γ. It is a matter of choice to locate the dividing line at any value of x in the range of Δx. When various different properties are plotted, we generally obtain different curve profiles so that, when we choose $x_{\gamma o}$ for equal shaded areas for a property (say for concentration), we will obtain differently (unbalanced) divided shaded areas of other properties (such as refractive index etc.) with the same location of the dividing plane, and this may confuse the surface thermodynamic treatment. To overcome this obvious difficulty, the property defined as having zero surface excess may be chosen at will, and $x_{\gamma o}$ is located according to this property, which is the most amenable to thermodynamic evaluation by the experimental or mathematical features of the problem at hand. On the other hand, the convention of assuming $\Gamma_1 = 0$ is mathematically unsymmetrical but it has great convenience in the case of binary dilute solutions. How this arbitrary assignment affects the magnitude of the calculated Γ_2 is a matter of great importance especially to any researcher who wishes to apply Equations (227)–(229) to their experimental data. It is important to understand clearly what conventions are used in the definitions of these quantities in any research paper.

There are some alternative methods for locating the dividing surface: Γ_2 can be experimentally measured at the liquid–vapor interface by using radioactive tracer methods or it may also be determined using ellipsometry so that the thickness of an adsorbed film is calculated from the ellipticity produced in light reflected from the film covered surface. On the other hand, the theoretical calculation of Γ_2 is also possible using Monte Carlo and the molecular dynamics methods.

For the adsorption of gases or volatile materials on liquid surfaces, if the gas or vapor obeys the ideal gas laws, Equation (227) may be written as

$$\Gamma_2 = -\frac{P_2}{RT^S}\left(\frac{d\gamma}{dP_2}\right)_T \tag{230}$$

where P_2 is the gas or vapor partial pressure (see Section 8.3.3).

3.3.2 Surface equation of state

The Gibbs adsorption isotherm shows the dependence of the extent of adsorption of an adsorbent on its bulk concentration or pressure. However, we also need to know the state of the adsorbate at the surface. These are interrelated because the extent of material adsorption on a surface depends on the state of the surface. The behavior of the molecules in the surface film is expressed by a *surface equation of state* which relates the *spreading pressure*, π, which is the difference between the solvent and solution surface tensions, $[\pi = \gamma_0 - \gamma]$ to the surface concentration of the adsorbent. This equation is concerned with the lateral motions and interactions of the molecules present in an adsorbed film. In general, the *surface equation of state* is a two-dimensional analogue of the three-dimensional equation of state of fluids, and since this is related to monomolecular films, it will be described in Sections 5.5 and 5.6. It should be remembered that on liquid surfaces, usually monolayers form, but with adsorption on solid surfaces, usually multilayers form (see Section 8.3).

3.4 Conditions of Equilibrium where Several Surfaces Intersect

When three different phases make contact with each other, where three surfaces intersect at a triple point, we obtain three contact angles and three interfacial tension values, as can be seen in Figure 3.7. We can obtain contact angle equilibria when we place an immiscible drop on a liquid or solid in air or a vapor phase; there are many applications of contact angle measurement in industry and surface science (see Chapter 9). In these conditions, the total excess surface internal energy can be written from Equation (201) so that

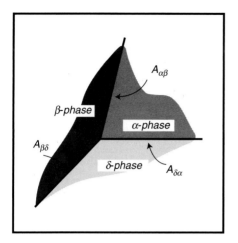
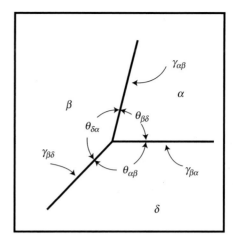

a. Three phases make contact in space at the triple point

b. The definitions of angles and surface tensions at the triple point

Figure 3.7   *a.* Schematic representation of the conditions of equilibrium where three phases (α, β and δ) intersect at a triple point, where $A_{\alpha\beta}$ is the contact area between phases (α and β); $A_{\beta\delta}$ is the contact area between phases (β and δ) and $A_{\delta\alpha}$ is the contact area between phases (δ and α). *b.* Similar schematic representation giving the interfacial tensions and angles between the three phases.

$$(dU^S)_{S,V,n_i} = \gamma_{\alpha\beta}dA^S_{\alpha\beta} + \gamma_{\beta\delta}dA^S_{\beta\delta} + \gamma_{\delta\alpha}dA^S_{\delta\alpha} \qquad (231)$$

and the total excess surface free energy can be written from Equation (206) so that

$$(dG^S)_{P,T,n_i} = \gamma_{\alpha\beta}dA^S_{\alpha\beta} + \gamma_{\beta\delta}dA^S_{\beta\delta} + \gamma_{\delta\alpha}dA^S_{\delta\alpha} \qquad (232)$$

At equilibrium, $(dU^S)_{S,V,n_i}$ and $(dG^S)_{P,T,n_i}$ must be a minimum, and from this requirement we can derive a relation for the angles of contact at the triple line:

$$\frac{\sin\theta_{\alpha\beta}}{\gamma_{\alpha\beta}} = \frac{\sin\theta_{\beta\delta}}{\gamma_{\beta\delta}} = \frac{\sin\theta_{\delta\alpha}}{\gamma_{\delta\alpha}} \qquad (233)$$

This equation is equivalent to the static force balance where each of the γ terms is considered as the force applied to the vertex. This condition of equilibrium is often called a *Neumann triangle*. Equation (233) is the basis of Young's equation, which is used in the contact angle determination of liquid drops on flat substrates (see Sections 5.5 and 9.1).

3.5   Relation of Thermodynamic Parameters with Intermolecular Forces

There is a need to find out the relationship between molecular pair potentials and thermodynamic energy functions, because all macroscopic bulk and surface properties arise from molecular interactions. For example, adsorption occurs when a molecule loses sufficient energy to the atoms in a surface by exciting them vibrationally, electrostatically or electronically to become effectively bound to the surface. In general, a wide variety of events may occur when a molecule impinges upon a surface. It may bounce back with no loss of energy, or it may suffer a redistribution of momentum and be diffracted by the surface, again with no loss of energy. Alternatively, it may lose a small amount of energy to the atoms or molecules in the surface by exciting them by some means, however this energy loss is not enough to bind that molecule to the surface and it may be inelastically reflected. Only after it loses sufficient energy may it be *accommodated* or *adsorbed* onto the surface. The magnitude of the energy loss (or the intermolecular interaction) depends upon the nature of the molecules (or atoms) involved.

It is very desirable to calculate the extent of this interaction in thermodynamical terms from the pair intermolecular potential values at hand, however it is a very difficult task at present. In general, quantum mechanics is used to determine the molecular properties from first principles. The link between thermodynamics and quantum mechanics is provided by *statistical mechanics*. When applied to equilibrium conditions, statistical mechanics is called *statistical thermodynamics* and is used to deduce the macroscopic properties of matter such as internal energy, entropy, heat capacity, surface tension, electrical conductivity, viscosity etc. from the molecular properties of the system, such as molecular volume, geometry, and intra- and intermolecular forces. The reverse is also true. Because of the very large number of molecules in a system, there are a huge number of different quantum states that are compatible with a given thermodynamic state, so that there are a huge number of different ways in which we can populate these energy levels and still end up with the same total internal energy. *Canonical ensembles* are used to apply statistical thermodynamics to ideal, non-interacting molecules. A canonical ensemble is a standard hypothetical collection of an indefinite number of non-interacting systems. The main assumption is that the time-average of a macroscopic property of a system is equal to the average value of that property in the canonical ensemble. As an example, in classical gas kinetic theory, the measured pressure is the time-average over the impacts of individual molecules on the walls of the container. In very simple cases, such time-averaging computing can be performed but it is impossible to do this for many complex systems. The above assumption on the canonical ensembles allows us to replace the difficult calculation of a time-average by the easier calculation of an average over the microstate systems in the ensemble at a fixed time. In order to achieve this, we should include the probability that a system in the ensemble has this property (i.e. energy for many cases) and thus *canonical partition functions* can be used in statistical mechanical calculations. The application of statistical thermodynamics is beyond the scope of this book. However, we should note that when real liquids and non-ideal gases with intermolecular interactions are considered, the canonical partition function must contain the potential energy of the interaction, $V(r)$; unfortunately the exact analytical solution is almost impossible. Moreover, it is a hopeless task to solve the Schrödinger equation for the entire system of Avogadro's number (6.02×10^{23}) of molecules to obtain the system quantum energies for only one mole of a material. In addition, there is a significant difference in the physical interpretation of intermolecular forces between gaseous molecules in free space, and those between liquid molecules in a dense fluid medium. Our quantitative knowledge of two-body intermolecular pair potentials is limited to simple systems under ideal conditions, so that we measure the interaction between two gas molecules, which are isolated from all others. In gases at low densities, this is applicable because there is little probability that three gas molecules will be simultaneously close together, and we can assume that the interaction of two gas molecules is unaffected by the nearby presence of a third molecule. However this approximation is not true for high-density gases, liquids and solids where the presence of the third molecule will polarize the interaction of the first two molecules. We have to consider many-body effects, molecular size and orientation effects, and repulsion interactions to find an exact analytical solution; this is almost impossible. As a result, instead of applying quantum mechanics directly, the use of approximate molecular interaction equations seems feasible at present. The experimental determination of virial coefficients from P–V–T data and transport properties of real gases, and the excess properties of liquid mixtures and some spectroscopic techniques such as molecular-beam scattering yield data for intermolecular pair potential values. During the treatment of these data, it is generally, assumed that the molecular vibrations are not substantially affected by intermolecular forces, and also the rotational and vibrational motions are independent of each other. Then, translational, rotational and intermolecular energies of molecules are treated classically, and the vibrational and electronic energies are treated quantum-mechanically. If the *configuration integral* can be evaluated by this means, then it is possible to apply the canonical partition function to determine the thermodynamic properties of the system.

3.5.1 Internal pressure and van der Waals constants

There are several examples where we can relate thermodynamically measurable parameters to the molecular properties. The link between the *internal pressure* and the van der Waals constants, a and b is a good example. The internal pressure of a fluid is defined from pure thermodynamics: if we consider the total differential of the internal energy of a fluid as a function of its entropy and volume, $U = f(S,V)$

$$dU = \left(\frac{\partial U}{\partial S}\right)_V dS + \left(\frac{\partial U}{\partial V}\right)_S dV \tag{234}$$

and, if we divide both sides of Equation (234) by dV, imposing the constraint of constant temperature, we have

$$\left(\frac{\partial U}{\partial V}\right)_T = \left(\frac{\partial U}{\partial S}\right)_V \left(\frac{\partial S}{\partial V}\right)_T + \left(\frac{\partial U}{\partial V}\right)_S \tag{235}$$

By combining Equation (235) with Equations (131) and (132), we obtain

$$\left(\frac{\partial U}{\partial V}\right)_T = T\left(\frac{\partial S}{\partial V}\right)_T - P \tag{236}$$

If we combine Equation (236) with the Maxwell equation, Equation (137), and rearrange, we have

$$P + \left(\frac{\partial U}{\partial V}\right)_T = T\left(\frac{\partial P}{\partial T}\right)_V \tag{237}$$

Now, the van der Waals equation of state which was derived for real gases (Equation (64)) contains a molecular attraction constant, a. In this equation, the (an^2/V^2) term is used to account for the attractive intermolecular forces. If we rearrange the van der Waals equation per mole of a gas similar to Equation (237), we have

$$P + \frac{a}{V^2} = \frac{RT}{V-b} \tag{238}$$

where P is the external pressure, V is the molar volume, R is the gas constant, and T is the absolute temperature. By comparing Equations (237) and (238), it is evident that there are two more pressure terms in the state equation, in addition to the external pressure P:

Internal pressure (attractive) $\qquad \left(\frac{\partial U}{\partial V}\right)_T \cong \frac{a}{V^2}$ $\qquad\qquad$ (239)

Thermal pressure (repulsive) $\qquad T\left(\frac{\partial P}{\partial T}\right)_V \cong T\frac{R}{V-b}$ $\qquad\qquad$ (240)

The internal pressure is due to the *cohesional* forces (see Section 3.4.3) between the molecules that contribute to the internal energy, U, and is equal to zero for ideal gases. The approximate equality of internal pressure values in liquids is a good criterion for the behavior of ideal liquid solutions obeying Raoult's law. In ideal liquid solutions, the molecules of the components are under similar forces in solution as in the pure liquids, and the approximate equality of internal pressures, especially for non-polar components, is the reason for their ideal behavior. On the other hand, the repulsive thermal pressure represents the tendency of a fluid to expand. The $(\partial P/\partial T)_V$ parameter is called an *isochore* and can be measured directly, or it is more often computed as the ratio of the *coefficient of thermal expansion*, $\alpha = 1/V(\partial V/\partial T)_P = (\partial \ln V/\partial T)_P$, to the *coefficient of compressibility*, $\beta = -1/V(\partial V/\partial P)_T = -(\partial \ln V/\partial P)_T$, so that $(\partial P/\partial T)_V = -\alpha/\beta$.

3.5.2   Relation of van der Waals constants with molecular pair potentials

We can write the van der Waals equation in terms of molecular (not molar!) parameters, (see Equation (77) in Section 2.6.2), where v is the volume occupied per gaseous molecule $v = (V_m/N_A)$ (v is not the volume of the molecule itself), and k is the Boltzmann constant. The number density of the gas (particles/m^3) is inversely proportional to the gas volume, and for a unit number it is given by $\rho = 1/v$. As we know, $V = (\partial G/\partial P)_T$, from Equation (133), and we then have the gas molecule volume in terms of the chemical potential of a molecule,

$$v = 1/\rho = \left(\frac{\partial \mu}{\partial P}\right)_T \tag{241}$$

By rearranging Equation (241) and applying the chain rule, we may write for the gas number density

$$\rho = \left(\frac{\partial P}{\partial \mu}\right)_T = \left(\frac{\partial P}{\partial \rho}\right)_T \left(\frac{\partial \rho}{\partial \mu}\right)_T \tag{242}$$

Then, by rearranging Equation (242) we can obtain the variation of the pressure by the change in the number density of the gas

$$\left(\frac{\partial P}{\partial \rho}\right)_T = \rho\left(\frac{\partial \mu}{\partial \rho}\right)_T \tag{243}$$

and we can calculate the pressure by integration of Equation (243) between 0 and ρ:

$$P = \int_0^\rho \rho\left(\frac{\partial \mu}{\partial \rho}\right)_T d\rho \tag{244}$$

If we can write an equation for the chemical potential of a molecule as a function of pair potentials, $\mu = f(\rho)$, and find the derivative of the chemical potential with the number density in terms of pair potential energies, we may solve the above integration. Since we know the $\mu_2 = \mu_2^* + kT\ln X_2$ expression per molecule from Equation (165), where X_2 is usually expressed as the mole fraction or volume fraction, we need to relate the chemical potential of pure gas, μ_2^*, to the molecular pair potentials and also the mole fraction, X_2, to the gas number density, ρ. As the chemical potential, μ_2, is the total free energy per mole, it includes the interaction energy, μ_2^*, as well as enthalpy (kT) and entropy of mixing ($k\ln X_2$) contributions.

Cohesive chemical potential, or *cohesive self-energy*, μ_{coh} is defined as the free energy of an individual gas or liquid molecule surrounded by the same molecules. This is not an intermolecular pair potential term itself, however it may be calculated from pair potentials by summing this molecule's interactions with all the surrounding molecules. Since we know that the intermolecular forces are not to extend over large distances but only interact with molecules in close proximity for gases, then, similar to the derivation of Equations (19) and (20), the pair potential for a molecule in the gas phase is given as

$$V(r) = -\frac{C_W}{r^n} \tag{245}$$

where C_W is the van der Waals interaction coefficient (see Section 2.6.2). If $n > 3$, we must sum all the pair potentials, $V(r)$, over all the space, by integrating them between the molecular *hard sphere* diameter, σ, and infinity (for $r > \sigma$). If the number of gas molecules in a region of space between r and $(r + dr)$ away is $\rho 4\pi r^2 dr$, then by integrating we have

$$\mu_{coh}^* = \int_\sigma^\infty V(r)\rho 4\pi r^2 dr = -4\pi C_W \rho \int_\sigma^\infty r^{2-n} dr = -\frac{4\pi C_W \rho}{(n-3)\sigma^{n-3}} = -K\rho \tag{246}$$

where $K = 4\pi C_W/(n-3)\sigma^{n-3}$ is a constant. Now, we must relate the gas number density, ρ, to the mole fraction, X_2. In order to do this, we have to consider the *excluded volume* of the gas molecule, B, which is not available for the molecules to move in. Since $(v - B)$ is the free space where the molecules move and σ is the closest distance that a molecule can approach another, we can write, $B = 4\pi\sigma^3/3$. Then, the mole fraction (or effective density) of the ideal and non-ideal molecules can be calculated

$$X_2 = \frac{1}{v - B} = \frac{\rho}{1 - B\rho} \tag{247}$$

By combining Equations (165), (246) and (247), we may write for the cohesive chemical potential of a gas (or cohesive self-energy of a gas)

$$\mu_{\text{coh}}^{\text{gas}} = -K\rho + kT \ln\left(\frac{\rho}{1-B\rho}\right) \tag{248}$$

By taking the derivative of the cohesive chemical potential of the gas with its number density at constant temperature, we have

$$\left(\frac{\partial \mu_{\text{coh}}^{\text{gas}}}{\partial \rho}\right)_{\text{T}} = -K + \frac{kT}{\rho(1-B\rho)} \tag{249}$$

By inserting Equation (249) into Equation (244), we are able to integrate Equation (244) analytically,

$$P = \int_0^\rho \rho\left[-K + \frac{kT}{\rho(1-B\rho)}\right]d\rho = \int_0^\rho\left[-K\rho d\rho + \frac{kT d\rho}{(1-B\rho)}\right] \tag{250}$$

giving

$$P = -1/2\,K\rho^2 - \frac{kT}{B}\ln(1-B\rho) \tag{251}$$

It is possible to expand the $[\ln(1 - B\rho)]$ term mathematically for the $(1 > B\rho)$ condition so that

$$\ln(1 - B\rho) = -B\rho - 1/2(B\rho)^2 + \ldots \approx -B\rho(1 + 1/2B\rho) \tag{252}$$

since $[1 - (1/2B\rho)^2] \approx 1$ is also valid for the $(1 > B\rho)$ condition, and then

$$\ln(1 - B\rho) \approx -\frac{B\rho}{(1-1/2B\rho)} \approx -\frac{B}{(v-B/2)} \tag{253}$$

By combining Equations (251) and (253) we have

$$P = -\frac{K}{2v^2} + \frac{kT}{(v-B/2)} \tag{254}$$

If we rearrange the van der Waals equation per molecule (Equation (77)) in order to make comparison with Equation (254), we may write

$$P = -\frac{a}{v^2} + \frac{kT}{(v-b)} \tag{255}$$

It is evident that van der Waals constants can be shown in terms of molecular parameters:

$$a = \frac{K}{2} = \frac{2\pi C_W}{(n-3)\sigma^{n-3}} \tag{256}$$

$$b = \frac{B}{2} = \frac{2\pi\sigma^3}{3} \tag{257}$$

As expected, the constant a depends on the attractive interactions, and b on the diameter of molecules, and thus b is the repulsive contribution to the pair potential (see also Sections 2.6.2 and 2.7).

3.5.3 *Cohesive energy and close-packed molecules in condensed systems*

Each molecule is in contact with several other molecules in a liquid or solid, and much of the behavior of a liquid or a solid depends upon how the shapes of the molecules determine their packing. Each molecule has a considerable negative potential energy in the condensed state, in contrast with vapor phase molecules which have negligible potential energy. Liquid and solid molecules are forced to overcome the attraction interactions, which are holding them in the condensed state, during vaporization and sublimation. Consequently, it is a good idea to use the experimentally determined macroscopic vaporization properties of liquids (and sublimation properties of solids when data are available) in order to link the thermodynamics with the molecular pair potentials.

In Section 2.1, we defined the term *cohesion* to describe the physical interactions between the same types of molecules, and the term *adhesion* between different types of molecules. The cohesion in a liquid or solid measures how hard it is to pull them apart. The *work of cohesion* (see Section 5.6.1), W_i^c, is the reversible work, per unit area, required to break a column of a liquid or solid into two parts, creating two new equilibrium surfaces, and separating them to such a distance that they are no longer interacting with one another. (Theoretically, this separation distance must be infinity, but in practice a distance of a few micrometers is sufficient.)

The cohesive energy of condensed matter (especially of a liquid), U^V, describes the molar internal energy of vaporization of the gas phase at zero pressure (i.e. infinite separation of the molecules). The term $-U^V$ is the energy of a liquid relative to its ideal vapor at the same temperature, assuming that the intramolecular properties are identical in liquid and gas states, which may not be true in the case of complex organic molecules (the minus sign shows that energy is required to vaporize liquids). The term consists of two parts: the energy required to vaporize the liquid to its saturated vapor, ΔU^V, plus the energy required isothermally to expand the saturated vapor to infinite volume,

$$-U^V = \Delta U^V + \int_{V_{vap}}^{\infty} \left(\frac{\delta U}{\delta V} \right)_T dV \tag{258}$$

For liquids at ordinary room temperatures (below the boiling point of the liquid) and low pressures, the second term is neglected:

$$-U^V \cong \Delta U^V = \Delta H^V - P\Delta V \tag{259}$$

ΔH^V is the enthaply of vaporization. At low pressures and ordinary temperatures, the vapor in equilibrium with the liquid is assumed to behave ideally and we may write

$$-U^V = \Delta U^V = \Delta H^V - RT \tag{260}$$

As a result, the cohesive energy is the enthalpy of evaporation plus the change in enthalpy to expand from the vapor pressure to the ideal gas state (which is usually neglected in liquids at ordinary temperatures and low pressures, see also Section 4.2.1) minus RT. The cohesive energy concept is used to derive *regular solution theory* and also the *solubility parameters* (see Section 5.3).

ΔH^V is a measure of the strength of intermolecular cohesive attractions in the liquid (see Equation (281) in Section 4.2.1). ΔH^V values vary between 20 and 50 kJ mol^{-1} for materials that are liquids at room temperature. However, ΔH^V does not give the molar energy of interaction between two liquid molecules, because each molecule in a liquid interacts with several other molecules, and the addition of all of these interactions sums to ΔH^V so that the molar energy of interaction between two liquid molecules is substantially less than ΔH^V.

It is possible to apply the cohesive chemical potential (or cohesive self-energy), μ_{coh} to the molecular packing in a liquid state (however we cannot apply Equation (248) for μ_{coh}^{gas} to liquids directly because it is derived for a gas state, and in general $\mu_{coh}^{liq} \gg \mu_{coh}^{gas}$, as we will see in Section 3.4.4). If we consider only spherical molecules for simplicity and try to calculate the energy change if we introduce a vapor molecule into its own liquid, we shall consider only the closest packing spherical molecules. It is well known that the intermolecular forces do not extend over large distances but only interact with molecules in close proximity (*short-range forces*), and we need to count the number of neighbor molecules interacting with this individual gas molecule in the liquid in order to calculate μ_{coh}^{liq}. If we apply the *close packing* conditions, in which each molecule in a condensed state can have up to 12 neighbor molecules in contact with it, six neighbor molecules surround the center molecule in the same layer, plus three above and three below, as shown in Figure 3.8 *a* from the vertical (plan) view, and 3.8 *b* from the horizontal view. A molecule in the surface may be in contact with as many as nine, or as few as three others. Spherical molecules in a liquid approach a close-packed structure but they do not contact completely; they are further separated to enable molecular motion. Consequently, if a vapor molecule is introduced into its own liquid medium from the vapor phase, then 12 liquid molecules must first separate from each other to form the cavity to accommodate the guest molecule. If we define $V(\sigma)$ as the pair energy of two molecules in contact where $r = \sigma$, then it is possible to calculate roughly the net free energy change. When 12 molecules are separated to form the cavity, the six *bonds* holding the 12 neighbor molecules together are broken and an energy

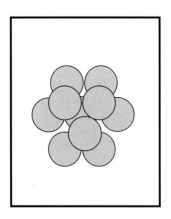

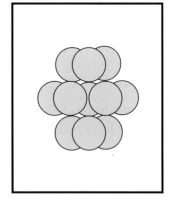

a. Plan view   *b. Horizontal view*

Close-packed spherical molecules

Figure 3.8 Schematic representation of *close-packed* spherical molecules with a central molecule having 12 neighbor molecules in contact with it: *a.* Plan view. *b.* Horizontal view.

of $-6\,V(\sigma)$ is expended. When the guest molecule is introduced, 12 new bonds are formed costing $+12V(\sigma)$, thus giving a net energy change of

$$\mu_{coh}^{liq} \approx -[6V(\sigma) - 12V(\sigma)] = 6V(\sigma) \tag{261}$$

which is half the total interaction energy of the guest molecule with its 12 nearest neighbors. Then the molar cohesive energy of a liquid in a constant volume may be expressed as

$$G_{coh}^{liq} = N_A\mu_{coh}^{liq} \approx 6N_AV(\sigma) \tag{262}$$

where N_A is Avogadro's constant. If we assume that the molecular entropy can be neglected at constant temperature, then we obtain $U_{coh}^{liq} \approx 6N_AV(\sigma)$ from Equation (124) (see also the explanations below Equation (211)). In practice, the cohesive energy of a molecule in a pure liquid or solid is determined experimentally to be somewhere between four and six times the pair potential energy, the lower value being applicable to complex molecules that deviate from simple spherical molecules that can condense to close-packed structures. However, the accurate calculation of μ_{coh}^{liq} for a liquid from pair potentials is extremely difficult because the average number of molecules surrounding any particular molecule is not known exactly; it may be between 4 and 12. In addition, the density of the neighbor molecules is not uniform locally; it is rather a function of the distance r, which is known as the *density distribution function*, $\rho = f(r)$ and can only be determined approximately. Calculation of the *adhesive free energy*, μ_{adh} between two dissimilar materials will be demonstrated in Chapters 7 and 10, in a similar manner. However, in many cases, the above approach is inadequate and the effect of hydrogen bonding, formation of dimers and trimers, and the presence of hydrophobic and hydrophilic forces may be dominant in determining thermodynamic properties and our fundamental understanding of these forces is insufficient. Future progress in surface and phase-equilibrium thermodynamics will be possible with increased knowledge of intermolecular forces.

3.5.4   Derivation of Trouton's rule

Nevertheless, there are some good applications of intermolecular pair potentials to macroscopic properties; we may estimate *Trouton's rule* from pair potentials. Trouton's rule provides an indication of the strength of cohesive forces in most liquids. This rule states that at 1 atm pressure, the enthalpy (or latent heat) of vaporization divided by its boiling point gives roughly $87\,\mathrm{J\,K^{-1}\,mol^{-1}}$, which corresponds to a cohesive energy, μ_{coh}^{liq}, of about 9–$10\,kT$ per molecule. In general, $(\Delta H^V/T_B)$ falls in the range of 75–$95\,\mathrm{J\,K^{-1}\,mol^{-1}}$ for most liquids; the higher values (up to 139 for hydrogen bonding liquids) are due to the cooperative association of liquid molecules arising from mostly hydrogen bonding, while the lower values are due to weak attraction or dimerization. The physical meaning of Trouton's rule is that the entropy of vaporization, $\Delta S^V = \Delta H^V/T$, is approximately the same for all non-associated liquids when evaporated to the same molar volume in the gas phase.

We know experimentally that vapor molecules will condense when their cohesive energy with all the other surrounding molecules in the condensed phase exceeds about $9\,kT$. Since $\mu_{coh}^{liq} \approx 6V(\sigma)$ for close-packed structures, from Equation (261), then if we equalize these two quantities, we will find $V(\sigma) \approx (9/6)kT \approx (3/2)kT$, and we may conclude that when the pair attraction energy potential of two molecules exceeds $(3/2)kT$ then the attraction energy is strong enough to condense them into a liquid (or solid) phase. That is why $(3/2)kT$ is used as a standard reference to gauge the cohesive strength of an attraction potential. (However we should be aware that $(3/2)kT$ is not the kinetic energy consumed during condensation, because the kinetic energy of a gas does not disappear when condensed into a liquid (or solid), the only change is the restriction of molecular motion in a narrower region of space around a potential energy minimum, without considering the very complex rotational and vibrational energies in this argument.)

This idea may be tested by examining the evaporation of liquids. We can try to calculate how strong the intermolecular attraction should be if it will allow vapor (gas) molecules to condense into a liquid at a particular temperature and pressure. Since we know from the Gibbs phase equation that the chemical potentials for gas and liquid molecules in equilibrium with each other are equal, we may write

$$\mu_{coh}^{gas} + kT\ln X_{gas} = \mu_{coh}^{liq} + kT\ln X_{liq} \tag{263}$$

Now, the volume of an ideal gas at standard atmospheric pressure (1 atm) and temperature (273.15 K) is approximately 22 400 cm^3, and the same gas occupies approximately 20 cm^3 when condensed; it is obvious that the cohesive energy of the liquid greatly exceeds the cohesive energy of the gas, $\mu_{coh}^{liq} \gg \mu_{coh}^{gas}$, so we may neglect, μ_{coh}^{gas}, and rearrange the above equation so that

$$-\mu_{coh}^{liq} = kT\ln\left(\frac{X_{liq}}{X_{gas}}\right) = kT\ln\left(\frac{22\,400}{20}\right) \approx 7kT \tag{264}$$

From the ideal gas equation, it is determined that the "ln" term changes by only 13% in the range $T = 100$–500 K. Since this range includes most boiling temperatures of liquids, we may write approximately for the cohesive energy of the liquid

$$-\mu_{coh}^{liq} \approx 7kT_B \tag{265}$$

where T_B is the boiling temperature. If we want to use the gas constant, R, instead of k, we may multiply both sides of the equation by Avogadro's number,

$$-\frac{N_A\mu_{coh}^{liq}}{T_B} \approx 7N_Ak = 7R \tag{266}$$

This equation shows that the boiling point of any liquid is simply proportional to the energy needed to evaporate the molecule. As the internal energy of vaporization is given by

$$\Delta U^V \cong -N_A\mu_{coh}^{liq} \cong 7RT_B \tag{267}$$

and if we want to calculate the enthalpy of vaporization, by combining Equations (260) and (267), we obtain *Trouton's rule* in a crude manner:

$$\frac{\Delta H^V}{T_B} = \frac{\Delta U^V}{T_B} + R \approx 7R + R = 8R \approx 70\,\mathrm{J\,K^{-1}\,mol^{-1}} \tag{268}$$

This 70 J K^{-1} mol^{-1} value is close to the 87 J K^{-1} mol^{-1} which is given by the original Trouton rule. The $(\Delta H^V/T_B)$ ratio may be replaced by $(\Delta H^{sub}/T_{SB})$ for solids which can sublime, where ΔH^{sub} is the enthalpy of sublimation and T_{SB} is the sublimation temperature. (We must be aware of the fact that Trouton's rule holds only on the earth because of the particular value of the atmospheric pressure on the earth (1 atm) which determines that one mole of gas molecules occupies a constant volume, according to the ideal gas law.)

3.5.5 Molecular interactions at the surface

It is important to understand the basic effects of pair potential energies of molecules on any liquid surface before investigating more complex interactions: in a homogeneous condensed liquid consisting of molecules of type A, in equilibrium with its vapor, the pair-potential energy between two molecules is shown as $V_{AA}(r)$. If we can determine the number of interacting pair molecules, n_{AA} experimentally, then we can calculate the Gibbs free energy in the bulk liquid so that

$$G_{A,\,bulk} = \frac{1}{2} n_{AA} V_{AA}(r) \qquad (269)$$

where $1/2 V_{AA}(r)$ is the energy per molecule deducted from the pair potentials. Now, if we assume that only nearest-neighbor interactions are operative in this condensed phase, we need to determine the number of interacting pair molecules, n_{AA} in the shell of these nearest neighbors. If we simply assume that the separation distance between the surface molecules is equal to the separation distance between the bulk molecules, and the number of interacting pair molecules in the surface nearest neighbors is only half of n_{AA}, the number of interacting pair molecules in the bulk nearest neighbors (because of the presence of air above the liquid surface) then we may write

$$G_{A,\,surface} = \frac{1}{4} n_{AA} V_{AA}(r) \qquad (270)$$

We may recall that the attractive $V_{AA}(r)$ is negative, so that work must be done to create a new surface, so there is an increase in free energy when a molecule is taken from the bulk and placed in the surface, as we have already discussed in Section 3.2.4. Unfortunately the experimental determination of n_{AA} is extremely difficult and we have to rely on density distribution functions and statistical mechanics or some other assumptions. It is well known that the effects of molecular structure and shape are often large for any condensed system, but since we have no adequate tools for describing such effects in a truly fundamental way, the best we can do is to estimate these effects by molecular simulation using computers. As we have already mentioned in Section 3.4.3, the density of the neighbor molecules is not uniform locally; it is rather a function of the distance r from the guest molecule, $\rho = f(r)$. This function is known as the *density distribution function* which can be approximately modeled and used in computation (see Section 4.1).

References

1. Pitzer, K.S. and Brewer L. (1961) *Thermodynamics* (2nd edn). McGraw-Hill, New York.
2. Atkins, P.W. (1998) *Physical Chemistry* (6th edn). Oxford University Press, Oxford.
3. Levine, I.N. (1990) *Physical Chemistry* (3rd edn). McGraw-Hill, New York.

4. Adam, N.K. (1968) *The Physics and Chemistry of Surfaces*. Dover, New York.
5. Lyklema, L. (1991) *Fundamentals of Interface and Colloid Science* (vols. I and II). Academic Press, London.
6. Israelachvili, J. (1991) *Intermolecular & Surface Forces* (2nd edn). Academic Press, London.
7. Prausnitz, J.M., Lichtenthaler, R.N. and Azevedo E.G. (1999) *Molecular Thermodynamics of Fluid-Phase Equilibria* (3rd edn). Prentice Hall, Englewood Cliffs.
8. Adamson, A.W. and Gast, A.P. (1997) *Physical Chemistry of Surfaces* (6th edn). Wiley, New York.
9. Scatchard, G. (1976) *Equilibrium in Solutions & Surface and Colloid Chemistry*. Harvard University Press, Cambridge.
10. Aveyard, R. and Haydon, D.A. (1973) *An Introduction to the Principles of Surface Chemistry*. Cambridge University Press, Cambridge.
11. Erbil, H.Y. (1997) Interfacial Interactions of Liquids. In Birdi, K.S. (ed.). *Handbook of Surface and Colloid Chemistry*. CRC Press, Boca Raton.
12. Murrell, J.N. and Jenkins, A.D. (1994) *Properties of Liquids and Solutions* (2nd edn). Wiley, Chichester.
13. Hirschfelder, J.O., Curtiss, C.F. and Bird, R.B. (1954) *Molecular Theory of Gases and Liquids*. Wiley, New York.

PART II

Liquids

Chapter 4
Pure Liquid Surfaces

4.1 What is a Liquid State?

A *liquid* is defined as "a medium, which takes the shape of a container without necessarily filling it". The term *fluid* is used to describe both gas and liquids phases, and the term *condensed phases* refers collectively to solids and liquids. Gases mix in all proportions, but certain liquids are partially and sometimes completely immiscible. The liquid state is usually distinguished from the gaseous state by its high density and small compressibility. From the standpoint of kinetic theory, a liquid may be considered as a continuation of the gas phase into the region of small volumes and very high molecular attractions. The cohesive forces in a liquid must be stronger than those in a gas even at high pressures and strong enough to keep the molecules confined to a definite liquid volume. The structure of a liquid refers to the spatial arrangement of molecules relative to one another (static) and the convection motions of the molecules in the liquid (dynamic). The molecules within a liquid have some freedom of motion, but this motion is considerably restricted, and when compared with gases the mean free path in liquids is much shorter than in the gas phase. In general, the distance between molecules in a liquid is so small that it is roughly equal to the molecular diameter, and the effect of intermolecular forces is correspondingly so large that the properties of liquids depend on the forces acting between the molecules. A liquid under normal pressures has a density, which is close to that of a solid; this is true over the whole liquid range from the melting point to the boiling point. A liquid is usually less dense than a solid at the melting point (but there are a few exceptions such as water, silicon, germanium, tin etc. due to the loose structure of their solids). Liquids have a low *compressibility* because there is not a great deal of free space between the molecules in a normal liquid, similarly to solids. On the other hand, a liquid may be distinguished from a solid by referring to the time-response to an applied force. The response of solids to a force is mainly *elastic*, at least for low strain; they return to their original shape following the application then removal of a force. The response of liquids to a force is mainly *inelastic*; they are permanently distorted by the applied force. This classification is useful especially in identifying the state of some materials having intermediate properties between liquids and solids, such as synthetic polymers, biological macromolecules and some glasses etc. If they flow to the shape of the container over time, they will be defined as liquids, otherwise as solids.

There is no complete theory for liquid state from first principles. Gases at high temperature and low pressure exhibit the properties of perfect chaos, whereas ideal solid crystals exhibit the properties of perfect order. Since perfect chaos and order are both simple to treat mathematically, their theories were advanced first. Liquids have neither the long-range structural order of solids nor the measurable intermolecular pair potential energies of gases and they are much harder to deal with theoretically than solids or gases. They exhibit properties between gases and solids and they have so far defied comprehensive theoretical treatment.

There are several other reasons that prevent us from deriving a successful theoretical description of liquids: although cohesive forces are sufficiently strong in liquids to lead to a condensed state, they are not strong enough to prevent considerable translational energy of the individual molecules. Consequently, thermal motions introduce a partial disorder into the liquid structure. This fact prevents evaluation of a partition function to describe a liquid in statistical mechanics. In order to model a liquid structure, we need to describe the spread of possible separation distances of molecules from a central molecule using a probability function, which defines mathematically the chance of finding two molecules separated by r. The *radial distribution function*, $g(r)$, is used for this purpose; this shows the variation in the average density of molecules with distance, r, from a given central molecule. Some knowledge of this function may be obtained from X-ray diffraction data. In most theoretical treatments, liquids are represented by random close packing of almost hard spheres. If we assume a liquid is composed of spherical, non-polar molecules having no long-range order characteristics of crystalline solids, the only ordering that exists is a short-range order resulting primarily since molecules cannot overlap. Then the probability that a volume element dV in a liquid of volume V will contain the center of a particular specified molecule is (dV/V). For a given pair of molecules, the probability of their occupying two such volume elements will be $(dV/V)^2$, provided that the distance r between them is sufficiently large so that their intermolecular pair potential can be neglected. However, in real liquids, the distance r between the molecular pairs is small, approaching σ, where σ is the diameter of the spherical molecule, and as a result, the intermolecular pair potentials are very effective in determining the positions of molecules. Then the probability that a given molecular pair will occupy two specified volume elements will be

$$P_{\text{pair}} = g(r)\left(\frac{dV}{V}\right)^2 \tag{271}$$

Equation (271) shows that $g(r)$ is the correction factor applied to the random probability $(dV/V)^2$ for ideal molecules, if no intermolecular interactions takes place. In a similar manner, we can calculate the probability of finding a particular molecule in the dV volume element, at a distance r from the center of a fixed molecule

$$P_{\text{single}} = g(r)\left(\frac{dV}{V}\right) \tag{272}$$

The volume element, dV, can be deduced from the volume of molecules in a spherical shell of thickness, dr, given as $dV = 4\pi r^2 dr$ for a spherical mathematical model, and the bulk molecular density $= N_m/V$, where N_m is the total number of molecules in a given volume V. To obtain the probability of finding N_m molecules in volume element dV, at a distance r from the center of a fixed molecule, we may write

$$P_{Nm} = N_m g(r)\left(\frac{dV}{V}\right) = \frac{N_m}{V} g(r) 4\pi r^2 dr \qquad (273)$$

The radial distribution function, $g(r)$, can be determined experimentally from X-ray diffraction patterns. Liquids scatter X-rays so that the scattered X-ray intensity is a function of angle, which shows broad maximum peaks, in contrast to the sharp maximum peaks obtained from solids. Then, $g(r)$ can be extracted from these diffuse diffraction patterns. In Equation (273) there is an enhanced probability due to $g(r) > 1$ for the first shell around the specified molecule at $r = \sigma$, and a minimum probability, $g(r) < 1$ between the first and the second shells at $r = 1.5\sigma$. Other maximum probabilities are seen at $r = 2\sigma$, $r = 3\sigma$, and so on. Since there is a lack of long-range order in liquids, $g(r)$ approaches 1, as r approaches infinity. For a liquid that obeys the Lennard–Jones attraction–repulsion equation (Equation (97) in Section 2.7.3), a maximum value of $g(r) = 3$ is found for a distance of $r = \sigma$. If $r < \sigma$, then $g(r)$ rapidly goes to zero, as a result of intermolecular Pauli repulsion.

On the other hand, more explicitly, the radial distribution function can also be shown as the ratio of local molecular density to the bulk molecular density,

$$g(r) = \frac{\rho_{m,local}(r)}{\rho_{m,bulk}(r)} \qquad (274)$$

where $\rho_{m,local}(r)$ is the local molecular density in the thin spherical shell from r to $r + dr$ around a given spherical molecule, and $\rho_{m,bulk}(r)$ is the bulk molecular density $= N_m/V$ where N_m is the total number of molecules in a given volume V.

Although the above analysis has been limited to pure liquids containing spherical molecules, the same ideas can be applied to liquids having non-spherical molecules or liquid mixtures. For non-spherical molecules, the radial distribution function depends on the directional angles θ and ϕ from the central molecule as well as on r. However, when we consider non-spherical molecules, not only does the mathematical complexity increase but also much more detailed information on liquid structure and intermolecular forces is required.

In 1933, Hildebrand related the internal energy of a mole of non-polar liquid to the radial distribution function and pair potentials

$$U = \frac{2\pi N_A^2}{V} \int_0^\infty g(r) V(r) r^2 dr \qquad (275)$$

It was later determined that the presence or absence of the attractive term in the Lennard–Jones potential does not alter the calculated $g(r)$ for spherical liquid molecules, indicating that the structure of most liquids is determined mainly by the intermolecular repulsive forces, with attractive forces playing only a relatively minor role. (The main exceptions are H-bonding liquids, electrolyte solutions and molten salts.) However, attractive forces (vectors) are important in determining thermodynamic properties such as internal energy, molar volume and molar energy of vaporization (see Section 5.3), although the attractive forces roughly cancel each other on all sides in a non-associated liquid in bulk. The attractive part of the potential energy is not a vector; it is a scalar quantity and does not cancel, but it is additive. In contrast, repulsive forces do not cancel each other because they are operative only when molecules are in contact. Thus, the complexity of repulsive and attractive forces, the large number of degrees of freedom of liquid molecules, and their convection movement make the application of the canonical partition functions of statistical mechanics very difficult to apply for a liquid to estimate the thermodynamical parameters from the molecular properties. Because of this, only approximate statistical–mechanical theories of liquids could be developed using radial distribution functions instead of canonical functions. (Some older theories, which assume liquids to be similar to solid crystals, were abandoned due to the poor approximation between these two states.) The most successful theories of liquids are *perturbation* theories, which initially calculate $g(r)$ while the intermolecular attractions are omitted, and then take them into account by treating them as a perturbation on the results obtained from repulsive forces only. If the correct intermolecular potentials between the molecules of simple liquids are known, then the perturbation approach is quite successful for calculating thermodynamic and transport properties of these liquids, as well as the behavior of simple liquid mixtures. (Unfortunately, the accurate intermolecular potentials are not known for most liquids of interest.) There have been several attempts to calculate the surface tension of fluids by using the radial distribution function and also a potential function such as that of Lennard–Jones, these have not been very successful at present. However, with the development of computers and new numerical calculation methods, such attempts may be successful in the future.

There are two computer methods for estimating $g(r)$: computer simulation approaches for liquids using *molecular dynamics* (MD) and *Monte Carlo* (MC) methods provide valuable insights into structure of liquids and allow calculation of some liquid properties. In the MD method, the history of an initial molecular arrangement is followed by calculating the trajectories of all the molecules under the influence of pair potentials. Computations on a system of 100–10 000 molecules (well below the 10^{25} molecules^{-1} for a standard liquid, chosen to allow for restrictions in computation capabilities) are carried out, which are kept in a theoretical "box" whose volume corresponds to the density of the chosen liquid. Newton mechanics are applied to predict where each molecule will be after a short time interval, and the calculation is then repeated for millions of such steps. The molecules are assigned initial positions, orientations and momenta; the momenta are chosen to be consistent with some desired temperature. The net force on the molecule arising from all the other molecules in the system is computed. An intermolecular potential energy function is used which is taken as the sum of pair-wise interactions, $V(r)$; the computer then solves the classical-mechanical equation of motion to find new configurations of the molecules at successive small intervals of time (about 10^{-15} sec, which is shorter than the average time between collisions). The MD method provides a "movie" of molecular motions; typically for a time interval of 10^{-15} sec and the system is followed for 10^{-11} sec. Molecular dynamics calculations allow $g(r)$ to be determined by averaging over successive configurations and they give information on structural details not accessible by experimentation. It is also possible to calculate the internal energy, U, of the liquid under examination, relative to that of a corresponding ideal gas having molecules without any intermolecular interactions, by taking the time average over the MD-calculated motions. Molecular dynamics methods are not restricted to equilibrium conditions and can be used to calculate some transport properties such as the viscosity of a liquid, or melting or evaporation phase transitions, and also the solution properties of polymers and proteins.

In the MC method, successive configurations of the system are not found by solving equations of motion. Instead, one molecule is picked at random and the computer gives small random changes in its position and orientation. If the potential energy of the new configuration is less than that of the original configuration, the new configuration is accepted. This recipe produces a sequence of configurations such that the probability of a configuration with potential energy $V(r)$ appearing in the sequence is proportional to the Boltzmann factor $e^{-V/kT}$. It is then possible to average over a sequence of typically 10^5–10^6 configurations to find $g(r)$ and the thermodynamic properties of the liquid. The MC method is restricted to equilibrium properties unlike the MD method. The MC method uses ensemble averages over classical-mechanical microstates to determine thermodynamical properties, unlike the time-averaging MD method.

4.2 Phase Transition of Pure Liquids

We can represent the regions of stability of gases, liquids and solids under various temperature and pressure conditions using a *phase diagram* showing at which phase each substance is the most stable. As we know from thermodynamics, the most stable phase of a pure substance at a particular temperature and pressure is the one with the lowest chemical potential. A *phase transition* is the spontaneous conversion of one phase into another phase, which occurs at a characteristic temperature at a given pressure. For example, as seen in Figure 4.1, under 1 atm external pressure, above 0°C, the chemical potential of

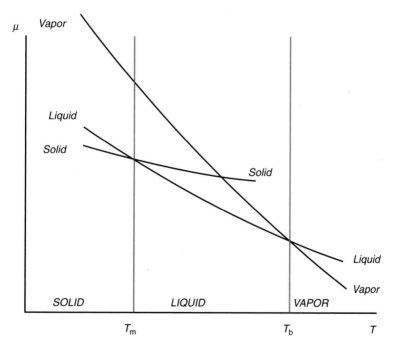

Figure 4.1 Variation of the chemical potential, μ, of a material such as water with the temperature, T, showing the phase transition between solid, liquid and vapor phases. A *phase transition temperature*, such as melting point, T_m and boiling point T_b is a temperature at which the two phases are in equilibrium and the two chemical potentials are equal.

liquid water is lower than that of ice, and thus the stable phase at these conditions is the liquid water; but below 0°C ice is more stable because its chemical potential is lower than the liquid water. The *phase transition temperature* is the temperature at which the two phases are in equilibrium and the two chemical potentials are equal. Two phases can coexist in equilibrium only at pressures and temperatures defined by the lines in the phase diagram, such as the liquid–vapor, solid–liquid and the solid–vapor lines. These lines are called *phase boundaries,* as shown in Figure 4.2. The pressure is a function of temperature along a phase boundary in the phase diagram (or vice versa). The high-temperature end of the liquid–vapor phase boundary in a phase diagram terminates at the critical point, T_c. All three phases can coexist in equilibrium only at the *triple point,* which is the intersection of the three two-phase boundaries. This is an invariant point, which can be determined experimentally for most materials with great accuracy.

If a liquid is sealed in an evacuated glass tube, a certain amount will evaporate to form vapor. This vapor will exert a pressure on the walls of the tube and also on the liquid surface as any gas does, and provided constant temperature is maintained, an equilibrium will be established between the liquid and vapor phases. The vapor pressure, which is a characteristic constant at a given temperature for each liquid, is known as the *saturated vapor pressure,* P_{vap}, of the liquid. In this sealed glass tube containing a liquid in equilibrium with its vapor, a *meniscus* is present showing the interface between the phases. If the tempera-

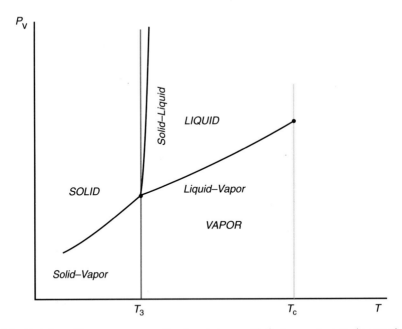

Figure 4.2 Variation of the vapor pressure, P_v, of a substance with the temperature, T, showing the phase transition between solid, liquid and vapor phases. Two phases can coexist in equilibrium only at pressures and temperatures defined by the phase boundary lines in the phase diagram, such as liquid–vapor, solid–liquid and solid–vapor lines. The liquid–vapor phase boundary terminates at the critical point, T_c. All three phases can coexist in equilibrium only at the *triple point,* T_3, which is the intersection of the three *two-phase* boundaries.

ture is raised, the density of the liquid will decrease and that of the gas increase, due to evaporation of the liquid molecules, and the saturated vapor pressure increases continuously with temperature until eventually a temperature and pressure is reached at which the densities of both liquid and vapor phases will be equal. At this point, the meniscus disappears and there is no distinction between the liquid and its vapor; the temperature and pressure at which this occurs are the *critical temperature, T_c,* and *critical pressure, P_c.* No liquid phase exists above the T_c and the application of pressures higher than the P_c only makes the vapor much denser than would normally be considered typical for gases, but no liquid condensation takes place, in contrast to compression below T_c. The term *supercritical fluid* is used for this homogeneous, dense phase of materials above their T_c. Supercritical fluids such as CO_2 and Xe are increasingly used in industry such as for the extraction of coffee caffeine and other materials, and also as a solvent. The critical phenomena are reversible and when the gas in the sealed tube is cooled below T_c, and if the pressure is sufficiently high, the meniscus reappears.

4.2.1 Liquid–vapor boundary: vapor pressure change by temperature: Clausius–Clapeyron equation

As stated above, the saturated vapor pressure, P_{vap}, of a liquid is the pressure at which that liquid is in equilibrium with its vapor at a given temperature. In a P–T phase diagram, the liquid–vapor phase boundary gives the vapor pressure change of the liquid as a function of temperature, as seen in Figures 4.2 and 4.3 *a.* In thermodynamics, the temperature dependence of the Gibbs energy in a closed system is expressed in terms of the entropy of this system by Equation (134). It follows that, by using chemical potential notation

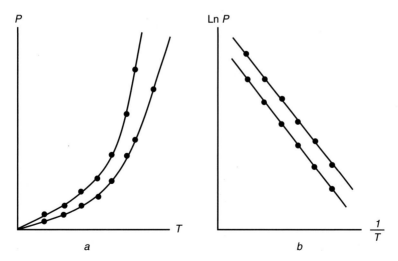

Figure 4.3   a. Experimental vapor pressure, P_v versus temperature T plot for a liquid–vapor interface. **b.** plot of ln P_v versus $(1/T)$ to calculate ΔH_m^v from the slope of the straight line, as given in Equation (284).

$$\left(\frac{\partial \mu}{\partial T}\right)_P = -S_m \qquad (276)$$

where S_m is the molar entropy. Since S_m is positive, Equation (276) indicates that the chemical potential, μ, decreases with the increase in temperature. The molar entropy of a vapor is much larger than its liquid state, $[S_m(\text{vap}) \gg S_m(\text{liq})]$ because of the larger disorder in the vapor phase; as a result Equation (276) implies that the slope of μ against T is steeper for vapors (gases) than liquids; consequently the liquids are more stable at temperatures below their normal boiling points (under 1 atm pressure).

The variation of Gibbs free energy with temperature and pressure in a closed system was given in Equation (126), and similarly, we can write $[d\mu = -S_m dT + V_m dP]$ for each phase. Since the chemical potentials are equal for two phases at equilibrium, it follows that

$$-S_m^1 dT + V_m^1 dP = -S_m^2 dT + V_m^2 dP \qquad (277)$$

where the V_m s are the molar volumes. Equation (277) may be rearranged as

$$\left(\frac{dP}{dT}\right)_\mu = \frac{\Delta S_m}{\Delta V_m} \qquad (278)$$

where $\Delta S_m = S_m^2 - S_m^1$ and $\Delta V_m = V_m^2 - V_m^1$. This is the *Clapeyron equation*, which is applicable to any phase transformation of a pure substance. Since its derivation contains no assumptions or approximations, it is an exact result for a one-component system. For the liquid–vapor boundary, the entropy of vaporization at a constant temperature is given as $\Delta S_m^V = \Delta H_m^V/T$, by the definition of entropy (Equation (111)). Then Equation (278) becomes

$$\frac{dP}{dT} = \frac{\Delta H_m^V}{T\Delta V_m^V} \qquad (279)$$

for vaporization of a liquid, where ΔH_m^V is the *molar enthalpy of vaporization*, the amount of heat required to evaporate a mole of liquid at a constant temperature. The term ΔH_m^V is often called the *latent heat* of vaporization. (This form of the Clapeyron equation can also be derived from the Carnot cycle.) The term ΔH_m^V is the difference in the enthalpies of vapor and liquid respectively; for one mole of a material, $\Delta H_m^V = H_m^{vap} - H_m^{liq}$. For an evaporation process, ΔH_m^V is always positive, i.e. heat is always taken from the surroundings, while for a condensation process ΔH_m^V is always negative and equal numerically to the heat taken in the vaporization. From the relation between the internal energy and the enthalpy (Equation (104)) we may also write

$$\Delta H_m^V = \Delta U_m^V + P(V_m^{vap} - V_m^{liq}) \qquad (280)$$

The ΔU_m^V term is the difference between the internal (intermolecular interaction) energies of the vapor and the liquid, $\Delta U_m^V = U_{intermol}^{vap} - U_{intermol}^{liq}$. The value of $P\Delta V_m$ is generally substantially smaller than that of ΔU_m^V. If the vapor pressure is low (well below its critical-point pressure), then $U_{intermol}^{vap} \approx 0$ can be assumed, and Equation (280) becomes

$$\Delta H_m^V \cong \Delta U_m^V \cong -U_{intermol}^{liq} \qquad (281)$$

Therefore ΔH_m^V is a measure of the strength of intermolecular cohesive attractions in the liquid. Equation (281) is the basis of the experimental determination of cohesive energies of liquids (see Section 3.5.3) and also solubility parameters (see Section 5.2). Values of ΔH_m^V vary between 20 and 50 kJ mol^{-1} for materials that are liquids at room temperature. (Values of ΔH_m^V are substantially lower than chemical bond formation energies, which are 150–800 kJ mol^{-1})

The vapor pressure of a liquid, though constant at a given temperature, increases continuously with increase in temperature, up to the critical temperature, T_c, of the liquid, because as the temperature increases, a greater proportion of the molecules acquire sufficient energy to escape from the liquid, and consequently a higher pressure is necessary to establish equilibrium between vapor and liquid. The value of P_{vap} increases slowly at the lower temperatures, and then quite rapidly with a steep rise in the P–T curve. Above T_c, the concept of a saturated vapor pressure, P_{vap}, is no longer valid. If we rewrite the Clapeyron equation for vaporizing liquids, from Equation (279), we have

$$\frac{dP_{vap}}{dT} = \frac{\Delta H_m^V}{T(V_m^{vap} - V_m^{liq})} \tag{282}$$

Now, at temperatures far from T_c, $[V_m^{vap} \gg V_m^{liq}]$, so that V_m^{liq} can be neglected. If we assume that the vapor behaves essentially as an ideal gas, $V_m^{vap} = RT/P_{vap}$ and Equation (282) becomes

$$\frac{dP_{vap}}{dT} \cong \frac{\Delta H_m^V P_{vap}}{RT^2} \tag{283}$$

This is the *Clausius–Clapeyron equation*. Actually, ΔH_m^V is reasonably constant over a short temperature range only, and if we assume that ΔH_m^V remains constant over the temperature range in question, we can integrate Equation (283) so that

$$\ln P_{vap} = -\frac{\Delta H_m^V}{R}\left(\frac{1}{T}\right) + \text{integration constant} \tag{284}$$

When $\ln P_{vap}$ is plotted versus $(1/T)$, we may calculate ΔH_m^V from the slope of the straight line which we can derive from Equation (284), as shown in Figure 4.3 b. In practice, ΔH_m^V may also be measured directly using a calorimeter, by condensing a definite weight of vapor and observing the temperature rise of the calorimeter, or by supplying the liquid with a definite amount of electrical energy and measuring the weight of liquid vaporized thereby. In general, ΔH^V decreases with increasing temperature and becomes zero at the critical temperature.

4.2.2 Liquid–solid boundary

A liquid freezes to give a solid at the freezing phase transition temperature under a specified pressure, which is usually taken as 1 atm (or conversely a solid melts at the melting temperature which is equal to the freezing temperature). For this phase transition, the Clapeyron equation (Equation (279)) becomes

$$\frac{dP}{dT} = \frac{\Delta H_m^M}{T\Delta V_m^M} \tag{285}$$

where ΔH_m^M is the molar heat of melting $[\Delta H_m^M = H_m^{liq} - H_m^{solid}]$ and $\Delta V_m^M = V_m^{liq} - V_m^{solid}$ is the change in molar volume that occurs on melting. The term ΔH_m^M is nearly always positive;

ΔV_m^M is usually positive (but negative for a few cases such as water, Ga, Bi) and always small so that we cannot neglect V_m^{solid} when compared with V_m^{liq}. Thus, we can integrate Equation (285) by just assuming that $(\Delta H_m^M/\Delta V_m^M)$ is a constant. Since ΔH_m^M can be approximated as being constant experimentally, unless $(P{-}P_M)$ is huge, and ΔV_m^M is also essentially constant, due to the incompressibility of solids and liquids, their ratio also changes little with the change of pressure and temperature, then we may write

$$\int_{P^M}^{P} dP = \left(\frac{\Delta H_m^M}{\Delta V_m^M}\right)\int_{T^M}^{T} \frac{dT}{T} \qquad (286)$$

The integration gives

$$P - P^M = \left(\frac{\Delta H_m^M}{\Delta V_m^M}\right)\ln\left(\frac{T}{T^M}\right) \qquad (287)$$

When T is close to T^M, we can approximate the logarithmic term in Equation (287) by using the approximation in the mathematical series expansion, so that $[\ln(1 + x) \cong x]$ and thus, $\ln(T/T^M) = \ln[1 + (T/T^M) - 1] \cong [(T/T^M) - 1]$ giving

$$P \approx P^M + \frac{\Delta H_m^M(T - T^M)}{\Delta V_m^M T^M} \qquad (288)$$

Equation (288) results in a straight line when P is plotted against T, which fits the experimental pressure–temperature data usually obtained during melting of solids or freezing of liquids.

4.3   Curved Liquid Surfaces: Young–Laplace Equation

Liquid surfaces and interfaces are usually well defined and easier to treat than solid surfaces. The determination of the surface and the interfacial tension of pure liquids and solutions is one of the most important aspect of surface science, and this is closely interrelated with the properties of curved liquid surfaces (see also Chapter 6). The formation of curved liquid surfaces such as spherical liquid drops in air, or curved liquid meniscuses in thin capillary glass tubes, is the consequence of the surface area minimization process due to the existence of liquid surface free energy. There are exceptions: if two phases are in hydrostatic equilibrium, they can be separated by a flat curvature-free interface. However, this exceptional case is rarely encountered and if a liquid interface is curved, this means that the pressure is greater on the concave side (the inside of a bubble, for example) than on the convex, by an amount, ΔP, which depends on the liquid surface tension and on the magnitude of the *curvature* (see Figure 4.4a for the formal description of concave and convex surfaces). *Curvature* is defined as the amount by which a geometric object deviates from being *flat*. The word *flat* might have different meanings depending on the object under consideration (for two-dimensional curves it is a straight line, and for three-dimensional surfaces it is a Euclidean plane). The pressure inside is larger than that outside for a spherical drop of liquid (or a soap bubble) in air, because an area increase is needed to displace the three-dimensional curved surface, parallel to itself, as the surface moves towards the convex side. The pressure difference, ΔP, does the work to increase this area.

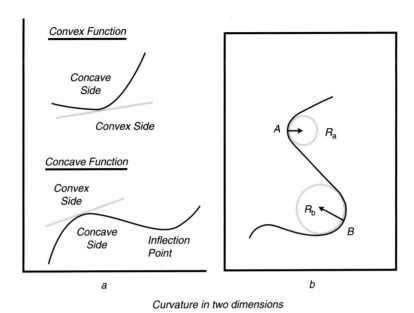

Curvature in two dimensions

Figure 4.4 *a.* Formal definition of *curvature* in two dimensions, for concave and convex curves. *b.* Definition of *curvature* in two dimensions, depending on the curve's *osculating circle* which is drawn at point *B* on the curve, merging as much as possible with the section of the curve around *B*. The value of radius of curvature, R_b is sufficient to characterize the shape of the curve around *B*. The same procedure leads to R_a at point *A*. The curvature at *A* is larger than the curvature at B.

(It should be noted that the pressure is always greater on the concave side of the interface irrespective of whether or not this is a condensed phase.) The phenomena due to the presence of curved liquid surfaces are called *capillary phenomena*, even if no capillaries (tiny cylindrical tubes) are involved. The *Young–Laplace equation* is the expression that relates the pressure difference, ΔP, to the curvature of the surface and the surface tension of the liquid. It was derived independently by T. Young and P. S. Laplace around 1805 and relates the surface tension to the curvature of any shape in capillary phenomena. In practice, the pressure drop across curved liquid surfaces should be known from the experimental determination of the surface tension of liquids by the capillary rise method, detailed in Section 6.1.

4.3.1 Young–Laplace equation from Newton mechanics

We can derive the simplest form of the Young–Laplace equation for a spherical vapor bubble in equilibrium with liquid in a one-component system (or a liquid drop in air) from Newton mechanics. In the absence of any external field such as gravitational, magnetic or electrical fields, the bubble will assume a spherical shape, and the force acting towards the boundary of the bubble (or liquid drop) from the interior of the bubble is given as

$$F_{interior} = A_{bubble}P_{interior} = 4\pi R_{sph}^2 P_{interior} \tag{289}$$

where R_{sph} is the radius of the spherical bubble. The force acting on the boundary of the bubble from outside must contain an additional surface tension force term, F_γ, which tries to diminish the surface area

$$F_{exterior} = A_{bubble}P_{exterior} + F_\gamma = 4\pi R_{sph}^2 P_{exterior} + F_\gamma \tag{290}$$

Since, the work to diminish the radius of the spherical bubble is, $dW_\gamma = F_\gamma dR_{sph}$ from Newton mechanics, and $dW_\gamma = \gamma dA$, from Equation (189), and $A = 4\pi R_{sph}^2$, from spherical geometry, then we have

$$F_\gamma = \gamma \left(\frac{dA}{dR_{sph}} \right) = \gamma (8\pi R_{sph}) \tag{291}$$

Since, $F_{interior} = F_{exterior}$ at equilibrium, then by combining Equations (289)–(291), we obtain,

$$4\pi R_{sph}^2 P_{interior} = 4\pi R_{sph}^2 P_{exterior} + 8\pi\gamma R_{sph} \tag{292}$$

By rearrangement of Equation (292), we may write for the pressure difference between the inside and outside of the spherical bubble (or liquid drop)

$$\Delta P = P_{interior} - P_{exterior} = \gamma \left(\frac{2}{R_{sph}} \right) \tag{293}$$

This simple form of the Young–Laplace equation shows that if the radius of the sphere increases, ΔP decreases, and when $R_{sph} \to \infty$, $\Delta P \to 0$, so that when the curvature vanishes and transforms into a flat Euclidean plane, there will be no pressure difference, and the two phases will be in hydrostatic equilibrium as stated above.

4.3.2 Young–Laplace equation from curvature

It is clear that the derivation of the Young–Laplace equation for a spherical shape is just a special case of a more general pressure difference–shape relationship, and we need to derive this equation for any shape having a different type of curvature. This is not a simple task and first of all we should define what is *curvature* mathematically. In two-dimensional terms, *curvature* is the rate of change of the slope of a curve with arc length. We see curves in plane graphs some of them are *concave* and some of them are *convex*. It is usual to define these concepts by analogy with a circle, so that concave curves are described as "similar to the curve seen from the interior of a circle", and convex curves as "similar to the curve seen from the exterior of a circle", but these descriptions are only schematic and can be applied to any curve. In mathematical terms, if the curve lies below each of its tangents, it has a *concave function*; and if the curve lies above each of its tangents, it has a *convex function*, these definitions also fit the schematic description (see Figure 4.4 *a*). It often happens that a graph is concave at certain intervals and convex downwards in others; the transition points are called *inflection points*. On the other hand, *curvature* may also be described by analogy with a moving car: if we imagine a car moving on a curved path at a specific rate, we need a parameter to give the rate of change of direction along the path. This may be given by an *inclination angle*, φ, which will vary by time or path length. Then we may define the absolute value of the *curvature*, κ, as

$$\kappa = \left| \frac{d\varphi}{ds} \right| \tag{294}$$

where, s is the arc length. The reciprocal of the curvature is the *radius of curvature* R_i

$$R_i = \frac{1}{\kappa} = \left| \frac{ds}{d\varphi} \right| \tag{295}$$

Curvature therefore has units of inverse distance (m^{-1}) and is sometimes reported in *diopter* units. As the car moves along the path, φ varies with the path length, and the more slowly we are changing direction, the smaller the value of κ; the less curved the path. In the extreme case, where $\kappa = 0$, the car travels along a straight line, and $R_i = \infty$. (This special case is applicable also to any inflection point in the graph.) The tangent of the inclination angle, φ, is the slope of the change in y due to the change in x, and for any coordinate point, we may write, $\tan \varphi = dy/dx$, and for the arc length, s, between two points, A and B,

$$s_{AB} = \int_{x=B}^{x=A} \sqrt{1 + \left(\frac{dy}{dx} \right)^2} \, dx, \text{ from analytical geometry. It has been proved that the curvature in}$$

two-dimensions can be calculated from Equation (294) for any curve written in the form of a $y = f(x)$ function, where f has continuous first and second derivatives in rectangular Cartesian coordinates so that

$$\kappa = \frac{\left| \dfrac{d^2 y}{dx^2} \right|}{\left[1 + \left(\dfrac{dy}{dx} \right)^2 \right]^{3/2}} \tag{296}$$

For example, for a circle having its center at the origin, the equation is $x^2 + y^2 = R^2$. By applying this circle equation to Equation (296) we can easily calculate the curvature of a circle, as $\kappa = 1/R$ and thus $R_i = R$ for a circle.

There is another method to quantify the curvature of a two-dimensional curve: it is dependent on the radius of the curve's *osculating circle* (a circle that *kisses* or closely touches the curve at a given point), a vector pointing in the direction of the center of the circle. In Figure 4.4 *b*, a circle is drawn at point B on the curve, merging as much as possible with the section of the curve around B by applying the best circular approximation to the curve path having a radius of R_b. Here R_b is called the *radius of curvature* and $\kappa_b = 1/R_b$ is the *curvature* of this portion of the curve. The value of R_b is sufficient to characterize the shape of the curve around B. The same procedure leads to R_a and κ_a at point A on the same curve. Since $R_b > R_a$, then the curvature at A is stronger ($\kappa_a > \kappa_b$) than at B. If the mathematical functions of the curves are known, the coordinates of the centers (*of curvature*) can be calculated for any curve which is written in the form $y = f(x)$, where f has continuous first and second derivatives

$$x_{center} = x - \left(\frac{dy}{dx} \right) \left[\frac{1 + \left(\dfrac{dy}{dx} \right)^2}{\left(\dfrac{d^2 y}{dx^2} \right)} \right] \tag{297}$$

$$y_{center} = y + \left[\frac{1 + \left(\dfrac{dy}{dx}\right)^2}{\left(\dfrac{d^2 y}{dx^2}\right)} \right] \tag{298}$$

Then, the radius of curvature, R_i, between the center of curvature and the tangent points (for example R_a in Figure 4.4 *b*) can be found from the simple distance formula of analytical geometry:

$$R_a = \sqrt{\left(x_a - x_{center}\right)^2 + \left(y_a - y_{center}\right)^2} \tag{299}$$

However, we are mostly dealing with three-dimensional objects such as bubbles, drops etc. in the real world. In order to describe the curvature of three-dimensional objects, two radii of curvature are needed, and things get a bit more complicated. This is because the curvature can appear different when in different directions. Curvatures may be positive or negative, and here we adopt the convention that a curvature is taken to be positive if the curve turns in the same direction as the surface's chosen normal; otherwise it is negative. In Figure 4.5 we see how to obtain these curvatures on a surface shown by the *mnpr* layer.

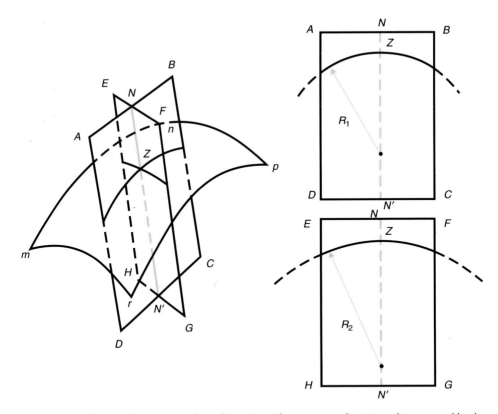

Figure 4.5   Definition of *curvature* in three dimensions. The curvature of *mnpr* can be expressed by the two *two-dimensional curvatures*, $\kappa_1 = 1/R_1$ and $\kappa_2 = 1/R_2$, where R_1 and R_2 are the radii of curvatures in the *ABCD* and *EFGH* planes, which are perpendicular to each other and are shown at the right of the figure. $R_1 \neq R_2$, unless the material has a spherical shape.

We can define a normal, NN', to this surface at point Z and pass a plane $ABCD$ through the surface containing the normal NN' (shown as a gray line). The line of intersection is curved on this plane and has a two-dimensional curvature, $\kappa_1 = 1/R_1$, where R_1 is the radius of the osculating circle tangent to the line at point Z, as we see in the $ABCD$ plane at the top right of Figure 4.5. The second curvature, κ_2, can be obtained by passing a second $EFGH$ plane through the surface, also containing the normal NN', but perpendicular to the first plane. This gives a second line of intersection and a second osculating circle tangent to the line at point Z, having a radius of R_2, as we see in the $EFGH$ plane at the lower right of Figure 4.5. $\kappa_2 = 1/R_2$, and $R_1 \neq R_2$, unless the figure is a sphere.

Now, if we vary the direction of the plane through the NN' axis, this curvature will of course change, and as κ varies, it achieves a minimum and a maximum (which are perpendicular to each other) known as the *principal curvatures*, κ_1 and κ_2, and the corresponding directions are called *principal directions*. The principal curvatures measure the maximum and minimum bending of a surface at each point. However, it has been shown mathematically that the sum of the curvatures ($\kappa_1 + \kappa_2 = 1/R_1 + 1/R_2$) is independent of how the first plane is oriented when the second plane is always at right angles to it, and consequently the direction of the planes is a matter of choice.

In order to describe the curvature of three-dimensional objects in terms of principal curvatures, there are two methods: *mean curvature* and *Gaussian curvature*. The *mean curvature*, H, is much more used in surface science and defined as the arithmetic mean of curvatures

$$H \equiv \frac{1}{2}(\kappa_1 + \kappa_2) \tag{300}$$

It has the dimension of m^{-1}. (In some surface textbooks, mean curvature is given as $H \equiv (\kappa_1 + \kappa_2)$, which is wrong in mathematical definition terms.) Let R_1 and R_2 be the radii corresponding to the curvatures the mean curvature H is then given by the multiplicative inverse of the harmonic mean,

$$H = \frac{1}{2}\left(\frac{1}{R_1} + \frac{1}{R_2}\right) = \frac{R_1 + R_2}{2R_1R_2} \tag{301}$$

We can also calculate the mean radius of curvature from analytical geometry: since there are x-, y- and z-axes in a three-dimensional Cartesian coordinate system, without proof, we may write the mean radius of curvature as

$$H = \frac{\left[1+\left(\frac{\partial z}{\partial y}\right)^2\right]\left(\frac{\partial^2 z}{\partial x^2}\right) - 2\left(\frac{\partial z}{\partial x}\right)\left(\frac{\partial z}{\partial y}\right)\left(\frac{\partial^2 z}{\partial x \partial y}\right) + \left[1+\left(\frac{\partial z}{\partial x}\right)^2\right]\left(\frac{\partial^2 z}{\partial y^2}\right)}{2\left[1+\left(\frac{\partial z}{\partial x}\right)^2 + \left(\frac{\partial z}{\partial y}\right)^2\right]^{3/2}} \tag{302}$$

If we apply this formula to the simplest case, for a sphere having its center at the origin whose equation is $x^2 + y^2 + z^2 = R^2$, then we will find $H_{sph} = \kappa_{sph} = 1/R_{sph}$. The angle between the normal at point (x, y) and the z-axis is θ and can be calculated from

$$\cos\theta = \left[\left(\frac{\partial z}{\partial x}\right)^2 + \left(\frac{\partial z}{\partial y}\right)^2\right]^{-1/2} \tag{303}$$

and the area of the interface can be found from

$$A = \int\int_{x\ y} \left[1 + \left(\frac{\partial z}{\partial x} \right)^2 + \left(\frac{\partial z}{\partial y} \right)^2 \right]^{1/2} dxdy \tag{304}$$

The other curvature term, *Gaussian curvature*, K, determines whether a surface is locally convex (when it is positive), or locally saddle (when it is negative), and also the intrisicity of the curvature. It is defined as

$$K \equiv \kappa_1 \kappa_2 = \frac{1}{R_1 R_2} \tag{305}$$

It has the dimension of m^{-2} and is positive for spheres, negative for one sheet hyperboloids and zero for planes. The Gaussian curvature, K, and mean curvature, H, are related to each other by a quadratic equation $[\kappa^2 - 2H\kappa + K = 0]$, which has solutions of $\kappa_1 = H - \sqrt{H^2 - K}$ and $\kappa_1 = H - \sqrt{H^2 - K}$. We may write the mean curvature in terms of the Gaussian curvature

$$H = \frac{1}{2}(R_1 + R_2)K \tag{306}$$

In analytical geometry terms, the Gaussian curvature can be given as

$$K = \frac{\left(\frac{\partial^2 z}{\partial x^2} \right) \left(\frac{\partial^2 z}{\partial y^2} \right) - \left(\frac{\partial^2 z}{\partial x \partial y} \right)^2}{\left[1 + \left(\frac{\partial z}{\partial x} \right)^2 + \left(\frac{\partial z}{\partial y} \right)^2 \right]^2} \tag{307}$$

Now, we can comment on the radii of curvature of some geometrically defined objects. For a plane, $\kappa_{pl} = 0$ and $H_{pl} = K_{pl} = 0$. For a sphere, the two radii of curvature that are the same $(R_1 = R_2 = R_{sph})$, and thus $\kappa_1 = \kappa_2 = \kappa_{sph} = 1/R_{sph}$, give $H_{sph} = \kappa_{sph} = 1/R_{sph}$, and $K_{sph} = \kappa_{sph}^2 = 1/R_{sph}^2$. However, there are two possible alternatives for a sphere surface in the real world. For a liquid drop in a gas, the two radii of curvature are positive and so the pressure difference is positive, which implies that the pressure inside the liquid is higher than outside. The other alternative is a gas bubble in a liquid environment where the two radii of curvature are negative so that ΔP is negative, and the pressure inside the liquid is lower than inside the bubble. For a cylinder, there is a single radius of curvature, the curvature of the base circle – there is no curvature in the other direction, so the other radius is infinite, thus giving $\kappa_1 = \kappa_{cyl} = 1/R_{cyl}$, $\kappa_2 = 0$ so that $H_{cyl} = (1/2)\kappa_{cyl} = 1/(2R_{cyl})$, and $K_{cyl} = 0$. We can imagine an ellipsoid with two different, finite radii of curvature at the same point and Equations (300) and (305) apply. For a drop hanging between the ends of two cylinders in a gas, one radius of curvature is negative and the other is positive. The value of ΔP depends on the specific values of κ_1 and κ_2, so that the Gaussian curvature is negative. Similarly a saddle actually has two radii of curvature of opposite signs – one positive, the other negative.

From another mathematical point of view, the mean radius of curvature, H, is half of the derivative of the curved surface area over the volume of a material. We can test it by simple geometric shapes, such as with a sphere:

$$H_{sph} = \frac{1}{2}\left(\frac{dA_{sph}}{dV_{sph}}\right) = \frac{1}{2}\left(\frac{dA_{sph}}{dR_{sph}}\right)\left(\frac{dR_{sph}}{dV_{sph}}\right) = \frac{1}{2}(8\pi R_{sph})\left(\frac{1}{4\pi R_{sph}^2}\right) = \frac{1}{R_{sph}} \tag{308}$$

For a cylinder, since $\kappa_2 = 0$, we need only consider the curved area $= 2\pi RL$ in the same calculation

$$H_{cyl} = \frac{1}{2}\left(\frac{dA_{cyl}}{dV_{cyl}}\right) = \frac{1}{2}\left(\frac{dA_{cyl}}{dR_{cyl}}\right)\left(\frac{dR_{cyl}}{dV_{cyl}}\right) = \frac{1}{2}(2\pi L)\left(\frac{1}{2\pi R_{cyl}L}\right) = \frac{1}{2R_{cyl}} \tag{309}$$

In general, numerical methods are applied to solve the curvatures of complex figures using computers. However, when the interfaces of objects of revolution are considered, the three-dimensional curvature equations are then simplified and standard calculus methods are applied for surface area calculations from the revolution of a two-dimensional half profile. Examples are the formation of a sessile liquid drop on a substrate and the formation of a meniscus in a wide capillary tube. We will see the application of such methods in Chapter 6.

Now we can derive the general version of a Young–Laplace equation which can be applied to any shape having different types of curvature. For flat interfaces, the surface excess functions were derived in Section 3.2.5 by applying the Gibbs dividing interface approach. Thus, we can write the following expression for the total internal energy of two phases, α and β, separated by a flat interface, through combining Equations (196) and (201)

$$dU = T^\alpha dS^\alpha - P^\alpha dV^\alpha + \sum_i \mu_i^\alpha dn_i^\alpha + T^\beta dS^\beta - P^\beta dV^\beta + \sum_i \mu_i^\beta dn_i^\beta$$
$$+ T^S dS^S - P^S dV^S + \gamma dA^S + \sum_i \mu_i^S dn_i^S \tag{310}$$

In the Gibbs treatment of the dividing interface, we assume that there is no volume for the interphase, $(V^S = 0)$ and the two phases are enclosed in a fixed volume, $V = V^\alpha + V^\beta$ containing a fixed amount of substance $(\Sigma n_i = \text{constant})$, and the system is in thermal equilibrium with its surroundings, $T^\alpha = T^\beta = T^S$. At equilibrium, the difference in internal energy, $dU = 0$, and for a given constant set of values of S and of n_i, Equation (310) reduces to,

$$0 = -P^\alpha dV^\alpha - P^\beta dV^\beta + \gamma dA^S \tag{311}$$

(the rigorous thermodynamic derivation of Equation (311) has been performed elsewhere but it is outside the scope of this book). Now, if we consider curved instead of flat interfaces, then the two curvatures κ_1 and κ_2 must also be taken into account for the equilibrium condition, so that Equation (311) becomes

$$0 = -P^\alpha dV^\alpha - P^\beta dV^\beta + \gamma dA^S + C_1 d\kappa_1 + C_2 d\kappa_2 \tag{312}$$

where C_1 and C_2 are two constants having their units of force multiplied by area to maintain the dimensional balance in Equation (312). Mathematically, Equation (312) may also be written as

$$0 = -P^\alpha dV^\alpha - P^\beta dV^\beta + \gamma dA^S + 1/2(C_1 + C_2)\, d(\kappa_1 + \kappa_2) + 1/2(C_1 - C_2)\, d(\kappa_1 - \kappa_2) \tag{313}$$

Since the actual effect on P^α and P^β must be independent of the location chosen for the Gibbs dividing surface, a condition may be predetermined for C_1 and C_2 and this may arbitrarily be taken as $C_1 + C_2 = 0$. The application of this condition gives a particular location of the Gibbs dividing surface and we may then define the *interfacial tension* between phases α and β as a result of this location. For the $C_1 + C_2 = 0$ condition, Equation (313) becomes

$$0 = -P^\alpha dV^\alpha - P^\beta dV^\beta + \gamma dA^S + 1/2(C_1 - C_2)d(\kappa_1 - \kappa_2) \tag{314}$$

Now, we may check the results of this approach on some interfaces with defined geometric shapes. For flat interfaces, $\kappa_1 = \kappa_2 = 0$ and only Equation (311) applies so that $dV^\alpha = -dV^\beta$, $dA^S = 0$ and thus, $P^\alpha = P^\beta$, as expected. For a spherical interface such as a liquid drop in air, since $\kappa_1 = \kappa_2$, $d(\kappa_1 - \kappa_2) = 0$ and again Equation (311) applies, but $dA^S \neq 0$ due to the formation of the curvature. For this case we may rearrange Equation (311) so that,

$$(P^\alpha - P^\beta)dV^\alpha = \gamma dA^S \tag{315}$$

We cannot apply Euler's reciprocity rule to Equations (313) and (315) because they are not homogeneous first-order differential equations. So we need to analyze the curvature dependence of the pressure in Equation (315): if the interface moves outwards by a distance $d\lambda$ and phase α is considered to be at the concave side, then $dV^\alpha = -dV^\beta = A^S d\lambda$ applies. Geometrically, it has been proved that the increase in the interfacial area can be given as

$$dA^S = (\kappa_1 + \kappa_2)A^S d\lambda \tag{316}$$

By combining Equations (315) and (316) one obtains

$$(P^\alpha - P^\beta)A^S d\lambda = \gamma(\kappa_1 + \kappa_2)A^S d\lambda \tag{317}$$

By simplifying Equation (317) and combining with Equations (300) and (308), for a spherical interface we have

$$\Delta P_{sph} = (P^\alpha - P^\beta) = \gamma(\kappa_1 + \kappa_2)_{sph} = 2\gamma H = \gamma\left(\frac{2}{R_{sph}}\right) \tag{318}$$

This is the derivation of the Young–Laplace equation for a spherical interface from surface thermodynamics. Now, if we consider the general case, where $\kappa_1 \neq \kappa_2$ and $d(\kappa_1 - \kappa_2) \neq 0$, for any curved figure, Equation (314) applies for this condition and it has been proved that

$$\Delta P = (P^\alpha - P^\beta) = \gamma(\kappa_1 + \kappa_2) - \frac{(C_1\kappa_1^2 + C_2\kappa_2^2)}{A^S} \tag{319}$$

This is the derivation of the general form of the Young–Laplace equation for any curved interphase, and the pressure difference can be calculated upon measurement of the curvatures and predetermination of the C constants. As noted in the thermodynamics of adsorption in Section 3.3, the Γ_1 and Γ_2 parameters are the *excess moles of the components 1 and 2 adsorbed at the interface per unit area* of an interphase, and are defined relative to an arbitrarily chosen Gibbs dividing surface as a plane of infinitesimal thickness. In order to obtain physically meaningful quantities for flat interfaces, the most widely used convention in dealing with binary solutions is to locate the Gibbs dividing plane so that the surface excess of the solvent is zero, $\Gamma_1 = 0$, and the Γ_2 quantity of the solute in the interface can be calculated by Equation (226). In analogy, the location of the dividing surface by assuming $C_1 + C_2 = 0$ corresponds to similar conditions for curved interphases where the properties of the bulk phase continue up to the Gibbs dividing surface, and even for a single pure substance, there will be a nonzero Γ that can be positive or negative. Nevertheless, this convention, while mathematically convenient, is not pleasing intuitively, and other conventions have been offered to locate the dividing surface for curved interphases by several scientists. There is not however a consensus on this complex matter in surface science to this today, and the debate continues.

4.3.3 Young–Laplace equation from plane geometry

The radii of curvature of the surface (*mnpr*) are determined by cutting it with two perpendicular planes (see Figure 4.6). Each of the planes contains a portion of the arc, where it intersects the curved surface; the radii of curvature (the radii of the osculating circles) are designated R_1 and R_2 and the arc lengths are designated as x and y, respectively. If we assume that the curved surface is moved outward by a small amount, dz, to a new position (*m'n'p'r'*) with a larger surface area than (*mnpr*). This moves increases the arc lengths to $(x + dx)$ and $(y + dy)$ since the corners of the surface continue to lie along the extension of the new diverging radial lines, as shown in Figure 4.6. This increase in area is given by

$$dA = (x + dx)(y + dy) - xy = xdy + ydx + dxdy \approx xdy + ydx \tag{320}$$

since it is convenient to neglect second-order differential quantities. The increase in Gibbs free energy associated with this increase in area at constant pressure and temperature is given (from Equation (205)) by

$$dG = \gamma(xdy + ydx) \tag{321}$$

The work required to increase the surface area must be supplied by a pressure difference, ΔP across the element of surface area. If the ordinary PV work is responsible for this, $(dG = \Delta PdV)$, then we may write

$$\Delta PdV = \Delta Pxydz = \gamma(xdy + ydx) \tag{322}$$

From plane geometry, the arc lengths are related to radian angles α_1 and α_2 so that, $x = R_1\alpha_1$ and $(x + dx) = (R_1 + dz)\alpha_1$, and also $y = R_2\alpha_2$ and $(y + dy) = (R_2 + dz)\alpha_2$, giving

$$\frac{x + dx}{R_1 + dz} = \frac{x}{R_1} \quad \text{simplifying into} \quad \frac{dx}{xdz} = \frac{1}{R_1} \tag{323}$$

$$\frac{y + dy}{R_2 + dz} = \frac{y}{R_2} \quad \text{simplifying into} \quad \frac{dy}{ydz} = \frac{1}{R_2} \tag{324}$$

By substituting Equations (323) and (324) into Equation (322) we may write

$$\Delta P = \gamma\left(\frac{1}{R_1} + \frac{1}{R_2}\right) \tag{325}$$

which is the same as the Young–Laplace equation (Equation (319)) where the C_1 and C_2 terms are neglected. When equilibrium is reached, ΔP is constant in the liquid, and the surface of the liquid has the same curvature everywhere, otherwise there would be a flow of liquid to regions of low pressure.

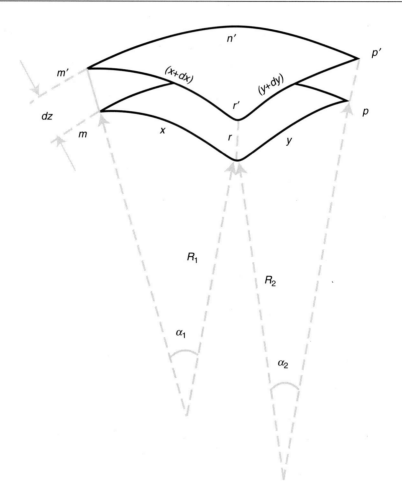

Figure 4.6 Description of three-dimensional curvature using plane geometry concepts. Each of the perpendicular planes contains a portion of the arc that intersects the curved surface. The radii of curvature (the radii of the osculating circles) are designated R_1 and R_2 and the arc lengths are designated as x and y, respectively. The radian angles are α_1 and α_2. If the curved surface is moved outwards by a small amount dz, the surface area increases by increasing the arc lengths to $(x + dx)$ and $(y + dy)$.

As a general rule, if we want to consider the placement of the Gibbs dividing plane in the interfacial region, then we should use C_1 and C_2 constants and apply Equation (319) instead of Equation (325).

4.4 Capillarity

When a solid capillary tube is inserted into a liquid, the liquid is raised or depressed in the tube, and the height of the liquid can be determined. A glass capillary is most commonly used for this purpose because it is transparent and is completely wettable by most liquids.

(However, other solid materials can also be used as capillary tubes when required.) It has been observed experimentally that there is an inverse proportionality between the height of the liquid present in the capillary tube and the radius of the tube (see also Section 6.1). Capillary rise was found to result from the adhesion interactions between the liquid and the capillary wall, which are stronger than the cohesion interactions within the liquid. This is a method used to measure the surface tension of pure liquids. During the measurement, the capillary tube must be very clean, placed completely vertical and be circular in cross section with accurately known and uniform radius.

A mechanism of capillary rise has been suggested (but not yet proved experimentally) for liquids such that a thin film of liquid forms inside the capillary wall during the first contact of the capillary tube with the liquid, due to the attraction force between the liquid and solid, giving a new large interfacial area. Then, very quickly, the liquid rises in the capillary column, because the surface free energy of the liquid must decrease the total interfacial contact area, but this rise in liquid is restricted by the opposing hydrostatic (gravity) forces, thus approaching an equilibrium between the capillary and hydrostatic pressures within the column at a definite height of the liquid. The attractive forces that raise the liquid in the capillary tube are exerted only along the edge at which the upper surface of the liquid meets the tube, where the capillary wall exercises an attraction. This leads to Jurin proportionality where a constant force, γ (surface tension), acting through the capillary perimeter determines the mass of the liquid to rise,

$$2\pi r\gamma \propto (m = \pi r^2 h\rho) \tag{326}$$

where r is the inner radius of the capillary tube, h is the height of the cylindrical liquid column, and ρ is the density of the liquid, so that $h \propto 2\gamma/r\rho$ can be experimentally observed for every liquid. It is not necessary for the tube to be of the same radius throughout. Only the radius at the upper surface is important and the capillary radius can be wider lower down. The capillary rise explains the transport of water to the tops of very tall trees, and many other processes in the nature.

On the other hand, the liquid surface in the capillary tube mostly takes the form of a concave spherical cap, as seen in Figure 4.7. In other terms, we can attribute the rise of a liquid in a capillary tube as simply the automatic recording of the pressure difference, ΔP, across the meniscus of the liquid in the tube, the curvature of the meniscus being determined by the radius of the tube and the angle of contact, θ, between the liquid and the capillary wall. If the capillary tube is circular in cross section and not too large in radius, then the meniscus will be completely hemispherical, that is $\theta = 0°$ and $r = R_1 = R_2$ in the Young–Laplace equation (Equation (325)) giving

$$\Delta P = \frac{2\gamma}{r} \tag{327}$$

In Figure 4.7, the pressure at the liquid level and at point a just above the meniscus is atmospheric. The height of the column of liquid, h, will be such that the pressure at point c in the column is also atmospheric, due to the balanced effects of hydrostatic and capillary pressures in the liquid column. The pressure at point b, just below the meniscus, will be less than atmospheric by an amount $(2\gamma/r)$. At equilibrium, ΔP is also equal to the hydrostatic pressure drop in the liquid column in the capillary. Thus,

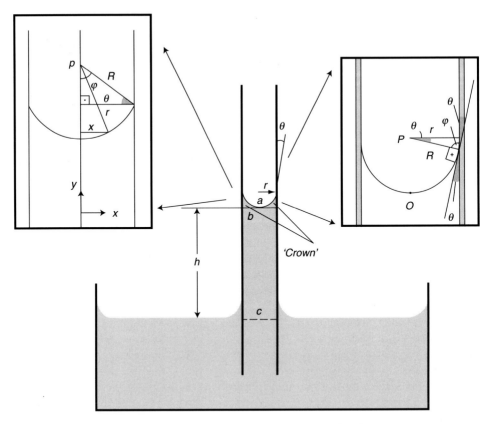

Figure 4.7 Concave liquid meniscus in a capillary tube during a liquid surface tension measurement: θ is the angle of contact between the liquid and the capillary wall; c is the point at the liquid level, a is the point just above, and b is just below, the meniscus level and h shows the height of the liquid column. The *crown* of the concave meniscus is the liquid between the top and the lower end of the meniscus. The term φ is the *inclination angle*, where ($\varphi = 90° - \theta$) from plane geometry.

$$\Delta P = \Delta\rho g h = \frac{2\gamma}{r} \tag{328}$$

where $\Delta\rho$ denotes the difference in density between the liquid and gas phase and g is the acceleration of gravity. By rearrangement, for completely hemispherical menisci we may write

$$rh = \frac{2\gamma}{\Delta\rho g} = a_o^2 \tag{329}$$

where a_o^2 is defined as the *capillary constant* for $\theta = 0°$. The square root of the capillary constant, a_o, has units of length.

The above treatment can be applied to the simplest ideal case, where $\theta = 0°$ and the liquid completely wets the capillary wall. However, occasionally a liquid meets the circularly cylindrical capillary wall at some contact angle, θ, between the liquid and the wall, as shown in Figure 4.7. The fluidity of the liquid permits the molecules to move about until they rest at this stable contact angle. Thus, the height to which the liquid will rise in the capillary tube is determined by the curvature of the meniscus. When $\theta \neq 0°$ and if the radius of the tube is sufficiently small, gravitational distortion of the curvature may be neglected, and since $\cos \theta = r/R$ in Figure 4.7, the radius of curvature of the meniscus can be given as $R_1 = R_2 = r/\cos \theta$ and we can write

$$rh = \frac{2\gamma \cos\theta}{\Delta\rho g} = a_\theta^2 \qquad (330)$$

The curvature of the interface depends on the relative magnitudes of the adhesive forces between the liquid and the capillary wall and the internal cohesive forces in the liquid. When the adhesive forces exceed the cohesive forces, θ lies in the range $0° \leq \theta \leq 90°$; when the cohesive forces exceed the adhesive forces, $90° \leq \theta \leq 180°$. When $\theta > 90°$, the $\cos\theta$ term is negative, resulting in a convex meniscus towards the vapor phase and the liquid level in the capillary falling below the liquid level in the container (capillary depression). This occurs with liquid mercury in glass where $\theta \cong 140°$ and also with water in capillary tubes coated internally with paraffin wax. Thus, liquid mercury is used in the evaluation of the porosity of solid adsorbents in the *mercury injection porosimetry* technique (see Section 8.5).

When deriving Equations (329) and (330), we neglected the weight of the liquid in the *crown* of the concave meniscus (the liquid between the top and the lower end of the meniscus), as shown in Figure 4.7. Consequently, surface tensions calculated by these equations are only approximate. More commonly, one measures the height, h, to the bottom of the meniscus and adds some correction factors. Jurin added a rough correction factor of $(r/3)$, to Equation (329) to account for the weight of the meniscus, assuming it to be spherical:

$$a^2 = r\left(h + \frac{r}{3}\right) \qquad (331)$$

and in a more refined approach, Rayleigh treated the above equation in a series extension for $h \gg r$,

$$a^2 = r\left(h + \frac{r}{3} - 0.1288\frac{r^2}{h} + 0.1312\frac{r^3}{h^2} + \ldots\right) \qquad (332)$$

the last two terms provide some rough corrections for deviation of the liquid surface from sphericity.

When the radius of the capillary tube is appreciable, the meniscus is no longer spherical and also $\theta > 0°$. Then, Equation (329) requires correction in terms of curvatures and it should give better results than those from the rough corrections given in Equations (330)–(332) for almost spherical menisci. Exact treatment of the capillary rise due to the curved meniscus is possible if we can formulate the deviation of the meniscus from the spherical cap. For this purpose, the hydrostatic pressure equation, $\Delta P = \Delta\rho g z$ (Equation (328)), must be valid at each point on the meniscus, where z is the elevation of that point above the flat liquid surface (see Figure 6.1 in Chapter 6). Now, if we combine the Young–Laplace equation (Equation (325)) with Equation (328), we have

$$\Delta \rho g h = \gamma \left(\frac{1}{R_1} + \frac{1}{R_2} \right) \tag{333}$$

which is applicable to any capillary rise situation. In Figure 4.7, if we define the radius of curvature as d at the apex (the lowest point O of the meniscus), then $R_1 = R_2 = d$, since the capillary tube is cylindrical, and point O is on the axis of revolution. For menisci of the axes of revolution of half profiles, the modification of Equation (329) is simple: in these conditions, $\theta \neq 0°$, $r \neq d$ and we must consider the gravitational distortion on the curvature. Then the capillary constant can be written as

$$dh = \frac{2\gamma}{\Delta \rho g} = a_d^2 \tag{334}$$

Since it is a very difficult task to measure d experimentally in a capillary tube, we need a relation between d and the experimentally accessible radius of the capillary tube, r. This relation can be derived by considering gravity and surface tension effects by applying fundamental Newton mechanics; the complete proof is given in Section 6.1. In the case of a figure of revolution, where $R_1 = R_2 = d$, when the elevation of a general point on the surface is denoted by z, the fundamental equation is given as

$$\Delta \rho g z + \frac{2\gamma}{d} = \gamma \left(\frac{1}{R_1} + \frac{1}{R_2} \right) \tag{335}$$

Equation (335) is the basis for calculating interfacial and surface tension of pure liquids from capillary rise, from drop profiles, and the height of a meniscus at a solid wall (see Sections 6.1–6.4). Unfortunately, this equation cannot be solved analytically and the application of numerical methods using computers is required. The inclusion of gravity correction into the Young–Laplace equation is feasible only for capillary tubes having appreciable diameters, and for large pendant or hanging liquid drops formed on solids. The determination of what is *large* and what is *small* can be done by simply comparing the radius of curvature of the meniscus (or the radius of the pendant or hanging drop) with the square root of the capillary constant, a_o (or a_d) of the liquid. If it is much smaller (say more than 10 times) than the square root of the capillary constant, then the influence of gravitation can be neglected.

4.5 Liquid Surface Tension Variation by Temperature

When the temperature is increased, the kinetic agitation of the molecules and the tendency to evaporate increases. As a result, the net inward pull may be expected to become less, thus weakening the molecular interactions, and the surface tension almost invariably decreases (some fused metals are the exception to this). For most liquids, the decrease in surface tension as the temperature rises is approximately linear, over long ranges. The surface tension of liquids varies between 5 and $74 \, mN \, m^{-1}$ for the 0–100°C temperature range. Fused salts have surface tensions of about $50–500 \, mN \, m^{-1}$ and fused metals $500–1000 \, mN \, m^{-1}$ (for much higher temperatures than 100°C). At constant pressure and surface area, the variation of surface tension with temperature is given in Equations (213)–(215) in Section 3.2.5. The $[-T^S \partial \gamma / \partial T^S_{A,n_i}]$ part of Equation (216) is often called the *latent heat of the surface*, and it is the amount of heat that has to be added to the surface to maintain its temperature constant during an isothermal expansion. Since $\partial\gamma/\partial T^S_{A,n_i}$ is almost always negative, the latent heat of the surface is positive, and the total surface (internal) energy, U^S, as given by Equation (216) is generally larger than γ. As Kelvin showed, there is absorption of heat during the extension of most surfaces because molecules must be dragged from the interior against an inward attractive force to form a unit area of new surface. Their motion is retarded by this inward attraction as they leave the interior to reach the surface, so the temperature of the surface layers is lower than that of the interior, unless heat is supplied from outside. However, the total (internal) energy of the surface, U^S is less affected by temperature changes than is γ, and U^S is found to be nearly temperature-independent for most liquids. The total energy is more easily related to molecular models describing surface. Total energy values vary between 4×10^{-2} and $18 \times 10^{-2}\,\mathrm{J\,m^{-2}}$ for most organic liquids and water, but it is possible to compare the internal energy values on a per mole basis. Now, the volume of a single liquid molecule $V_i = V_M/N_A$, where V_M is the molar volume and N_A is Avogadro's number. For a spherical molecule, the surface area of the molecule can be found from simple geometry as, $A_i = 4\pi(3V_M/4\pi N_A)^{2/3}$, and the total surface area per mole can be obtained by multiplying the surface area per molecule by Avogadro's number as $A_T = 4\pi N_A(3V_M/4\pi N_A)^{2/3}$. However, it was found that only about a quarter of spherical surface molecules are exposed to the interface giving, $A_S = \pi N_A(3V_M/4\pi N_A)^{2/3}$ so this equation reduces to a semi-empirical expression, $A_S = fN_A^{1/3}V_M^{2/3}$ where f is a factor near unity. If we multiply U^S by the area per mole at the surface, A_S, then values of 6×10^3 to $12 \times 10^3\,\mathrm{J\,mol^{-1}}$ are obtained for the internal surface energy, which are close to each other for a large number of materials, showing the strong dependence of U^S on the number of molecules at the surface.

As we know $(\partial\gamma/\partial T)_{A,n_i} = -(\partial S^S/\partial A)_{T,n_i}$, from Equation (213), the temperature variation of surface tension can be related to the differential surface excess entropy, and since the left-hand side of the equation is almost always negative, there is an increase in the interfacial entropy with the increase in surface area. For a constant unit area of $A^S = 1\,\mathrm{m^2}$, if we want to compare the surface excess entropy of a one-component pure liquid, with the entropy of its bulk liquid, we have to use Equation (219), $S^S dT^S + d\gamma + \sum_i \Gamma_i^S d\mu_i^S = 0$

giving $[S^S_{spec}dT^S + d\gamma = 0]$ when the Gibbs dividing plane is located at a place where $\Gamma_1^S = 0$. Then we have $[S^S_{spec} = -(d\gamma/dT)]$ where S^S_{spec} is the *specific surface excess entropy*, that is the entropy of a unit area of surface liquid less the entropy of the same amount of bulk liquid. Since, $(d\gamma/dT)$ is negative, S^S_{spec} is positive, indicating that molecules in the surface have more freedom of movement or are more disordered than in the bulk.

If we further increase the temperature towards the critical temperature, T_c, the restraining force on the surface molecules diminishes, and the vapor pressure increases, and when T_c is reached, the surface tension vanishes altogether ($\gamma = 0$). There are several empirical approaches using critical properties and molar volume to predict the surface tension of pure liquids. By comparing the surfaces on the basis of the number of similarly shaped and symmetrically packed molecules per unit area, Eötvös derived an equation in 1886,

$$\gamma(V_M)^{2/3} = k(T_c - T) \tag{336}$$

where V_M is the molar volume of the liquid and k is a constant ($V_M = M_W/\rho_L$ where M_W is the molecular mass of the compound and ρ_L is the liquid density). The term $(V_M)^{2/3}$ is called the *molar surface area* of the liquid and is a proportionality constant for the number of similarly shaped and symmetrically packed molecules per unit area. The left side of Equation (336) is also called the *molar surface free energy*. By applying simple graphical procedures to the data obtained by measuring surface tensions at different temperatures, the constant k has been found for a great number of non-polar and non-hydrogen bonding liquids, all close to an average value of 2.12. H-bonding liquids in which the molecules are associated give a lower value for k; 0.7–1.5 for alcohols; 0.9–1.7 for organic acids; and for water, k is not a constant but varies between 0.9 and 1.2, according to the measurement temperature range. It has been shown that molar volumes, V_M are higher than the calculated value from the (M_W/ρ) formula, for the associating liquids. However, for the dissociating chain molecules, k is larger than 2.12 and high values of k up to 7.0 were found, for example, for tristearin.

Later, Katayama replaced the density of the liquid by the difference in density between the liquid and saturated vapor in the Eötvös equation:

$$\gamma \left(\frac{M_W}{\rho_L - \rho_V} \right)^{2/3} = k(T_c - T) \tag{337}$$

which gives a better agreement with the experimental data at high temperatures. In 1893, Ramsay and Shields proposed that Equation (336) should be corrected as

$$\gamma (V_M)^{2/3} = k(T_c - T - 6) \tag{338}$$

because they experimentally determined that the surface tension of most liquids reaches zero 6 degrees before the critical temperature, T_c. The reason for this behavior is not clear.

Simple expressions were offered for the nearly linear variation of surface tension with temperature, such as $\gamma = \gamma^o(1 - bT)$, where, γ^o, is a constant for every liquid. Since $\gamma \cong 0$ at T_c, by denoting $b = 1/T_c$, this linear plot may be expressed as

$$\gamma = \gamma^o \left(1 - \frac{T}{T_c} \right)^n \tag{339}$$

for $n = 1$. The accuracy of Equation (339) is good for some liquids. When $n = 1$ is assumed, by comparing Equations (336) and (339) one obtains $\gamma^o = \dfrac{kT_c}{(V_M)^{2/3}}$, which is approximately correct for many liquids. However, for some non-polar and non-hydrogen bonding liquids the temperature–surface tension plot is not linear, and the curves are concave upward showing the power dependence. Van der Waals proposed $n = 1.5$ for Equation (339), but most of the surface tension experiments indicated that $n = 1.23$. Guggenheim proposed $n = 11/9$, which was derived from the theory of close-packed non-polar and non-hydrogen bonding liquids.

4.6 Parachor

In 1923, McLeod assumed $n = 6/5$ in Equation (339), and by combining with Equation (337) and eliminating $(T_c - T)$, he found an expression to relate the surface tension to the density:

$$\gamma = \left(\frac{k^6 T_c^6}{\gamma_o^5 M_w^4} \right) (\rho_L - \rho_V)^4 = K(\rho_L - \rho_V)^4 \tag{340}$$

where K is a constant which is different for every liquid. This fits very well with the experimental results for the majority of organic liquids over a large range of temperatures. In 1924, Sugden modified McLeod's equation to derive a new empirical equation to neutralize the effect of temperature:

$$P = \gamma^{1/4} \left(\frac{M_W}{\rho_L - \rho_V} \right) = \left(\frac{k^6 T_c^6}{\gamma_o^5} \right)^{1/4} \tag{341}$$

where P is called a *parachor*. It is well known that the molecular volume of organic compounds, V_M, depends on chemical constitution so that specific molecular groups have characteristic sizes and shapes, and probably occupy similar volumes in the liquid state. However, since the volume of a liquid changes with the variation in temperature, because the thermal motion of the molecules gradually overcomes the cohesional interactions between them, the use of the V_M term in any molecular theory seems erroneous. Sugden showed that the parachor parameter, P, does not vary with temperature and is a much better parameter than V_M for a molecular theory. The P value is comparable for different substances under similar conditions of surface tension. He proposed that the parachor may be used as a means of determining structure and he dissected the P value into parachors for different atoms, such as H = 17.1, C = 4.8, O = 20.0, double bond = 23.2, triple bond = 46.6, closed six ring = 6.1 etc.; these *atomic* or *group* parachors, when added up, reproduce experimentally observed parachors very accurately. It was later found that there is a rough correspondence between atomic parachors and atomic volumes. The weakest argument for the semi-emprical *parachor* approach is its dependence on $n = 6/5$ for all liquids; this cannot be true for many polar and hydrogen-bonding types.

It should be noted that Eötvös, Ramsay–Shields and Sugden's *parachor* equations are empirical in nature and their theoretical foundations are rather obscure. There have been several attempts to associate these equations with strict thermodynamical terms, but none have been successful.

4.7 Liquid Surface Tension Variation by Pressure: Kelvin Equation

The vapor pressure of a liquid increases with increase in temperature, which can be calculated by applying the Clausius–Clapeyron equation (Equation (283) in Section 4.2.1) and tables showing vapor pressure–temperature variation are given in many textbooks and handbooks. These vapor pressures are reported for vapors in thermodynamical equilibrium with liquid of the same material with a flat surface. However, as we see in the Young–Laplace equation, the vapor pressure inside a liquid drop is higher than that of a planar, flat surface because of the curvature present. Lord Kelvin derived an expression in 1870 showing how the vapor pressure depends on the curvature of the liquid. As we know from above, for a spherical liquid drop in a gas having a radius of r, the two radii of curvature are positive and the same ($R_1 = R_2 = r$) and thus $\kappa_1 = \kappa_2 = \kappa_{sph} = 1/r$, giving $H_{sph} = \kappa_{sph} = 1/r$; the pressure inside the liquid drop is higher than outside, and this positive pressure difference, ΔP, causes the liquid molecules to evaporate more easily than from a flat liquid surface.

However, for a gas bubble in a liquid environment where the two radii of curvature are also the same but negative for the liquid, the resultant ΔP is negative for the liquid; the vapor pressure inside the gas bubble is lower than for the flat liquid surface and it is easier for the liquid molecules to evaporate in the bubble, causing vapor condensation within the gas bubble.

Now, conversely, if we consider a spherical liquid drop in air, having a radius of r, the vapor pressure of a drop, $P_v^c > P_v$, that is P_v^c is higher than that of the same liquid with a flat surface, P_v (the superscript c indicates a curved surface). If dn mol of liquid evaporates from the drop and condenses onto the bulk flat liquid under isothermal and reversible conditions, the free-energy change of this process can be written by differentiating Equation (155) as

$$dG = dnRT \ln \frac{P_v}{P_c^v} \tag{342}$$

Since $P_v^c > P_v$, dG is negative and the process is spontaneous. This free-energy change can also be calculated from the surface free-energy change of the droplet, which results from the surface area decrease due to the loss of dn mol of the liquid having a molar mass of M_W. This evaporation process produces a volume decrease of $-dn(M_W/\rho_L)$ in the liquid drop. As a result of this volume decrease, a spherical shell from the drop surface whose volume is $= 4\pi r^2 dr$ is lost from the total drop volume. Then we can write

$$-dn\left(\frac{M_W}{\rho_L}\right) = -4\pi r^2 dr \tag{343}$$

and by rearranging Equation (343), we can calculate the decrease in drop radius as, $dr = (M_W/4\pi r^2 \rho_L)dn$. Now, the decrease in surface free-energy is γ times the decrease in the surface area of the drop that results from the decrease of dr in the droplet radius, so that

$$dG = \gamma dA = \gamma[4\pi(r - dr)^2 - 4\pi r^2] \cong -8\pi\gamma r dr \tag{344}$$

Inserting dr into Equation (344) gives

$$dG = -8\pi\gamma r(M_W/4\pi r^2 \rho_L)dn = -\frac{2\gamma M_W}{r\rho_L} dn \tag{345}$$

By equating Equations (342) and (345) for the free-energy change, one obtains

$$dnRT \ln \frac{P_v}{P_c^v} = -\frac{2\gamma M_W}{r\rho_L} dn \tag{346}$$

and

$$\ln \frac{P_v^c}{P_v} = \frac{2\gamma M_W}{r\rho_L RT} \tag{347}$$

Equation (347) is the Kelvin equation for spherical drops. The presence of r in the denominator shows the dependence of the vapor pressure on the drop size. For spherical drops, the vapor pressure can be calculated by

$$P_v^c = P_v e^{-\frac{2\gamma V_M}{RTr}} \qquad (348)$$

The constant, $[2\gamma V_M/RT]$ is 2.024 nm for benzene ($\gamma = 0.0282$ N/m, $V_M = 88.9 \times 10^{-6}$ m^3 mol^{-1}), 2.772 nm for octane ($\gamma = 0.0211$ N m^{-1}, $V_M = 162.5 \times 10^{-6}$ m^3 mol^{-1}), 1.058 nm for water ($\gamma = 0.072$ N m^{-1}, $V_M = 18 \times 10^{-6}$ m^3 mol^{-1}), and 1.035 for ethanol ($\gamma = 0.022$ N m^{-1}, $V_M = 58.4 \times 10^{-6}$ m^3 mol^{-1}) at 25°C. From Equation (348), it can be seen that curvature has little effect on the vapor pressure until the radius is of the order of 10 nm. In addition, the above derivation of the Kelvin equation assumes that the surface tension is unaffected by the curvature, and it is invalid for very small droplets whose radii are smaller than 10 nm, where the number of molecules will be so small that some variation of surface tension may take place. The surface tension of a water droplet of radius 1 nm was calculated to be approximately 55 mN m^{-1} at 20°C, instead of 72.8 mN m^{-1} for a plane surface.

The Kelvin equation can be combined with the *relative humidity, RH*, if water is involved as the fluid; relative humidity indicates how moist the air is. The amount of water vapor in the air at any given time is usually less than that required to saturate the air. The relative humidity is the percentage of saturation humidity, generally calculated in relation to the saturated vapor density. Relative humidity may be defined as the ratio of the water vapor density (mass per unit volume) to the saturation water vapor density, usually expressed in percent. Relative humidity is also approximately equal (exactly equal when water is assumed as an ideal gas) to the ratio of the actual water vapor pressure to the saturation water vapor pressure, $RH = P_v/P_v^o$. The P_v^o values corresponding to each temperature are given in tables which can be found in handbooks. If RH is measured in an experiment, then P_v can be calculated by using the saturation water vapor pressure tables and can be inserted into the Kelvin equation.

For drops, which are not spherical, we must consider two radii of curvature, and Equation (347) becomes

$$\ln\frac{P_v^c}{P_v} = \frac{\gamma M_W}{\rho_L RT}\left(\frac{1}{R_1} + \frac{1}{R_2}\right) \qquad (349)$$

Equation (349) is the general form of the Kelvin equation. There are some important consequences of liquid surface tension variation by pressure in nature; for example, fogs are unstable and always disappear: fogs are aerosols of water droplets; some droplets are larger than others in a fog. As we know $P_v^c > P_v$ for a liquid drop in a gas, small droplets have a higher vapor pressure than that of the large droplets from Equation (347) and hence more liquid evaporates from the surface of small droplets. They will therefore become smaller, and their vapor pressure will additionally increase, so that evaporation will continue increasingly rapidly. The evaporated vapor condenses on the large drops, whose vapor pressure will decrease, and they will continue to grow further. In summary, the bigger drops grow at the expense of the smaller drops. This process is called *Ostwald ripening*. After achieving a large radius, these large drops fall to the ground through the effect of gravity, and finally the fog disappears completely, with wet ground. This shows that the equilibrium is unstable in a medium where a number of droplets are present, which are initially in equilibrium with a surrounding vapor.

If a vapor is cooled rapidly, its vapor pressure will be larger than the saturation vapor pressure at this cold temperature, so that it becomes super-saturated. Under these conditions, condensation to the liquid state cannot occur spontaneously unless some nucleation sites are present which are large enough to continue to grow at the prevailing vapor pressure. If no nucleation sites are present, every newly formed drop will instantly evaporate again and the super-saturated medium cannot condense.

There is a critical drop radius, $r_{critical}$, for equilibrium with the surrounding vapor pressure, because smaller drops have a higher vapor pressure and will spontaneously evaporate, and all drops larger than this size will grow at the expense of smaller (and unstable) droplets. For a water drop in air and also an air bubble in water, the effect of radius of curvature on equilibrium vapor pressures is given in Table 4.1. It can be seen that below a droplet size of 10 nm, there is a considerable vapor pressure increase over the vapor pressure of flat surfaces due to the presence of curvature.

As we stated above, for a gas bubble in a liquid environment where the two radii of curvature are negative, the resultant ΔP is also negative and the pressure inside the gas bubble is lower than the flat liquid surface. For this case, we again use the Kelvin equation (Equation (347)) with a minus sign on the right-hand side. At the boiling point of the liquid, the vapor pressure over the plane surface of the liquid, P_v, equals the external pressure, P_{ext}. If a bubble were to be formed in these conditions, the pressure inside the vapor bubble, P_{int}, would have to be greater than P_{ext}, from the definition, $[\Delta P = P_{int} - P_{ext}]$ however, this is not permitted because $\ln(P_v^c/P_v)$ is negative in Equation (347) for this case. (This is equivalent to $P_v = P_{ext} > P_v^c = P_{int}$). Consequently vapor bubbles cannot exist at the boiling point so the Kelvin equation explains why we must heat liquids above their normal boiling point (the process of *superheating*).

In practice, if air is dissolved previously in the water, it forms very small air bubbles while the water is heated up to its boiling point. These tiny air bubbles are released from the liquid phase as the solubility of the air decreases with the increase in the temperature. Those bubbles form the nuclei into which water can evaporate because it is easier for the liquid molecules to evaporate into the air bubble due to the negative ΔP, and with the increase in the bubble size, boiling starts. However, if the water had been previously boiled so that it no longer contains dissolved air, the next boiling process quickly becomes much more difficult, and the liquid may pass to a superheated stage where it boils very abruptly causing *bumping*, which can sometimes be very dangerous on an industrial scale.

At temperatures well above the normal boiling point (around 200°C for water at atmospheric pressure), the temperature is high enough that the equilibrium vapor pressure, even

Table 4.1 Relative vapor pressures of curved surfaces in equilibrium at 25°C, for water drops in air and air bubbles in water

Radius r (nm)	$\dfrac{P_v^c}{P_v}$ (drop)	$\dfrac{P_v^c}{P_v}$ (bubble)
1000	1.001	0.999
100	1.011	0.989
50	1.021	0.979
10	1.112	0.899
5	1.236	0.809
1	2.880	0.347
0.5	8.298	0.121

inside a small bubble, is large enough to allow it to continue to grow. In this case the formation of bubbles of pure vapor (not air) occurs. This is called *homogenous nucleation*, as we will see later in Section 4.9.

On the other hand, the Kelvin equation has been extensively used in research on gas adsorption onto porous solids (see Sections 8.4 and 8.5) and capillary condensation.

4.8 Capillary Condensation

Lord Kelvin realized that, instead of completely drying out, moisture is retained within porous materials such as plants and vegetables or biscuits at temperatures far above the dew point of the surrounding atmosphere, because of capillary forces. This process was later termed *capillary condensation*, which is the condensation of any vapor into capillaries or fine pores of solids, even at pressures below the equilibrium vapor pressure, P_v. Capillary condensation is said to occur when, in porous solids, multilayer adsorption from a vapor proceeds to the point at which pore spaces are filled with liquid separated from the gas phase by menisci. If a vapor or liquid wets a solid completely, that is the contact angle, $\theta = 0°$, then this vapor will immediately condense in the tip of a conical pore, as seen in Figure 4.8 *a*. The formation of the liquid in the tip of the cone by condensation continues until the cone radius, r, reaches a critical value, r_c, where the radius of curvature of the vapor bubble reaches the value given by the Kelvin equation $(r = r_c)$. Then, for a spherical vapor bubble, we can write

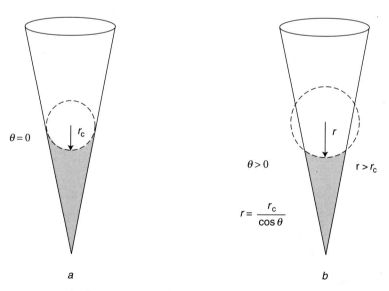

$$\theta = 0 \qquad r_c$$

$$\theta > 0 \qquad r \qquad r > r_c$$

$$r = \frac{r_c}{\cos \theta}$$

a   *b*

Figure 4.8   *a.* For completely wetting materials in a conical pore, the formation of the liquid in the tip of the cone by condensation continues until the cone radius, r, reaches a critical value, r_c $(r = r_c)$. *b.* For partially wetting materials, where $\theta > 0°$, the radius of curvature increases, so that $(r > r_c)$ and for a first approximation, $r = r_c \cos \theta$.

$$\ln \frac{P_v^c}{P_v} = -\frac{2\gamma M_W}{r_c \rho_L RT} \tag{350}$$

where r_c is the capillary radius inside the pore at the point where the meniscus is in equilibrium, and since $P_v > P_v^c$, due to the negative curvature of the vapor bubble, the vapor pressure of the liquid inside the pore decreases to P_v^c at this point. In practice, many solids cannot be completely wetted by contacting (or condensing) liquids so that $\theta > 0°$, as seen in Figure 4.8 *b*. For this case, the radius of curvature increases and for a first approximation, $r = R_1 = R_2 = r_c/\cos\theta$ is used in Equation (350). If $\theta < 90°$ the meniscus is concave, $P_v > P_v^c$, and the vapor will condense in the capillary surface first. If $\theta > 90°$, $P_v^c > P_v$, then the vapor will condense on the plane surface first at P_v, and a pressure greater than P_v is required to force the liquid to enter into the capillary. This is the basis of *mercury injection porosimetry* (see Section 8.3.4). It is necessary to apply a positive pressure to mercury liquid, which must be larger than the present positive ΔP, in order to make the mercury enter into a porous solid. Mercury injection porosimetry is a useful method of determining the pore size distribution of a solid by measuring the volume of mercury taken up by this solid as a function of the increase in the pressure. Smaller pores will be entered by successive increases in the applied pressure, and the method is very successful for comparing different samples of the same or similar materials where θ does not vary too much.

If a fissure or crack is present in any pore, the pore is no longer assumed to be spherical, and if r_c is chosen as the radius of curvature perpendicular to the fissure direction, then Equation (349) is used, where $R_1 = r_c$ and $R_2 = \infty$ for these conditions.

Capillary condensation explains why liquids are strongly absorbed into porous materials. This situation occurs occasionally in nature, such as the wetting of mineral and clay particles present in the ground by water. In many instances, capillary condensation determines the strength of adhesion between fine particles and the flow, and other behaviors, of powders. The capillary force, F_γ, arising from capillary condensation of a liquid between particles may be treated in several ways, the simplest approach being calculation of the capillary force between two contacting spherical particles having identical radius, R_S, as seen in Figure 4.9. If water (or another liquid) completely wets their surfaces ($\theta = 0°$), the liquid will condense into the gap around the contact zone. The meniscus of the liquid is negatively curved and the first radius of curvature is given by $R_1 = -r$. The second radius of curvature is $R_2 = z$ by definition. The Young–Laplace pressure in the liquid is negative and consequently the particles attract each other. In most practical situations, $z \gg r$, so that Equation (333) can be written as

$$\Delta P = \gamma\left(-\frac{1}{r} + \frac{1}{z}\right) \approx -\frac{\gamma}{r} \tag{351}$$

and the pressure, ΔP, is lower in the liquid than in the outer vapor phase. This attractive capillary force can be expressed as $F_\gamma = \Delta P A_c$, where A_c is the cross-sectional area of the liquid between two spherical particles. Therefore, $A_c = \pi z^2$ can be written, from simple geometry, but we need to express z in terms of r. If we apply Pythagoras' theorem of plane geometry so that $(R_S + r)^2 = R_S^2 + (z + r)^2$ giving $2 R_S r = z^2 + 2 zr$, and if we assume that $z \gg 2r$, then we may write $2R_S r \approx z^2$. Thus, the attractive capillary force due to capillary condensation between two identical spherical particles can be expressed as

$$F_\gamma = 2\gamma\pi R_S \tag{352}$$

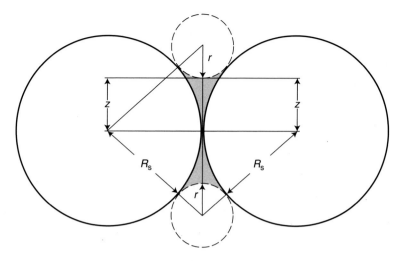

Figure 4.9 Capillary force calculation between two contacting spherical particles having identical radius of R_S. If a liquid completely wets the surfaces of the particles ($\theta = 0°$), then the liquid will condense into the gap around the contact zone, $(-r)$ is the first radius of curvature, and the second radius of curvature is $R_2 = z$ by definition, where the distance, z, is as shown in the figure.

Equation (352) shows that the capillary force depends mostly only on the size of the particle and the surface tension of the liquid. However, in reality, the surfaces of most of the particles are rough and they touch only at some points, having much smaller liquid menisci between them (and giving much smaller z values), so that the resultant capillary force, F_γ, is much smaller than the value calculated by Equation (352).

4.9 Nucleation

The production of a new phase such as a liquid from vapor, or a solid from liquid, even though thermodynamically favorable in terms of the chemical potentials of the bulk phases, requires a *nucleation* mechanism. Nucleation is the formation of an embryo or nucleus of a new phase in another phase. This topic covers a wide range of processes in industry such as the condensation of a vapor, the precipitation of a solute from a solution and the crystallization of a liquid; and in nature, rain formation in the clouds and the formation of opals from colloidal silica sediments. The nucleation mechanism helps to overcome the free energy barriers involved, which prevent the start of new phase formation. For example, during freezing, all the molecules of a liquid cannot suddenly adopt the positions required to form the solid, but when nucleation starts, an interface is produced between solid and liquid by forming tiny clusters with an increase in free energy. Another example is the formation of liquid droplets in a vapor phase. Tiny molecular clusters form in the vapor phase which will grow by condensation and transform into droplets. The nucleation of liquid droplets from a supersaturated vapor depends on the pressure difference, ΔP, between the actual and saturation vapor pressures and also on the curvature of the droplet, which is generally spherical in shape. Nucleation processes can be divided into two subsections, *homogeneous* and *heterogeneous* nucleations.

4.9.1   Homogeneous nucleation during a phase transition

The nucleation process is termed *homogeneous nucleation* when only a single pure component or a phase of uniform composition is involved during a phase transition, in the absence of any external surfaces and particles. Many natural phenomena, such as cloud formation as water droplets from water vapor or hail formation as ice, involve homogeneous nucleation (and also heterogeneous nucleation if foreign particles and surfaces are present) mechanisms. In general, the Kelvin equation (Equation (347)) can be applied to super-saturation, super-heating and sub-cooling processes that are sometimes observed in phase transitions. In this section we will focus on the nucleation and growth of liquid drops from their vapors. Small molecular clusters are always present in any vapor in the absence of participating foreign surfaces and particles. These clusters consist of only a few vapor molecules, some of them being dimers, trimers and mostly n-mers. They are continually forming and disintegrating in a dynamic mechanism. If the vapor pressure, P_v, is significantly above the equilibrium saturation vapor pressure with the flat liquid surface, P_v^o, then there is a natural tendency to form small clusters of molecules. The (P_v/P_v^o) ratio is called the *super-saturation ratio*. If the super-saturation ratio is high (say more than 4) then large clusters occur more frequently. These clusters then grow by the condensation of other new molecules. Ostwald ripening occurs in these conditions and the clusters grow further, sometimes aggregating to form even larger recognizable droplets. If this process continues, many macroscopic liquid drops form and finally coalesce to yield large amounts of the liquid phase. However, if the system is just beyond the saturation pressure, this liquid drop growth sequence does not occur, that is, the super-saturation ratio must be high to realize the above droplet nucleation process. When we consider the precipitation of solids from a super-saturated solution, the same rule applies so that the solution concentrations must be much higher than the saturation concentration. Similarly, the temperature of the liquids should be much lower than the crystallization temperature to form solid crystals from their liquids. For example, very pure, dust-free liquid water can be cooled down to −48°C before spontaneous freezing occurs (sub-cooling process). Chemists are familiar with the use of seed crystals to initiate the crystallization in super-cooled solutions. Dust-free water can be heated considerably above 100°C before it boils.

The nucleation and growth of liquid water drops from water vapor is a very important process in nature and also in industry. If we assume that the water obeys the ideal gas law, the difference in the Gibbs energies to form a liquid water drop in its vapor phase by the phase change process can be written as

$$\Delta G = G_L - G_v = -nRT \ln \frac{P_v}{P_v^o} \tag{353}$$

Since $P_v > P_v^o$, the higher the super-saturation ratio, the higher the ΔG value. The number of moles, $n = m/M_W$ where m is the mass of the droplet, M_W is the molecular mass ($m = 4\pi r^3 \rho_L/3$), r is the radius of the droplet and ρ_L is the density of the liquid. The value of n is very small because n represents the number of moles of vapor that condense into a tiny nucleus (liquid droplet) at the vapor pressure, P_v. Then, we can write

$$\Delta G = -\frac{4\pi r^3 \rho_L}{3M_w} RT \ln \frac{P_v}{P_v^o} \tag{354}$$

In order to find the total free energy difference to form the drop, ΔG_T, we also need to consider the surface free energy effects:

$$\Delta G_T = -\frac{4\pi r^3 \rho_L}{3M_w} RT \ln \frac{P_v}{P_v^o} + 4\pi r^2 \gamma \tag{355}$$

The term ΔG_T represents the free-energy barrier that must be overcome to form a liquid drop. When we check Equation (355) in terms of vapor pressures, if the super-saturation ratio is negative so that $P_v^o > P_v$, then ΔG_T is positive, so that any nucleus that is formed by some clustering vapor molecules will evaporate again and no droplet forms spontaneously. For the normal case, where the super-saturation ratio is positive so that $P_v > P_v^o$, the first term on the right-hand side of this equation is always negative and varies as r^3, and the second term varies as r^2. Since the second term is always positive, the formation of droplets from the vapor phase by homogenous nucleation is radius-dependent. Equation (355) can be most effectively checked for the size of drop radius between 0 and 2.5 nm because of the tiny size of the possible droplets. The numerical results obtained by applying Equation (355) are plotted in Figure 4.10. In this figure, the plots of ΔG_T versus the drop radius, r, are shown for water at 20°C for different super-saturation ratios

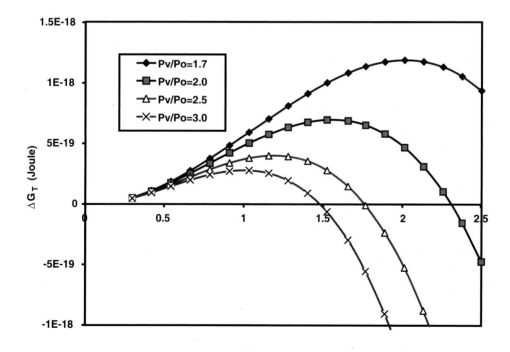

Figure 4.10 Variation of the free-energy change, ΔG_T versus the drop radius, r, for water at different super-saturation ratios, ranging from 1.7 to 3.0 at 20°C.

ranging from 1.7 to 3.0. At the beginning, the ΔG_T value increases with increasing r, but as r increases, the first negative term on the right-hand side of Equation (355) also increases rapidly because of the domination of the r^3 term, imparting a greater negative effect; and after a maximum value, ΔG_T starts to decrease. Consequently, a maximum ΔG_T value occurs at a specific critical radius, r_c, depending on the super-saturation ratio (P_v/P_v^o), as can be seen in Figure 4.7. At the maximum point, where $r = r_c$, the derivative $[d(\Delta G_T)/dr]$ is equal to zero. For molar quantities, at $r = r_c$ differentiating Equation (355) and equating to zero results in

$$\ln \frac{P_v^c}{P_v^o} = \frac{2\gamma M_W}{r_c \rho_L RT} \tag{356}$$

Equation (356) is identical to the Kelvin equation (Equation (347)) for saturation vapor pressures $(P_v \Rightarrow P_v^o)$ at $r = r_c$. The P_v^c parameter is the critical vapor pressure, which corresponds to the vapor pressure when the drop radius, r equals r_c, which is also the radius of curvature for a spherical drop having the critical size. We should note that Equation (356) is valid when two phases coexist in equilibrium.

According to the above analysis, droplets having a radius of r_c will re-evaporate for pressures below P_v^c, and only after the vapor pressure exceeds P_v^c, although all sizes of droplet would exist; the large ones are preferred, and then the droplets having a radius of r_c will start growing indefinitely by vapor condensation. In these conditions, only after most of these droplets exceed this critical radius, r_c, is it possible for a new phase to grow spontaneously. On the other hand, droplets with a smaller radius require a greater degree of super-saturation to exist in this metastable equilibrium, as can be seen in Figure 4.7, which shows that r_c becomes smaller with the increase in super-saturation ratio.

It would be a mistake to apply Equation (356) to $r = 0$ because, if we assume that it is possible, this means that an infinite super-saturation condition would occur and would make the presence of a new phase impossible. As we stated above, Equation (356) is valid for the coexistence of two phases, so we cannot apply it to $r = 0$. If $r > 0$, but becomes extremely small, the size of the cluster may approach the dimensions of individual molecules, but of course there is a limit to applying the Kelvin equation to these extremely small clusters because of spherical shape distortions. It is usually accepted that the Kelvin equation can be applied down to a radius of curvature that corresponds to about seven times the molecular diameter, but this figure was found to be much larger for water, having a critical radius of 0.8 nm and containing about 90 molecules inside. This is due to its complex cluster formation mechanism, containing more molecules than any of the organic liquids, this arising from its high hydrogen bond formation properties.

Equation (356) may be applied to the equilibrium solubility of a solid in a liquid. In this case the ratio (P_v^c/P_v^o) is replaced by the ratio (a_i^c/a_i^o), where a_i^o is the activity of the dissolved solute in equilibrium with a flat surface, and a_i^c is the corresponding quantity for a spherical surface. Then we may write

$$\ln \frac{a_i^c}{a_i^o} = \frac{2\gamma M_W}{r_c \rho_L RT} \tag{357}$$

However, the calculation of the solid surface tension from Equation (357) does not give quantitatively consistent results. This may be due to the non-uniformity of solid particles as a sphere; rather they are irregularly shaped and polydisperse. In addition, the effect of the presence of sharp points or protuberances on the solid particles has a much larger effect on the solid solubility than expected. Also, the linear dependence of the solubility of solid particles on their radius is not clear.

4.9.2 Rate of homogeneous nucleation

The essential problem in nucleation studies is to estimate the rate of formation of nuclei having the critical size. When $r = r_c$, the maximum free-energy barrier that must be overcome to form a liquid drop can be found by combining Equations (355) and (356)

$$(\Delta G_T)_{max} = -\frac{8\pi r_c^2 \gamma}{3} + 4\pi r_c^2 \gamma = \frac{4\pi r_c^2 \gamma}{3} \tag{358}$$

Equation (358) shows that the maximum free-energy barrier that must be overcome to form a liquid drop is equal to one-third of the surface energy of formation, as Gibbs stated. When this equation is applied for crystal formation, the numerical factor of 3 changes because of the non-spherical shape of crystals. If we substitute the critical drop radius, r_c, from Equation (356) into Equation (358), we obtain

$$(\Delta G_T)_{max} = \frac{16\pi\gamma^3 M_W^2}{3\rho_L^2 R^2 T^2 \left(\ln\dfrac{P_v^c}{P_v^o}\right)^2} \tag{359}$$

The rate of nuclei formation under particle-free, surface-free conditions is called the *rate of homogeneous nucleation, J*. In order to express the rate of homogeneous nucleation for nuclei of size, r_c, first we must consider the thermal fluctuations of the molecules to determine the number of molecules initially located in the embryonic nuclei, which can be found using Boltzmann statistics. Second, we must consider the gas phase collision frequency, f, which is the number of vapor molecules colliding with unit surface area per second, which determines the growth of the nuclei by adding the mass of arriving molecules to the nuclei. If a steady-state condition is assumed, such that the average number of nuclei consisting of 2, 3, . . . , N molecules does not change with time, then mathematical treatment of the subject is possible. The rate of homogeneous nucleation, J, for nuclei having a radius of r_c, is then given by

$$J = fe^{-\frac{(\Delta G_T)_{max}}{RT}} \tag{360}$$

By combining with the maximum free energy barrier, $(\Delta G_T)_{max}$, given in Equation (359), we obtain rate, J, as

$$J = f \exp\left[-\frac{16\pi\gamma^3 M_W^2}{3\rho_L^2 R^3 T^3 \left(\ln\dfrac{P_v^c}{P_v^o}\right)^2}\right] \tag{361}$$

For ideal gases, the gas phase collision frequency, f, which represents the number of molecules striking a unit area of surface per unit time can be expressed as

$$f = \frac{P_v^c N_A}{\sqrt{2\pi M_w RT}} \tag{362}$$

Other derivations of the f factor are also possible, but these are outside the scope of this book. Equation (361) shows that the rate of nuclei formation, J, increases with increasing vapor pressure. When J versus (P_v^c/P_v^o) is plotted it is seen that when the super-saturation ratio (P_v^c/P_v^o) is increased above 4, then J increases very sharply for water. The value of J is also very dependent on γ, which is the interfacial tension between the new and parent phases. For liquid drops formed in their own vapor, this is the surface tension of the liquid itself, since the surface tension of their vapor is neglected. Quantitative fits with Equation (361) and experimental expansion chamber results, where the density of the nuclei are measured by light scattering, are not good. Equation (361) predicts nucleation rates that are too low at low temperatures, and overly high rates for high temperatures; some empirical correction factors have to be added to obtain a good fit.

Equation (361) is a sample equation for the rate of formation of droplets in their vapor, and similar equations can be derived for the rate of crystal formation by freezing their liquids, or precipitate formation from supersaturated solutions, although diffusion controlled cluster formation kinetics in liquids, lattice strain and anisotropic growth in crystals must be considered whenever necessary.

4.9.3   Heterogeneous nucleation during a phase transition

If some solid particles and substrate surfaces are present in a phase transition, then *heterogeneous nucleation* takes place, where the maximum free energy barrier $(\Delta G_T)_{max}$ is lowered by energetically more favorable cluster formation on the solid surface. There are several examples of this phenomenon: water vapor condenses on a dust particle to form a liquid water drop whenever possible, and artificial clouds are seeded by silver iodide particles to start rain. The formation of air bubbles when pouring beer or water into a glass is another example. Air bubbles nucleate at the glass surface while pouring, then grow in size, rise in the liquid and leave the liquid at the surface. Capillary condensation, described in Section 4.8, is also an important example of heterogeneous nucleation.

Most systems in everyday life are not perfectly clean, and dusts, particulates or several substrates provide a template for initial growth of a new phase. If the value of the interfacial tension between the new condensing phase and the particle (or the substrate) surface is low, then this property considerably decreases the energy requirement of nuclei formation $(\Delta G_T)_{max}$ so that heterogeneous nucleation occurs as an easier and preferable path for phase transformations. In addition, phase transformations take place at or very close to the equilibrium temperature.

When heterogeneous nucleation occurs, usually a drop with a contact angle less than 90° forms on a substrate. This liquid drop is generally treated as a spherical cap on the substrate. Since the contact angle between the condensing liquid and the substrate is a measure of the interfacial tension between them, by both considering the liquid–solid and liquid–vapor interfacial tensions (see Section 9.1), the reduction in the free-energy barrier $(\Delta G_T)_{max}$ for heterogeneous nucleation, for a flat, ideal (hysteresis-free) system can be given as

$$[(\Delta G_T)_{max}]_{heterogeneous} = [(\Delta G_T)_{max}]_{homogeneous}\left[\frac{(2+\cos\theta)(1-\cos\theta)^2}{4}\right] \tag{363}$$

When $\theta = 0°$ this means that the liquid completely wets the solid, $[(\Delta G_T)_{max}]_{heterogeneous}$ equals zero, and spontaneous heterogeneous nucleation occurs with no super-saturation at the equilibrium temperature. When $\theta = 90°$ then $[(\Delta G_T)_{max}]_{heterogeneous}$ is half the homogeneous value, and when $\theta = 180°$ both homo- and heterogeneous nucleation energy barriers are equal.

References

1. Adam, N.K. (1968) *The Physics and Chemistry of Surfaces*. Dover, New York.
2. Hirschfelder, J.O., Curtiss, C.F. and Bird, R.B. (1954) *Molecular Theory of Gases and Liquids*. Wiley, New York.
3. Prausnitz, J.M., Lichtenthaler, R.N. and Azevedo E.G. (1999) *Molecular Thermodynamics of Fluid-Phase Equilibria* (3rd edn). Prentice Hall, Englewood Cliffs.
4. Scatchard, G. (1976) *Equilibrium in Solutions & Surface and Colloid Chemistry*. Harvard University Press, Cambridge.
5. Adamson, A.W. and Gast, A.P. (1997) *Physical Chemistry of Surfaces* (6th edn). Wiley, New York.
6. Erbil, H.Y. (1997) Interfacial Interactions of Liquids. In Birdi, K.S. (ed.). *Handbook of Surface and Colloid Chemistry*. CRC Press, Boca Raton.
7. Lyklema, L. (1991) *Fundamentals of Interface and Colloid Science* (Vols. I and II). Academic Press, London.
8. Aveyard, R. and Haydon, D.A. (1973) *An Introduction to Principles of Surface Chemistry*. Cambridge University Press, Cambridge.
9. Hiemenz, P.C. and Rajagopalan, R. (1997) *Principles of Colloid and Surface Chemistry* (3rd edn). Marcel Dekker, New York.
10. Murrell, J.N. and Jenkins, A.D. (1994) *Properties of Liquids and Solutions* (2nd edn). Wiley, Chichester.
11. Butt, H.J., Graf, K., Kappl, M. (2003) *Physics and Chemistry of Interfaces*. Wiley-VCH, Weinheim.
12. Abraham, F.F. (1974) *Homogeneous Nucleation Theory*. Academic Press, New York.

Chapter 5
Liquid Solution Surfaces

In Chapter 4, we dealt with the thermodynamic, physical and chemical properties of pure liquids. However, in most instances *solutions* of liquids are used in chemistry and biology instead of pure liquids. In Chapter 5, we will examine the surfaces of mainly non-electrolyte (ion-free) liquid solutions where a solid, liquid or gas solute is dissolved in a liquid solvent. A *solution* is a one-phase homogeneous mixture with more than one component. For a two-component solution, which is the subject of many practical applications, the major component of the solution is called the *solvent* and the dissolved minor component is called the *solute*. Liquid solutions are important in the chemical industry because every chemical reaction involves at least one reactant and one product, mostly forming a single phase, a solution. In addition, the understanding of liquid solutions is useful in separation and purification of substances.

The presence of a solute in a solution affects the entropy of the solution by introducing a degree of disorder that is not present in the pure solvent, so that many physical properties of the solution become different from that of its pure solvent. Furthermore, solute–solvent molecular interactions affect the total internal energy of the solution. The change in the vapor pressure or the surface tension of a solution from those of its pure solvent are examples of these solute effects. The concentration of the solute in the surface layer is usually different (or very rarely the same) from that in the bulk solution, and the determination of this concentration difference is very important in surface science.

The investigation of solution and surface film properties of two- or three-component liquid solutions is the subject of this chapter. In one extreme, the components in the liquid solution are completely miscible giving a one-phase solution, and in the other extreme, the components are almost completely immiscible, and an insoluble monomolecular film of one component forms on the surface of the other giving a two-phase solution. Between these two extremes, different kinds of films form on the solution surfaces depending on the extent of molecular interactions between the components. The theoretical approaches and experimental techniques that are applied to these solution types will be described in Chapters 5 and 6 respectively.

5.1 Equilibrium in Solutions

Initially we will start with completely miscible liquid solutions. A liquid solution may be in *equilibrium* or may be approaching equilibrium or in a non-equilibrium condition; the transport properties are only important for the last case. A system is in *equilibrium* if no further spontaneous changes take place in constant surroundings, and if the same state can be approached from different directions. When equilibrium occurs, the Gibbs phase equation, $\mu_i^\alpha = \mu_i^\beta = \ldots = \mu_i^\pi$ (Equation (175), given in Section 3.1.3) applies. When the temperature is constant, one method of determining the difference in chemical potential of a substance, i, in two liquid solutions is to measure the vapor pressure of each solution, that is, the partial pressure of the substance in a vapor that is in equilibrium with the solution. This utilizes the fact that at equilibrium the chemical potential is the same in the liquid as in the gas phase, given by the Gibbs phase rule, and we can determine the difference in the chemical potential of that substance in the two gas phases as detailed in Section 3.1.2. These relations hold for either solvent or solute in solutions of any concentration. For simplicity, the vapor is treated as an ideal gas and the vapor volume of the solute is neglected relative to that of solvent, and, similar to Equation (163), the difference in chemical potential in the gas phase may be written as

$$\mu_i = \mu_i^* + RT \ln \frac{P_i}{P_i^*} \tag{364}$$

where superscript $*$ denotes the quantities related to pure substances, P_i is the vapor pressure of the solution and P_i^* is the vapor pressure of pure i. F. Raoult measured the vapor pressures of the pure solvents and corresponding solutions by changing the concentrations of the solutes and found that only ideal liquid solutions obey Equation (164), named as Raoult's law, $X_i = (P_i/P_i^*)$, where X_i is the mole fraction of the solvent. Equation (364) can then be rearranged into $\mu_i = \mu_i^* + RT \ln X_i$ for ideal solutions. The molecular interpretation of Raoult's law is that the less volatile solute molecules present at the solution surface layer partially block the evaporation of the solvent molecules, thereby reducing the vapor pressure of the solvent.

Ideal liquid solutions are different from ideal gas mixtures where we assume that there is no intermolecular interaction. In a liquid, the molecules are close together and an assumption of negligible intermolecular interaction is not rational. Instead, in an ideal solution, we assume that intermolecular interactions between molecules are present, but the molecular properties of the solvent and the solute are so similar to one another that molecules of one component can replace molecules of the other component in the solution, without changing the total energy of intermolecular interaction and the spatial structure of the solution. The requirement of this assumption is that the solute molecules must be essentially of the same size and shape as the solvent molecules; and the intermolecular interaction energies should be essentially the same for solute–solute, solvent–solvent and solute–solvent pairs of molecules. Since it is nearly impossible to find any solute and solvent molecules having the above requirements, there is no absolutely ideal solution and Raoult's law is only approximately valid. Even for the best solute–solvent pairs chosen as examples of ideal solutions, there is a slight departure from ideal behavior. Benzene–toluene, cyclohexane–cyclopentane and ethyl chloride–ethyl bromide are good examples of roughly ideal solutions. The approximate equality of the internal pressure values in liquids, $(\partial U/\partial V)_T$, given by Equation (239), which indicates the attractive interactions between molecules, is a good criterion to be used in choosing ideal liquid solutions. When nonpolar liquid solutions are considered, a difference in internal pressure produces a positive deviation from Raoult's law for both components, decreasing the solubility of the solute in the solvent, and very large differences in internal pressures are necessary for the incomplete miscibility of two nonpolar liquid components. However, when a polar and nonpolar liquid solution are considered, they show strong positive deviations from Raoult's law, with accompanying decreases of solute solubilities irrespective of their internal pressure values; most immiscible liquids giving two separate phases belong to this class for this reason.

Apart from intermolecular interactions due to chemical dissimilarities, the presence of a solute in a solution changes the entropy of the solution by introducing a degree of disorder that is not present in the pure solvent, so that the physical properties of the solution (other than lowering the vapor pressure of the solvent) are also different from those of the pure solvent. The presence of a nonvolatile solute raises the boiling point of the solution, lowers the freezing point and gives rise to an osmotic pressure. These effects are collectively termed *colligative properties*, depending on the number of solute molecules (particles) present in the solution and not their chemical identity. These subjects are well documented in many standard physical chemistry textbooks and are beyond the scope of this book, but must be kept in mind when examining the surface properties of such solutions.

In thermodynamic terms, although there is no perfectly ideal solution, some liquid solutions roughly obey Raoult's law, over a wide range of concentrations; but most of the solutions deviate from the law and only obey it when the solution is very dilute. Thus, Raoult's law is a limited law in nature and is only strictly valid at the limit of zero concentration. Consequently, the properties of real liquid solutions should be discussed in terms of departures from ideal solution properties. In principle, the deviations from ideal-solution behavior are due to differing solute–solute, solvent–solvent and solute–solvent intermolecular interaction pair energies, and also to their differing sizes and shapes. For real liquid solutions that show strong deviations from Raoult's law, we use *activitity* parameters instead of mole fractions, as given in Equations (166)–(168). Activities may be related to the measurable molar fractions of the solutions by the use of *activity coefficients*, φ_i^X, as a constant for a given concentration, temperature and pressure, $\varphi_i^X = a_i/X_i$, as expressed in Equation (168). The φ_i^X parameter varies strongly when the solution concentration changes, and weakly when temperature and pressure change. We can usually neglect the effect of pressure on φ_i^X, unless the pressure is very high.

5.2 Mixing and Excess Thermodynamic Functions

5.2.1 Mixing of ideal gas and liquid solutions

When two ideal gases are mixed at a constant pressure, provided that no chemical reaction occurs, there is a difference between the free energy of the gases before and after the mixing. Before mixing, there are n_A moles of gas A at pressure P and n_B moles of gas B, also at pressure P. The total Gibbs free energy is

$$G_{\text{before}} = n_A(\mu_A^o + RT\ln P) + n_B(\mu_B^o + RT\ln P) \tag{365}$$

After mixing, the total pressure is still P, but there will be $(n_A + n_B)$ molecules. The partial pressure of gas A can be given from Dalton's law as

$$p_A = \left(\frac{n_A}{n_A + n_B}\right) P = X_A P \tag{366}$$

and a similar equation applies for gas B. After mixing, the total Gibbs free energy can be written as

$$G_{\text{after}} = n_A(\mu_A^o + RT \ln p_A) + n_B(\mu_B^o + RT \ln p_B) \tag{367}$$

Then the free energy change due to the mixing process is

$$\Delta G_{\text{mix}} = G_{\text{after}} - G_{\text{before}} = RT \left[n_A \ln\left(\frac{p_A}{P}\right) + n_B \ln\left(\frac{p_B}{P}\right) \right] \tag{368}$$

which can be simplified by combining with Equation (366), $(n = n_A + n_B)$

$$\Delta G_{\text{mix}} = nRT(X_A \ln X_A + X_B \ln X_B) = nRT \sum_i X_i \ln X_i \tag{369}$$

Since the pressure is constant during mixing, $(\partial \Delta G_{\text{mix}}/\partial P)_T = 0$, so that there is no volume change of mixing, $\Delta V_{\text{mix}} = 0$, as calculated from Equation (133), which can also be expected from an ideal gas. Since X_i is always positive, $\ln X_i$ is always negative, and it follows that ΔG_{mix} is always negative and gives a minimum point for $X_i = 0.5$. We can also calculate the entropy of mixing by using Equation (134) so that

$$\Delta S_{\text{mix}} = -\left(\frac{\partial \Delta G_{\text{mix}}}{\partial T}\right)_P = -nR \sum_i X_i \ln X_i = -nR(X_A \ln X_A + X_B \ln X_B) \tag{370}$$

Since $\ln X_i$ is always negative, it follows that ΔS_{mix} is always positive and gives a maximum point for $X_i = 0.5$. Then, by combining Equations (369) and (370), one obtains

$$\Delta G_{\text{mix}} = -T \Delta S_{\text{mix}} \tag{371}$$

giving $\Delta H_{\text{mix}} = 0$ for a constant temperature, for an ideal gas mixture from Equation (125), which can also be expected from the assumptions of an ideal gas provided that no chemical reaction occurs. The mixing of real gases can be determined relative to ideal gas behavior.

For an ideal solution, we use μ_i^* instead of μ_n^o, which was used for gases. When the mixing of non-electrolyte liquids is considered, the ideal system is defined as one whose thermodynamic mixing parameters obey the same equations (Equations (368)–(371)), similar to an ideal gas, and $\Delta H_{\text{mix}} = 0$ also.

5.2.2 Excess thermodynamic functions

When non-ideal liquid solutions are considered, we use *excess thermodynamic functions*, which are defined as the differences between the actual thermodynamic mixing parameters and the corresponding values for an ideal mixture. For constant temperature, pressure and molar fractions, excess Gibbs free energy is given as

$$G^E = G_{\text{actual}} - G_{\text{ideal}} = (\Delta G_{\text{mix}})_{\text{actual}} - nRT \sum_i X_i \ln X_i \tag{372}$$

and the excess entropy is given as

$$S^E = S_{actual} - S_{ideal} = (\Delta S_{mix})_{actual} + nR\sum_i X_i \ln X_i \tag{373}$$

and also $H^E = H_{actual} - H_{ideal} = (\Delta H_{mix})_{actual}$ and $V^E = V_{actual} - V_{ideal} = (\Delta V_{mix})_{actual}$ equations are valid. Thermodynamic relations between these excess functions are exactly the same as those between the total functions. Partial molar excess functions are defined analogously to those used for partial molar thermodynamic properties. Excess thermodynamic functions may be positive or negative depending on the conditions, and may also be nearly zero for liquid solutions that contain components very similar in molecular structure. When G^E is positive for a solution, it is said to have a positive deviation from Raoult's law. When H^E is negative, this shows there is an attraction between the mixing molecules of the two compounds, and an enthalpy must be supplied to separate them. When hydrogen bonds are present between the mixing molecules, H^E and S^E are negative. According to Equation (373) $(\Delta S_{mix})_{actual}$ will be positive because mixing increases the disorder, but S^E may be negative because, due to hydrogen bonding, there is more order in the actual mixture than for an ideal system.

Excess thermodynamic functions show the deviations from ideal solution behavior and there is of course a relation between G^E and the activity coefficients. Similar to Equation (369), if we write the actual Gibbs free energy of mixing $(\Delta G_{mix})_{actual}$ in terms of activities, we have

$$(\Delta G_{mix})_{actual} = nRT\sum_i X_i \ln a_i \tag{374}$$

By combining Equations (372) and (374) we obtain

$$G^E = nRT\sum_i X_i \ln a_i - nRT\sum_i X_i \ln X_i \tag{375}$$

Since the activity coefficient, φ_i^X is given in terms of molar fraction as $\varphi_i^X = a_i/X_i$ in Equation (168), then by inserting this expression in Equation (375), the total excess Gibbs free energy for the constant temperature, pressure and mole number of other constituents can be written as

$$(G^E)_{T,P,n_j} = nRT\sum_i X_i \ln \varphi_i^X \tag{376}$$

and for the chemical potential which is the partial molar Gibbs free energy, we may write

$$(\mu_i^E)_{T,P,n_j} = RT \ln \varphi_i^X \tag{377}$$

In practice, determination of the activity coefficients of a solvent in a solution is easy, if the solute is nonvolatile. The vapor pressure of the solution and the pure solvent are measured and $a_A = P_i/P_i^*$ (Equation (166) applies). However, if the solute is volatile, then the partial pressure of both the solute and the solvent should be determined.

Determination of the activity coefficients of the non-volatile solute in a solution is difficult. If electrolytes (ions) are present, the activities can be obtained from experimental electromotive force (EMF) measurements. However, for non-electrolyte and non-volatile solutes an indirect method is applied to find initially the activity of the solvent over a range of solute concentrations, and then the Gibbs–Duhem equation is integrated to find the solute activity. If the solution is saturated, then it is easy to calculate the activity coefficient of the solute because the chemical potential of the solute is equal to that of the solid, which can be found in thermodynamic tables. Colligative properties also lead to the calculation of solute activity; these directly measure the activity of the solvent and then the Gibbs–Duhem equation is used to calculate the solute activity.

5.3 Regular Solutions and Solubility Parameter Approach

One of the most important problems in the chemical industry is the prediction of which solutes dissolve in which solvents. Following the development of the polymer industry in the 1930s, solvent selection for the newly developed polymers was a necessity in order to find the best applications; a reliable method was required to replace the countless trial and error situations in the laboratory. The problem was attacked using various theoretical and empirical approaches but has not been solved rigorously to date; however several semi-empirical methods are presently in use. Hildebrand was the first to point out the importance of internal pressure, $(\partial U/\partial V)_T$, as given in Equation (239), of fluids for solubility predictions. As given above, internal pressure is due to *cohesional* attractive forces (see Section 3.5.1) between the molecules which contribute to internal energy, U, and which is zero for ideal gases. The approximate equality of internal pressure values in non-electrolyte liquids is a good criterion for their behavior as ideal liquid solutions obeying Raoult's Law. If the internal pressure values of two materials are too dissimilar from each other, this shows that the intermolecular forces within them are alike and they will not dissolve in each other. By rearranging Equation (237), the internal energy may be expressed as

$$\left(\frac{\partial U}{\partial V}\right)_T = T\left(\frac{\partial P}{\partial T}\right)_V - P \tag{378}$$

The $(\partial P/\partial T)_V$ parameter is called an *isochore* and can be measured directly from a P–T plot at a particular constant volume, or it is more often computed from the expression, $(\partial P/\partial T)_V = -\dfrac{\alpha}{\beta}$, where α is the *coefficient of thermal expansion*, $\alpha = \dfrac{1}{V}\left(\dfrac{\partial V}{\partial T}\right)_P$, and β is the *coefficient of compressibility*, $\beta = -\dfrac{1}{V}\left(\dfrac{\partial V}{\partial P}\right)_T$ of the fluids; these α and β parameters can easily be determined using simple experiments.

5.3.1 Cohesive energy density

In liquids, strong attractive forces exist between neighboring molecules, and as a result each molecule has a considerable negative potential energy in contrast to the vapor phase molecules, which have negligible potential energy. Liquid (and some solid) molecules are forced to overcome attraction interactions between them during vaporization (and sublimation), which are holding them in the condensed state. Consequently, it is a good idea to use the experimentally determined macroscopic vaporization properties of liquids (and sublimation properties of some solids when data are available) in order to estimate the physical intermolecular interaction energies within them. The term *cohesion* describes the physical interactions between the same types of molecule and *adhesion* between different types of molecule, as given in Sections 2.1 and 3.5.3. The cohesive energy of a liquid, U^V, gives the molar internal energy of vaporization to the gas phase at zero pressure (i.e. infinite separation of the molecules). In other terms, $-U^V$ is the molar internal energy of vaporization of a liquid relative to its ideal vapor at the same temperature (the minus sign shows that energy is required to vaporize liquids). For liquids at room temperature (generally well below their boiling point) and atmospheric pressure, it can be safely assumed that (from Equation (259)) $-U^V \cong \Delta U^V = \Delta H^V - P\Delta V$, where ΔH^V is the molar enthalpy of vaporization. If the vapor in equilibrium with the liquid behaves ideally then, $P\Delta V = RT$ so that $-U^V = \Delta U^V = \Delta H^V - RT$ (Equation (260)) is valid. The molar enthalpy of vaporization ΔH^V can easily be determined calorimetrically at any desired temperature. Values of ΔH^V vary between 20 and 50 kJ mol^{-1} for most liquids at room temperature, and thus are a measure of the strength of intermolecular cohesive attractions in the liquid.

van Laar proposed in 1906 that a simple theory of mixtures and solutions could be constructed if we neglect the excess mixing entropy, S^E_{mix} and excess volume of mixing, V^E_{mix} for special cases. He assumed a) the volume change during mixing, (ΔV_{mix}) at constant pressure is zero, so that the excess volume, $V^E_{mix} = 0$, and b) the molecules of mixtures are randomly distributed in both position and orientation, and the entropy of mixing corresponds to an ideal solution (that is the excess entropy, $S^E_{mix} = 0$). Since, at constant pressure, $G^E_{mix} = U^E_{mix} + PV^E_{mix} - TS^E_{mix}$, it follows from van Laar's simplifying assumptions that $G^E_{mix} = U^E_{mix}$. Later, in the 1920s, Hildebrand found that the experimentally determined solution properties of iodine in various nonpolar organic solvents fitted van Laar's simplifying assumptions, and he defined such nonpolar solutions as *regular solutions* where orienting and chemical effects are absent. Hildebrand had chosen iodine as a model solute in organic solvents because this gives violet solutions when dissolved *physically* in a solvent, whereas it gives red or brown solutions when it forms *chemical complexes* within a solution, so the violet solutions could be discriminated by naked eye and examined for the physical mixing rules. In addition, the concentration of iodine could be determined easily and accurately by titration, which was a popular reliable method in the 1920s. Hildebrand and co-workers determined the solubility of iodine in a great variety of organic solvents and reported the solubility variations in molar fractions with temperature. Hildebrand later generalized the definition of regular solutions having $S^E_{mix} = 0$ and $V^E_{mix} = 0$, "*a regular solution is one involving no entropy change when a small amount of one of its components is transferred to it from an ideal solution of the same composition, the total volume remaining unchanged*". Regular solutions have molecules randomly mixed by thermal agitation, regardless of their size differences. This model may also be applied to certain very unsymmetrical liquid–liquid solutions in which the molar volume of one component can be as much as six times that of the other.

Hildebrand also relied on van Laar's expressions using the equations derived by van der Waals in 1890 for the a and b constants in the van der Waals equation of state for binary fluid mixtures, when the interactions of three kinds of molecular pairs, 1–1, 2–2 and 1–2 are present:

$$a = X^2_{11}a_{11} + 2X_{11}X_{22}a_{12} + X^2_{22}a_{22} \tag{379}$$

$$b = X_{11}b_{11} + X_{22}b_{22} \tag{380}$$

where the X terms are the mole fractions. Equation (380) gives the generally accepted volume additivity rule in thermodynamics. As we see in Section 3.4.1, the internal pressure can be related to the van der Waals constant a, so that $(\partial U/\partial V)_T \cong a/V^2$ (Equation (239)), and its integration at a constant temperature gives, $U = -a/V$. The total internal energy of mixing, which is the change in the internal energy before and after mixing, can be written as

$$\Delta U_{mix} = U_{mix}^{after} - U_{mix}^{before} = U_{mix}^{after} - (X_{11}U_{11} + X_{22}U_{22}) = -\frac{a}{V} + X_{11}\frac{a_{11}}{V_{11}} + X_{22}\frac{a_{22}}{V_{22}} \quad (381)$$

where the $U_{mix}^{after} = -a/V$, and $U_i = -a_i/V_i$ expressions are used in the above derivation. If there is no volume change during mixing, the term, $\Delta V_{mix} = 0$; then by taking $V = b$, $V_{11} = b_{11}$, $V_{22} = b_{22}$, and by inserting into Equation (381) and combining with Equations (379) and (380), we may write

$$\Delta U_{mix} = -\frac{\left(X_{11}^2 a_{11} + 2X_{11}X_{22}a_{12} + X_{22}^2 a_{22}\right)}{\left(X_{11}b_{11} + X_{22}b_{22}\right)} + X_{11}\frac{a_{11}}{b_{11}} + X_{22}\frac{a_{22}}{b_{22}} \quad (382)$$

which can be simplified into

$$\Delta U_{mix} = -\frac{\left(X_{11}X_{22}\right)\left(a_{11}b_{22}^2 - 2a_{12}b_{11}b_{22} + a_{22}b_{11}^2\right)}{\left(X_{11}b_{11} + X_{22}b_{22}\right)b_{11}b_{22}} \quad (383)$$

since $\Delta V_{mix} = 0$, and then $\Delta U_{mix} = \Delta H_{mix}$. van Laar further simplified Equation (383) by combining this with Berthelot's well-known geometric mean equation for fluids $[a_{12} = (a_{11}a_{22})^{1/2}]$, which was derived for binary mixtures to apply to the van der Waals constant, a, giving

$$\Delta U_{mix} = \Delta H_{mix} = -\left(\frac{X_{11}X_{22}b_{11}b_{22}}{X_{11}b_{11} + X_{22}b_{22}}\right)\left(\frac{\sqrt{a_{11}}}{b_{11}} - \frac{\sqrt{a_{22}}}{b_{22}}\right)^2 \quad (384)$$

The b_i parameter represents the space occupied by the molecules in the densely packed fluids, and not very suitable for liquids; then van Laar and Lorentz modified Equation (384) by substituting the actual molar volumes, v for the b terms, in 1925:

$$\Delta U_{mix} = \Delta H_{mix} = -\left(\frac{X_{11}X_{22}v_{11}v_{22}}{X_{11}v_{11} + X_{22}v_{22}}\right)\left(\frac{\sqrt{a_{11}}}{v_{11}} - \frac{\sqrt{a_{22}}}{v_{22}}\right)^2 \quad (385)$$

Hildebrand showed that Equation (385) gives a good fit with some regular solutions but is not adequate when the liquids are appreciably expanded over their close-packed volumes, where $\Delta V_{mix} \neq 0$; and volume changes during mixing contribute to ΔH_{mix} considerably. Later Hildebrand defined the *cohesive energy density* parameter, *ced*, for one mole of a material, by using energy of vaporization parameters

$$ced \equiv -\frac{U^V}{V} \cong \frac{\Delta U^V}{V} \cong \frac{\Delta H^V - RT}{V} \quad (386)$$

where V is the molar volume of the material. In Equation (386), the cohesive energy density is approximately equal to the internal pressure of the material. Since the molar enthalpy of vaporization ΔH^V can easily be determined experimentally at any desired temperature, the *ced* value can be calculated from Equation (386). However, at higher vapor pressures, $P\Delta V \neq RT$, and real gas corrections should be applied, whereas the error is less than 10%

in most cases. If no calorimetric measurements are made to determine the ΔH^V parameter, then it may be calculated from the variation of vapor pressure with temperature using the Clausius–Clapeyron equation (given as Equation (283) in Section 4.2.1). Hildebrand also proposed an empirical equation in terms of the boiling point, T_b, to calculate the unknown ΔH^V parameter at 25°C:

$$\Delta H^V = -12.340 + 99.2T_b + 0.084T_b^2 \tag{387}$$

by using the Kelvin scale of boiling points; ΔH_{vap} is obtained in J mol^{-1} units. Later, in 1931, Scatchard developed Hildebrand's cohesive energy density approach for binary solutions. He assumed three points: a) the mutual interaction energy between two molecules depends only on the distance and on their relative orientation, b) the molecular distribution is random and c) $\Delta V_{mix} = 0$. With these assumptions and inserting the *ced* parameter for each of the components, so that $U_i^V = -ced_i v_i$, Scatchard proposed that the cohesive internal energy of a mole of liquid mixture can be written as

$$U_{mix}^V = -\frac{\left(ced_{11} v_{11}^2 X_{11}^2 - 2ced_{12} v_{11} v_{22} X_{11} X_{22} + ced_{22} v_{22}^2 X_{22}^2\right)}{\left(X_{11} v_{11} + X_{22} v_{22}\right)} \tag{388}$$

Since the total volume of a binary solution is

$$V = n_{11} v_{11} + n_{22} v_{22} \tag{389}$$

where v_{11} and v_{22} are the individual volumes per molecule, and then the volume fraction is

$$\phi_{11} = \frac{V_{11}}{V_{11} + V_{22}} = \frac{n_{11} v_{11}}{n_{11} v_{11} + n_{22} v_{22}} = \frac{X_{11} v_{11}}{X_{11} v_{11} + X_{22} v_{22}} \tag{390}$$

Inserting the volume fraction term, ϕ_i, in Equation (388) gives

$$U_{mix}^V = -(X_{11} v_{11} + X_{22} v_{22})(ced_{11}\phi_{11}^2 - 2ced_{12}\phi_{11}\phi_{22} + ced_{22}\phi_{22}^2) \tag{391}$$

Then, similarly to Equation (381), the total cohesive internal energy of mixing, which is the change of the cohesive internal energy before and after the mixing, can be written as

$$\Delta U_{mix}^V = U_{mix}^{after} - U_{mix}^{before} = U_{mix}^V - (X_{11} U_{11} + X_{22} U_{22}) \tag{392}$$

ΔU_{mix}^V is also the excess energy of mixing and by combining Equations (391) and (392), Scatchard obtained

$$\Delta U_{mix}^V = (X_{11} v_{11} + X_{22} v_{22})(ced_{11} - 2ced_{12} + ced_{22})\phi_{11}\phi_{22} \tag{393}$$

Scatchard also assumed the geometric mean rule for the cohesive energy density between the 1–2 molecules, similar to Berthelot and van Laar, in analogy with the result of London's dispersion force treatments for nonpolar molecules

$$(ced)_{12} = [(ced)_{11}(ced)_{22}]^{1/2} \tag{394}$$

By combining Equations (393) and (394), Scatchard obtained

$$\Delta U_{mix}^V = (X_{11} v_{11} + X_{22} v_{22})\left(\sqrt{ced_{11}} - \sqrt{ced_{22}}\right)^2 \phi_{11}\phi_{22} \tag{395}$$

5.3.2 Solubility parameter approach

Later Hildebrand defined the *solubility parameter*, δ, as the square root of the cohesive energy density, *ced*, after Scatchard derived Equation (395)

$$\delta \equiv (ced)^{1/2} = \left(\frac{\Delta U^V}{V}\right)^{1/2} = \left(\frac{\Delta H^V - RT}{V}\right)^{1/2} \tag{396}$$

where $\delta_i = (\Delta U_i^V/v_i)^{1/2}$ for all of the components. The unit of δ was $(cal\,cm^{-3})^{1/2}$ formerly, and now, in SI units $(MJ\,m^{-3} = MPa = megapascal)^{1/2}$, and it should be noted that the numerical values are different so that $(MPa)^{1/2}$ are 2.0455 times larger than $(cal\,cm^{-3})^{1/2}$. In addition, since $\Delta V_{mix} = 0$, for regular solutions, then $\Delta U_{mix} = \Delta H_{mix}$ and Equation (395) can be written as

$$\Delta U^V_{mix} = \Delta H^V_{mix} = (X_{11}v_{11} + X_{22}v_{22})(\delta_{11} - \delta_{22})^2 \phi_{11}\phi_{22} \tag{397}$$

Equation (397) is identical to the van Laar–Lorentz equation (Equation (385)) if we assume that $a_i = v_i \Delta U_i^V$ in this equation. On the other hand, if we use mole numbers instead of mole fractions (see Equation (390)), then Equation (397) can also be written as

$$\Delta U^V_{mix} = (n_{11}v_{11} + n_{22}v_{22})(\delta_{11} - \delta_{22})^2 \phi_{11}\phi_{22} \tag{398}$$

The partial molar internal energy of transferring a mole of component 2 from its pure liquid to the solution can be written as

$$\left(\frac{\partial \Delta U^V_{mix}}{n_{22}}\right)_{n_{11}} = v_{22}(\delta_{11} - \delta_{22})^2 \phi_{11}^2 \tag{399}$$

Since $(dU)_V = (dG + TdS)_V$ for constant volume conditions, we may write

$$\left(\frac{\partial \Delta U^V_{mix}}{n_{22}}\right)_{n_{11}} = \left(\frac{\partial \Delta G^V_{mix}}{n_{22}}\right)_{n_{11}} + T\left(\frac{\partial \Delta S^V_{mix}}{n_{22}}\right)_{n_{11}} = v_{22}(\delta_{11} - \delta_{22})^2 \phi_{11}^2 \tag{400}$$

The Gibbs free energy of mixing is $(\partial \Delta G^V_{mix}/n_{22})_{n_{11}} = RT \ln a_{22}$, and if we assume that the entropy transfer for such a process for regular solutions is given by $(\partial \Delta S^V_{mix}/n_{22})_{n_{11}} = -R \ln X_{22}$, then Equation (400) becomes

$$RT \ln a_{22} - RT \ln X_{22} = RT \ln \frac{a_{22}}{X_{22}} = v_{22}(\delta_{11} - \delta_{22})^2 \phi_{11}^2 \tag{401}$$

Since, $a_{22}/X_{22} = \varphi_{22}^X$, then by combining Equations (377) and (401) we obtain the excess chemical potential of the component, 2, for a regular solution

$$(\mu_{22}^E)_{T,P,n_{11}} = RT \ln \varphi_{22}^X = v_{22}(\delta_1 - \delta_2)^2 \phi_{11}^2 \tag{402}$$

Similarly, for component 1, Equation (402) can be written as

$$(\mu_{11}^E)_{T,P,n_{22}} = RT \ln \varphi_{11}^X = v_{11}(\delta_1 - \delta_2)^2 \phi_{22}^2 \tag{403}$$

Equations (402) and (403) were found to be applicable to many binary regular solutions containing nonpolar components. However, because of the various simplifying assumptions made in the course of their derivation, a complete quantitative agreement cannot be achieved. These equations always predict $\varphi_i \geq 1$, that is a regular solution can exhibit only positive deviations from ideality, which is not true experimentally. This is the direct consequence of the geometric-mean assumption. We know that assumptions of regular random mixing ($S^E = 0$) and no volume change ($V^E = 0$) are also not correct even for simple mixtures, but due to a cancellation of errors, these assumptions frequently do not seriously affect calculations of G^E. The solubility parameters δ_1 and δ_2 are functions of temperature but the difference ($\delta_1 - \delta_2$) is often almost independent of temperature.

The solubility parameter approach is a thermodynamically consistent theory and it has some links with other theories such as the van der Waals internal pressure concept, the Lennard–Jones pair potentials between molecules, and entropy of mixing concepts of the lattice theories. The solubility parameter concept has found wide use in industry for non-polar solvents (i.e. solvent selection for polymer solutions and extraction processes) as well as in academic endeavor (thermodynamics of solutions), but it is unsuccessful for solutions where polar and especially hydrogen-bonding interactions are operating.

5.3.3 Three-component solubility parameters

According to the solubility parameter approach, liquids having similar solubility parameters will be miscible, and polymers can be dissolved in solvents whose solubility parameters are not too different from their own. The lack of complete success in the chemical industry with this approach raises questions about the validity of the solubility parameter approach because deviation from the regular solutions is unexpectedly large for polar and hydrogen-bonding solvents, and solubility parameter equations cannot be used effectively. Later, in 1967, C. Hansen applied a multicomponent solubility parameter approach in order to solve this important solvent selection problem in the paint industry, and he progressively developed his approach up to the 1990s.

Hansen proposed that all types of bonds holding liquid molecules are broken during vaporization, and he divided the cohesive energy, ΔU^V into three parts:

$$\Delta U^V = \Delta U^V_d + \Delta U^V_p + \Delta U^V_h \tag{404}$$

where, ΔU^V_d are the cohesion contributions from nonpolar (mainly London dispersion) interactions; ΔU^V_p, the cohesion contributions from polar (permanent dipole–permanent dipole, mainly Keesom orientation) interactions and ΔU^V_h are the cohesion contributions from hydrogen-bonding interactions. Dividing Equation (404) by the molar volume of the solvent, V, gives

$$\frac{\Delta U^V}{V} = \frac{\Delta U^V_d}{V} + \frac{\Delta U^V_p}{V} + \frac{\Delta U^V_h}{V} \tag{405}$$

and introduction of the three-component solubility parameters is then possible by combining Equations (396) and (405),

$$\delta^2 = \delta^2_d + \delta^2_p + \delta^2_h \tag{406}$$

where, δ_d, δ_p, δ_h are dispersion, polar and H-bonding solubility parameters respectively. In order to evaluate the nonpolar dispersion contribution, δ_d, the *homomorph* concept was used, which was formerly invented by Brown. The homomorph of a polar molecule is a nonpolar molecule having very nearly the same size and shape as the polar molecule. In order to find the molecular properties of a homomorph, the experimentally determined vaporization enthalpy of a polar liquid is divided into two parts: one corresponds to the vaporization enthalpy of the nonpolar homomorph, which is assigned to the polar molecule at the same reduced temperature, so that the other polar part can then be deduced. The molar volumes of the polar molecule and the homomorph molecule were assumed to be equal for this comparison. The molar volumes of hydrocarbons were calculated according to their type, such as linear straight-chain, cyclic, aromatic etc., and the variation of internal vaporization energy is plotted against the molar volume. When, a polar molecule was investigated, δ_d of this polar molecule was found using the ΔU^V data of its corresponding homomorph. In an alternative method, Keller evaluated δ_d using the Lorentz–Lorenz refractive index, n_D ($_D$ for sodium light) function. The ratio

$$y = \frac{n_D^2 - 1}{n_D^2 + 2} \tag{407}$$

is used as an indication. If $y \leq 0.28$ then the linear relationship $[\delta_d = 62.6y]$ is used in (MPa)$^{1/2}$ units. If $y > 0.28$, then the linearity vanishes and a polynomial expression of $\lfloor \delta_d = -4.58 + 108y - 119y^2 + 45y^3 \rfloor$ is used.

The polar contribution, δ_p, was calculated using a slight modification of Böttcher's equation in terms of measurable dipole moment, μ, refractive index, n_D, molar volume, V and relative permittivity, ε.

$$\delta_p^2 = \frac{12108}{V^2} \left(\frac{\varepsilon - 1}{2\varepsilon + n_D^2} \right) (n_D^2 + 2)\mu^2 \tag{408}$$

For many solvents the above physical data are not available and Beerbower developed a much simpler and somewhat reliable equation for such cases:

$$\delta_p = \frac{37.4\mu}{\sqrt{V}} \tag{409}$$

Alcohols, glycols, carboxylic acids, amines and other hydrophilic compounds have high hydrogen-bonding parameters. In the earlier times of the three-parameter solubility parameter approach, the hydrogen-bonding contribution, δ_h, was calculated by subtracting the ΔU_d^V and ΔU_p^V data from the total energy of vaporization and then applying Equations (405) and (406). In the absence of any reliable latent heat and dipole moment data, some empirical group contribution methods can be applied. Hansen formerly calculated δ_h for alcohols directly from $\delta_h = (20\,900 \, N_{OH}/V)^{1/2}$ in (MPa)$^{1/2}$ units, where N_{OH} is the number of alcohol groups in the molecule, V is the molar volume and the numerical value of 20 900 arises from the fact that the average hydrogen-bond energy of the —OH . . . O— bond is 20 900 cal mol^{-1}. Crowley applied the effect of hydrogen bonding on the shift formation of the infra-red spectrum to calculate δ_h. Bagley calculated the δ_h parameter from the residual solubility parameter which was evaluated from internal pressure–molar volume and molar vaporization energy data. However, all these approaches did not consider the property that hydrogen bonding is an unsymmetrical interaction, involving a donor and an acceptor with different roles, rather than two equivalent mutually H-bonding species. This unsymmetrical property may be qualitatively estimated by multiplying the hydrogen-bond accepting capacity with the hydrogen-bond donating capacity. This may be shown to calculate the enthalpy of hydrogen bonding as

$$\Delta H_h = \phi_1 \phi_2 (A_1 - A_2)(D_1 - D_2) \tag{410}$$

where A is the H-bond acceptor capability and D is the H-bond donor capability. It is apparent that the maximum interaction occurs when $A_1 = D_2 = 0$, or $A_2 = D_1 = 0$. On the other hand, hydrogen bonding was also thought of as a quasi-chemical bond rather than a physical interaction between molecules. For example, alcohols may be considered as linear polymer complexes in non-hydrogen-bonding solvents. If chemical H-bonding forces are strong then dispersion and polar forces may be neglected but careful study has shown that, for accurate work, both physical and chemical forces must be taken into account. Thus, the determination of δ_h is not very reliable in all the above methods and forms the weakest part of the solubility parameter theory. Some other semi-empirical approaches were also proposed, but any theory of solutions with a sufficient number of adjustable parameters can never be considered to be adequate unless supported by independent physicochemical experimental data. The need for so many parameters is due to our inadequate understanding of intermolecular forces. Since real liquid mixtures are much more complicated than the oversimplified solution models, we must not take any theory too seriously.

Three-component solubility parameters are used in coatings industries to select suitable solvents for polymers and other ingredients. Since ΔU^V cannot be measured for most of the solids, it is not possible to determine δ values for non-volatile solutes such as polymers directly, as is done for liquids. As a practical solution to this important industrial problem, it is assumed that a solute δ value has exactly the same value as a solvent δ value, when the solute dissolves very well, such that they can mix in all proportions without enthalpy or volume change, and without specific chemical interaction. Polymers are assumed to dissolve in solvents whose δ_d, δ_p, δ_h solubility parameters are not too different from their own. In practice, the usual procedure is that the polymer is dissolved in a limited number of solvents chosen specifically for this purpose at a given concentration, usually 10% by weight. The solvents are selected to maximize the information regarding all types of interaction. Then the solubility data of the solute are plotted in a three-dimensional system of δ_d, δ_p, δ_h. After defining a region of spherical volume in this three-dimensional solubility plot for the solute under consideration, one can proceed to use it in solving practical problems (see Reference 8, Chapter 10, written by C. Hansen in the *CRC Handbook of Surface and Colloid Chemistry*).

It is clear that a match in solubility parameters leads to zero enthalpy of mixing, and the entropy change should cause complete dissolution. Solvents with smaller molecular volumes promote lower heats of mixing, so that materials having smaller molecules will be thermodynamically better solvents than those with larger molecules. Entropy change is beneficial to mixing: the higher the excess entropy, the higher the dissolution. Higher temperatures will also contribute to a greater degree of dissolution when multiplied with the entropy change, giving negative free energy of mixing. For polymer chains, it is obvious that the entropy changes associated within a solvent will be smaller than the entropy of liquid–liquid solutions, since the monomer units of the polymer macromolecule are already bound into the configuration of the macromolecule. Thus it is difficult to achieve complete polymer–polymer miscibility in any medium. In spite of these facts, polymer compatibility and polymer interfacial tensions may also be examined, somewhat roughly, by using the three-component solubility parameter approach.

Beerbower related δ_d, δ_p, δ_h solubility parameters to the surface tension of organic solvents:

$$\gamma = 0.0715 V^{1/3}[\delta_d^2 + 0.632(\delta_P^2 + \delta_h^2)] \tag{411}$$

when δs are inserted in cal cm$^{-3 1/2}$ units, then surface tension can be found in ergs cm^{-2} units. It is interesting to note that the constant 0.07152 can also be derived from the number of nearest neighbors lost in the surface formation, assuming that the liquid molecules on average occupy the corners of a regular octahedron in the bulk liquid. When Equation (411) is applied to many organic liquids, the computer fit gives a constant of 0.07147; the similarity of these values is remarkable. However, this approach is not very successful for hydrogen-bonding liquids, and also cannot be applied to predict polymer surface tensions.

5.4 Solutions Containing Surface-active Solutes

Some solutes, which are called *surface-active solutes* have special chemical groups in their structure, and their solutions have unusual properties. Surface activity is the ability to alter (generally reduce) the surface tension of the solvent when the surface-active solute is dissolved in small concentrations. For example, the surface tension of pure water, which is 72.8 mN m^{-1} at room temperature, can be reduced down to 22 mN m^{-1} by the addition of a surface-active solute at concentrations of less than 0.1%. However, when nonaqueous solutions are considered, this effect is much smaller. Surface-active agents are also termed *tensides, soaps, detergents, wetting agents, emulsifiers* and *dispersants*, but the term *surfactant* is the most generally accepted. (The term *soap* indicates the alkali metal salts of long-chain fatty acids.) Surfactants have the property of adsorbing onto the surfaces (or interfaces when two immiscible phases are present) of the solution. They usually reduce the interfacial free energy of the solution but, very rarely, there are occasions when they can increase the interfacial free energy. A surfactant is therefore a substance that significantly decreases the amount of work required to expand these solution interfaces.

Surfactants have a characteristic molecular structure consisting of a chemical group that has very little attraction for the solvent (solvent-repellent), known as a *lyophobic* group, together with a group that has a strong attraction for the solvent (solvent-loving), called a *lyophilic* group. This is known as an *amphipathic* structure (combines both natures). When a molecule with an amphipathic structure is dissolved in a solvent, the lyophobic group, with little affinity for the solvent, may change the structure of the bulk solvent, increasing the free energy of the system. When this occurs, the system responds in some way in order to minimize the contact area between the solvent-repellent lyophobic group and the solvent. In the case of a surfactant dissolved in aqueous medium in air, the lyophobic (in water medium, it is called *hydrophobic*, water-repellent, tail) group breaks the hydrogen bonds between the water molecules and distorts the pre-structure of the pure water. As a result of this distortion, some of the surfactant molecules are expelled to the surface of the solution, with their hydrophobic, water-repellent groups oriented towards the air so as to minimize contact with the water molecules. On the other hand, the presence of the lyophilic (in water medium it is called a *hydrophilic*, water-loving, head) group prevents the surfactant from being expelled completely from the water solvent as a separate phase, since that would require dehydration of the hydrophilic groups. Then the surface of the water becomes covered with a monolayer of the surfactant molecules with their hydrophobic groups oriented predominantly towards the air, and the hydrophilic groups oriented predominantly towards the water. Since air molecules are essentially non-polar in nature, as are the hydrophobic groups, this decrease in the dissimilarity of the two phases contacting each other at the surface results in a decrease in the surface tension of the aqueous solution. This situation is energetically more favorable. Consequently, the amphipathic structure of the surfactant causes an increase in its concentration at the water–air surface over its concentration in the bulk solution, and reduces the surface tension of the water.

Most of the surfactants used as detergents, emulsifiers or wetting agents have hydrophilic and lipophilic (oil-loving, hydrophobic) groups in different proportions. The properties of these groups determine the behavior of the surfactant. Every surfactant occupies a place on a scale of hydrophile–lipophile balance (HLB), and this balance determines whether the surfactant is water- or oil-soluble (see Section 5.4.3). A long-chain hydrocarbon tail is a lipophile, is completely insoluble in water and will not spread on a water surface, but floats as a compact liquid lens. If a hydrogen-bonding, hydrophilic head group such as hydroxyl (—OH) or amino (—NH$_2$) is covalently attached to this hydrophobic group in the molecule, its water solubility increases although the new molecule is still practically insoluble, but it spreads spontaneously on the water surface and if the surface area is sufficiently large, a *monomolecular layer* forms. These *monolayers* are generally insoluble films. If the surface area is not large enough to form a monolayer, then surfactant multilayers form on the water surface.

If the hydrogen-bonding group is ionic, such as carboxylate (—COOH), sulfate (—OSO$_3$), sulfonate (—SO$_3^-$), or ammonium ions (R$_4$N$^+$), the corresponding surfactants are much more soluble in water but are still positively adsorbed at the water–air (or water–oil interfaces). These surfactants reduce the surface tension of the water because of their positive adsorption at the interface; this was quantitatively shown by the Gibbs adsorption isotherm equation (Equations (224)–(227)). When the interfacial region contains excess adsorbed solute that is thicker than the monolayer, then this is called a *soluble interfacial film* or *soluble multilayer film*, to distinguish it from the insoluble monolayers. For surfactants having the same hydrophilic groups, when the length of the hydrocarbon (lipophile) group decreases, the degree of hydrophilicity further increases. When fewer than six carbon atoms are present in the hydrocarbon part of the surfactant, it hardly differs from an ordinary strong electrolyte, comprising a small anion and a small cation. Then, water molecules surround these small ions in order to dissolve them, and thermodynamically it is more favorable for these surfactant molecules to be dissolved in the bulk water, rather than locating at the interfacial region. This leads to negative adsorption. Thus the sign of the adsorption depends on the HLB balance of the surfactants. It should be noted that achieving equilibrium during surfactant monolayer formation at the interface is not an instantaneous process, but is governed by the rate of diffusion of the surfactant through the solution to the surface. It might take several seconds for a surfactant solution to attain its equilibrium surface tension, especially if the solution is dilute and the solute molecules are large and unsymmetrical.

Surfactants are very important in the chemical industry, appearing in such diverse products as detergents, emulsion polymer-based adhesives, surface coatings, pharmaceuticals, cosmetics, motor oils, drilling muds used in petroleum prospecting, ore flotation agents, and more recently also in the electronics and computer industries, such as for printing, micro-electronics, magnetic recording, and also in supercritical carbon dioxide, biotechnology, nanotechnology and viral research. The use of surfactants is important when the phase boundary area is so large relative to the volume of the system that a substantial fraction of the total mass of the system is present at the boundaries, such as dispersions of solids in liquids, in emulsions, foams etc. Surfactants also play a major role when the phenomena occurring at phase boundaries are so unusual relative to the expected bulk phase interactions that the entire behavior of the system is determined by interfacial processes, such as heterogeneous catalysis, detergency, flotation, and corrosion etc.

5.4.1 Effect of hydrophilic and hydrophobic group types

The lyophobic and lyophilic structural groups of the surfactant molecule are chosen according to the nature of the solvent and the conditions of use. The hydrocarbon part of the molecule is responsible for its solubility in oil, while the polar carboxyl (—COOH) or hydroxyl (—OH) groups, which have sufficient affinity to water, are responsible for the solubility of the surfactant in water. The hydrophobic tail (lyophobic) group is usually a long-chain hydrocarbon [$CH_3(CH_2)_n$—] or fluorocarbon [$CF_3(CF_2)_n$—], and less often a halogenated or oxygenated hydrocarbon or siloxane —[$OSi(CH_3)_2]_n$— chain of proper length when the surfactant is used in water. Only fluorocarbon or siloxane chains may be used as the hydrophobic group when the surfactant is to be dissolved in a less hydrogen-bonding solvent such as polypropylene glycol. Materials such as short-chain fatty acids and alcohols are soluble in both oil and water.

Hydrophobic tail groups are generally long-chain hydrocarbon residues [$CH_3(CH_2)_n$—] including such different structures as:

1 Straight-chain, long alkyl groups (C_8–C_{20})
2 Branched-chain, long alkyl groups (C_8–C_{20})
3 Long-chain (C_8–C_{16}) alkylbenzene residues
4 Alkylnaphthalene residues (C_3 and greater-length alkyl groups)
5 High-molecular-weight propylene oxide polymers (polyoxypropylene glycol derivatives)
6 Long-chain perfluoroalkyl groups
7 Polysiloxane groups, H—[$OSi(CH_3)_2]_n$—OH
8 Rosin and lignin derivatives

When the length of the hydrophobic tail group increases in a surfactant dissolved in water, it decreases the solubility of the surfactant in water, generally causes closer packing of the surfactant molecules at the interface, increases the tendency of the surfactant to form micelles in water, increases the melting point of the surfactant, and, if it has ionic groups, increases the precipitation sensitivity by the counterions such as Ca^{2+} or Mg^{2+} present in the water. However, when the length of the hydrophobic group increases in a surfactant dissolved in an organic nonpolar liquid, conversely it increases the solubility of the surfactant in this solvent.

The introduction of branching or unsaturation into the hydrophobic tail group increases the solubility of the surfactant in water or in organic solvents (compared to the straight-chain, saturated isomer), decreases the melting point of the surfactant, causes looser packing of the surfactant molecules at the interface (the *cis* isomer is particularly loosely packed, the *trans* isomer is packed almost as closely as the saturated isomer) and may increase thermal instability and cause oxidation and color formation in unsaturated compounds. The presence of an aromatic nucleus in the hydrophobic group may increase the adsorption of the surfactant onto polar surfaces and cause looser packing of the surfactant molecules at the interface. Cycloaliphatic nuclei, such as those in rosin derivatives, are even more loosely packed. The presence of aromatics in the structure decreases the biodegradability of the surfactant. When polyoxypropylene units are present, they increase the hydrophobicity of the surfactant, its adsorption onto polar surfaces, and its solubility in organic solvents. The presence of the perfluoroalkyl or polysiloxane groups results in the largest reductions in the surface tension of water solutions. Perfluoroalkyl surfaces are both water- and hydrocarbon-repellent.

The hydrophilic head part of the most effective water-soluble surfactants is often an ionic group. Ions have a strong affinity for water owing to their electrostatic attraction to the water dipoles, and are capable of pulling fairly long hydrocarbon chains into solution with them. For example, palmitic acid, which is virtually non-ionized, is insoluble in water, whereas sodium palmitate, which is almost completely ionized, is soluble. It is possible to use non-ionic hydrophilic groups in surfactants, which also exhibit a strong affinity for water. For example, a poly (ethylene oxide) chain shows a strong affinity for water. As the temperature and use conditions (e.g. presence of electrolyte or organic additives) vary, modifications in the structure of the lyophobic and lyophilic groups may become necessary to maintain the surface activity at a suitable level.

5.4.2 Types of surfactant

Conventional surfactants are classified as *anionic, cationic, non-ionic* or *amphoteric*, according to the charge carried by the surface-active part of the molecule:

Anionic surfactants

Sodium stearate, sodium oleate, sodium dodecylsulphate and sodium dodecyl benzene sulphonate are examples of anionic surfactants. The surface-active (head) portion of the molecule has a negative charge, such as a sulfonate, sulfate, or carboxylate group. For example, in the surfactants sodium dodecylsulphate $[C_{12}H_{25}OSO_3^-Na^+]$, $[RC_6H_4SO_3^-Na^+]$ (alkylbenzene sulfonate), sodium dodecanoate $[C_{11}H_{23}COO^-Na^+]$ or sodium perfluorooctanate $[C_8F_{17}COO^-Na^+]$, the sodium atom (or other alkali metal if present) dissociates to give a sodium cation in water at neutral pH and room temperature, and the remaining surfactant becomes anionic. Anionic surfactants are cheap and the most widely used types. Because of their low cost, wide pH tolerance and lower precipitation properties when the hardness ions Ca^{2+} and Mg^{2+} are present, branched-chain alkylbenzene sulfonates were largely used until the 1960s. However, because of the environmental pollution they create in rivers and lakes, their production was considerably decreased and the production of biodegradable linear chain alkylbenzene sulfonates began for the same applications.

Cationic surfactants

Hexadecyl trimethyl ammonium bromide $[C_{16}H_{33}N^+(CH_3)_3Br^-]$ and dodecyl trimethyl ammonium chloride $[C_{12}H_{25}N^+(CH_3)_3Cl^-]$ are examples of cationic surfactants. The surface-active portion (head) bears a positive charge localized at the nitrogen atom, for example, $RN^+(CH_3)_3Cl^-$ (quaternary ammonium chloride) or $RN^+H_3Cl^-$ (salt of a long-chain amine). If the surface is to be made hydrophobic (water-repellent) with the use of a surfactant, then the best type of surfactant to use is cationic. This is because a cationic surfactant will adsorb onto the water surface with its positively charged hydrophilic group oriented toward the negatively charged surface (due to the electrostatic attraction), and its hydrophobic group oriented away from the surface, making the surface water-repellent. On the other hand, if the surface is to be made hydrophilic (water-wettable), then cationic surfactants should be avoided. Cationic surfactants are expensive, but their germicidal action makes them useful for many industrial applications. Cationic surfactants are incompatible with anionic surfactants, and since anionic surfactants are generally used in the manufacture of emulsion polymers, the use of cationic surfactants is mostly avoided in the latex industry.

Nonionic surfactants

These surfactants do not contain charged groups. Polyethylene oxides, alkyl glucosides, sorbitan esters, polyoxyethylene sorbitan esters are examples of nonionic surfactants. For example, polyoxyethylenated alcohols $[R(OC_2H_4)_nOH]$ such as dodecyl hexaoxylene glycol monoether $[C_{12}H_{25}(OC_2H_4)_6OH]$ belong to this class. Some aromatic derivatives such as polyoxyethylenated alkylphenol $[RC_6H_4(OC_2H_4)_nOH]$, and some monoglycerides of long-chain fatty acids $[RCOOCH_2CHOHCH_2OH]$ are also examples of nonionic surfactants. When nonionic ethylene oxide is block copolymerized with hydrophobic polypropylene oxide, $H—[OCH_2CH_2]_n—[OCH(CH_3)CH_2]_m—OH$ surfactant is obtained and its properties vary according to the chain lengths of the blocks. Polyethylene oxide is the hydrophilic part of this surfactant. When polyethylene oxide is combined with fatty-acid esters of glycerol or sorbitol, the resulting surfactant is used in the food industry, generally as a wetting agent in dehydrated milk and eggs, flour, cocoa etc. All polyethylene oxide-containing surfactants become insoluble in water at higher temperatures, as the extent of hydrogen bonding of water to the ether oxygen in polyethylene oxide is decreased by the supply of heat, and polyethylene oxide portions become hydrophobic in these circumstances.

An advantage of the nonionic surfactants is that both of their hydrophilic head and hydrophobic tail group lengths can be varied when required. Nonionics adsorb onto surfaces with either the hydrophilic or the hydrophobic group oriented towards the surface, depending upon the nature of the solvent. If polar groups in the solvent are capable of H-bonding with the hydrophilic group of the surfactant, then the surfactant will probably be adsorbed onto the surface with its hydrophilic head group oriented towards the surface, making the surface more hydrophobic. However, if such polar or H-bonding groups are absent in the solvent, then the surfactant will probably be oriented with its hydrophobic tail group towards the surface, making it more hydrophilic. Nonionic surfactants are widely used as detergents in laundry applications.

Amphoteric (or zwitterionic) surfactants

Both positive and negative charges are present in the amphoteric surfactant molecule, and the net charge is zero. For example, long-chain amino acids [$RN^+H_2CH_2COO^-$], sulfobetaine [$RN^+(CH_3)_2CH_2CH_2SO_3^-$] and dodecyl dimethyl propane sultaine [$C_{12}H_{25}N^+(CH_3)_2(CH_3)_2SO_3^-$] are amphoteric. Amphoteric surfactants may behave as either anionic or cationic in water depending on the pH of the medium. Since they carry both positive and negative charges, they can adsorb onto both negatively charged and positively charged surfaces without changing the charge of the surface significantly. Amphoteric surfactants are expensive and are only used for special tasks on charged surfaces.

As described above, the *conventional surfactants* have only one hydrophilic head group and only one hydrophobic tail group. However, there are special *polymeric surfactants* having more than one hydrophilic and hydrophobic group which have been used extensively in the chemical industry over the past decade. If two conventional surfactants are connected by a spacer chemical group, then a *dimeric surfactant* is obtained. There are two main types of these, as shown in Figure 5.1: *Gemini surfactants* are formed if two conventional surfactants are connected by a spacer group which is close to the hydrophilic head group, and *Bolaform surfactants* are formed if two conventional surfactants are connected by a spacer group close to the middle or the end of the hydrophobic tail group. By the same procedure, trimeric or tetrameric surfactants can be synthesized having superior surface-active properties to those of the conventional surfactants. *Polymeric surfactants* contain a large number of monomeric units having amphipathic structure, as shown in Figure 5.1, and sometimes behave much more effectively than conventional surfactants. Finally some *diblock copolymers* are used as surfactants; these have long hydrophilic —AAAA— and hydrophobic —BBBB— blocks connected in the middle of the copolymer. By applying suitable polymerization procedures it is possible to adjust the length of each block. In addition, it is also possible to use more than two types of polymeric block giving *triblock or tetrablock copolymers.*

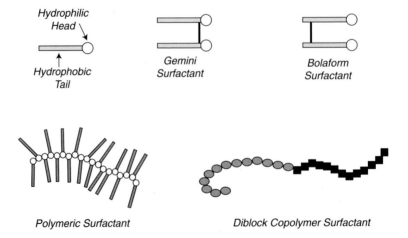

Figure 5.1 Types of monomeric and polymeric surfactants.

5.4.3 HLB method

If an oil is finely dispersed in water or the water is finely dispersed in an oil by the use of a surfactant, then the dispersion is called an *emulsion*. The size of the liquid droplets in an emulsion is generally larger than 50 nm. If the droplets are 10–50 nm in diameter, giving a transparent emulsion, it is called a *microemulsion*. Conventional emulsions (or *macroemulsions*) are unstable thermodynamically and separate into oil-rich and water-rich bulk phases rapidly if the surfactants are ineffective. Emulsions are very important in the chemical, cosmetic and pharmaceutical industries, and the selection of a suitable surfactant for an emulsification process should be carried out on the basis of sound scientific principles. Correlations between the chemical structure of surfactants and their emulsifying power are complicated, because any emulsifier selection method must be applied to both oil-in-water (O/W) and water-in-oil (W/O) emulsion processes. Nevertheless, there are several general guidelines that can be helpful in the selection of a surfactant as an emulsifying agent. It must have a tendency to migrate to the interface, rather than to remain dissolved in either one of the bulk phases. It must therefore have a balance of lyophilic and lyophobic groups so that it will distort the structure of both bulk phases to some extent, although not necessarily equally. Griffin introduced in 1949 a useful rating scheme to be used for this purpose, known as the HLB (hydrophile–lipophile balance) method. In this method, a number between 0 and 40 indicates the emulsification behavior in the arbitrary one-dimensional scale of surfactant behavior and relates it to the balance between the hydrophilic and lipophilic (hydrophobic) portions of the molecule. In this scale, the least hydrophilic materials have low HLB numbers, and increasing HLB corresponds to increasingly hydrophilic character, so the surfactants with a low HLB number generally act as W/O emulsifiers and those with high HLB numbers are O/W emulsifiers, and also act as solubilizers in water. In general a value of 3–6 is the recommended range for W/O emulsification; 8–18 is recommended for O/W emulsification. The *phase inversion temperature* (PIT) is the temperature at which, upon heating, an oil–water–emulsifier mixture changes from an O/W to a W/O emulsion. It has been shown that there is a linear dependence between the HLB number and the phase inversion temperature.

In water, the surfactant solubility behavior can be shown in more detail as: no dispersibility (HLB: 1–4); poor dispersibilty (HLB: 3–6, such surfactants may be applied as W/O emulsifiers); unstable milky dispersion after vigorous agitation (HLB: 6–8, these may be applied as wetting agents); stable milky dispersion (HLB: 8–10); from translucent to clear (HLB: 10–13, these may be applied as O/W emulsifiers up to HLB 15); clear solution (HLB: 14–18, these may be applied as detergents for HLB 13–15 and solubilizer for HLB 15–18).

The assignment of the HLB number for a new surfactant is generally based on the emulsification experience in the laboratory, rather than on the structural considerations. (However various methods have been proposed to calculate the HLB number from the structure of the surfactant molecule; these have not been very successful.) A similar range of HLB numbers has been assigned to various substances that are frequently emulsified, such as hydrocarbons, plant oils, lanolin, paraffin wax, xylene, chlorinated solvents etc. Then, an emulsifying agent (or preferably a combination of emulsifying agents) is selected whose HLB number is approximately the same as that of the ingredients to be emulsified. It has become apparent that, although the HLB method is useful as a rough guide to emulsifier selection, it has serious limitations. For example, paraffinic mineral oil has an HLB value of 11 for emulsification as the dispersed phase in an O/W emulsion, and a value of 4 as the continuous phase in a W/O emulsion. Since the HLB number depends on the particular type of phase, it cannot be used additively.

In summary, the HLB method is only an empirical approach which has made it possible to organize a great deal of rather messy information on emulsion preparation, in order to choose candidate surfactants in trial and error laboratory work for suitable emulsion selection. There are other surfactant selection methods such as the *phase inversion temperature* (PIT) and the *hydrophilic–lipophilic deviation* (HLD) methods used for the same purpose in the emulsion industry, but these are outside the scope of this book.

5.5   Gibbs Surface Layers of Soluble Materials on Liquid Solutions

5.5.1 Gibbs monolayers: thermodynamics of adsorption

When a soluble organic or inorganic material, a surfactant, polymer or biomaterial solute, vapor, liquid or solid, is mixed in a liquid solvent, depending on the molecular interactions between the solute and the solvent, the solute may be dissolved completely in the solvent giving a one-phase solution, or dissolve partially in the solvent giving a two-phase solution. If the solute molecules enrich at the surface in a one-phase liquid solution, they form a surface layer, and if this layer is monomolecular, then it is called a *Gibbs monolayer*, and the surface tension of the solution differs from that of the pure solvent. A *monolayer* is defined as a monomolecular layer of the solute having a thickness of only one molecule, where the gravitational effects are negligible. The resultant surface tension of the solution varies with the composition and, if the quantity of solute is appreciably small, the surface excess may be calculated by applying the Gibbs adsorption isotherm equation (Equations (224)–(230)) at constant temperature, provided that the surface and bulk of the system are in equilibrium.

On the other hand, if the solute is partially or completely immiscible with the sub-phase solvent, the solute molecules form a layer on the surface of the sub-phase (or in the interface between the two immiscible phases) which is different from the Gibbs monolayer, and if the layer is monomolecular, it is called a *Langmuir insoluble monolayer* (see Section 5.6).

The solute and solvent molecules present in any solution have different intensities of attractive force fields, and also have different molecular volumes and shapes. A concentration difference between the surface region and the bulk solution occurs because the molecules that have the greater fields of force tend to pass into the interior, and those with the smaller force fields remain at the surface. The Gibbs surface layer of a solution is more concentrated in the constituents that have smaller attractive force fields, and thus whose intrinsic surface free energy is smaller than the interior. As we stated in Section 3.3, this concentration difference of one constituent of a solution at the surface is termed *adsorption*. In qualitative terms, if the solution has a smaller surface tension than its pure solvent, the solute is concentrated in the surface layer indicating a *positive adsorption* according to

Gibbs' equation. The adsorption process involves the transport of molecules from the bulk solution to the interface, where they form a specially oriented monomolecular layer according to the nature of the two phases. When a *Gibbs monolayer* forms, it does not necessarily mean that the molecules are touching each other in this monolayer. Instead, if the anchoring from the sub-phase molecules is weak, the molecules may move freely in the two-dimensional interfacial area. Thus, the physically measurable monolayer area is sometimes much larger than the close-packed area where all the molecules touch each other. When any monolayer is fairly well populated with adsorbed molecules, it exerts a lateral spreading (film) pressure, π, which is equal to the depression of the surface tension (see Section 5.5.2).

For aqueous dilute solutions of many organic solutes, the adsorption equilibrium is attained quickly, and the results of changing the concentration of the system can very often be predicted by the *Gibbs adsorption isotherm* equations given in Section 3.3.1 (Equations (224)–(230)), depending on the availability of experimental surface tension and concentration data for the solution. In reality, the liquid surface is sharply defined, but thermal agitation of the molecules renders it indefinite to a thickness of one or two molecular diameters, and all the properties of the interfacial phase change gradually in this transition region until we reach the bulk phase. The Gibbs adsorption approach does not describe this gradual change; it only compares the actual system with a physically impossible system in which two phases touch each other without any transitional layer. Gibbs' method is consistent mathematically and very useful in understanding surface chemistry processes. The only problem arises in locating the *Gibbs dividing plane* in order to quantify the adsorbed surface excess. Nevertheless, some methods have been developed to eliminate the requirement to choose the location of the Gibbs dividing plane by introducing the *relative adsorption* concept given below:

For a two-component system, the total number of moles of component (1) can be written (from Equation (194) in section 3.2.5) as

$$n_1 = n_1^\alpha + n_1^\beta + n_1^S \tag{412}$$

and since $c_1^\alpha = n_1^\alpha/V^\alpha$, and $c_1^\beta = n_1^\beta/V^\beta$, from the definition of the concentration, the number of moles of component (1) in the interphase region can be given as

$$n_1^S = n_1 - (c_1^\alpha V^\alpha + c_1^\beta V^\beta) \tag{413}$$

From the Gibbs convention where $V^S = 0$ is assumed, we know that $V = V^\alpha + V^\beta$, where V is the total volume of the system (or solution). If we substitute $[V^\alpha = V - V^\beta]$ into Equation (413), we have

$$n_1^S = n_1 - c_1^\alpha V + V^\beta(c_1^\alpha - c_1^\beta) \tag{414}$$

and similarly

$$n_2^S = n_2 - c_2^\alpha V + V^\beta(c_2^\alpha - c_2^\beta) \tag{415}$$

In the above equations, n_1, n_2, V, c_1^α, c_1^β, c_2^α and c_2^β are all experimentally measurable quantities, but we cannot calculate n_1^S and n_2^S if we do not know the value of V^β, which depends on the location of the Gibbs dividing plane. In order to eliminate this V^β term, we multiply both sides of Equation (414) by $[(c_2^\alpha - c_2^\beta)/(c_1^\alpha - c_1^\beta)]$ giving

$$n_1^S \left[\frac{\left(c_2^\alpha - c_2^\beta \right)}{\left(c_1^\alpha - c_1^\beta \right)} \right] = n_1 \left[\frac{\left(c_2^\alpha - c_2^\beta \right)}{\left(c_1^\alpha - c_1^\beta \right)} \right] - c_1^\alpha V \left[\frac{\left(c_2^\alpha - c_2^\beta \right)}{\left(c_1^\alpha - c_1^\beta \right)} \right] + V^\beta \left(c_2^\alpha - c_2^\beta \right) \tag{416}$$

and if we subtract Equation (416) from Equation (415), then we obtain

$$n_2^S - n_1^S \frac{\left(c_2^\alpha - c_2^\beta \right)}{\left(c_1^\alpha - c_1^\beta \right)} = n_2 - c_2^\alpha V - \left(n_1 - c_1^\alpha V \right) \left[\frac{\left(c_2^\alpha - c_2^\beta \right)}{\left(c_1^\alpha - c_1^\beta \right)} \right] \tag{417}$$

Equation (417) is important in practice because, the right side of the equation does not depend on the location of the Gibbs dividing plane and can be calculated from the experimental data. If there are more than two components in a system, then Equation (417) can be written as

$$n_i^S - n_1^S \frac{\left(c_i^\alpha - c_i^\beta \right)}{\left(c_1^\alpha - c_1^\beta \right)} = n_i - c_i^\alpha V - \left(n_1 - c_1^\alpha V \right) \left[\frac{\left(c_i^\alpha - c_i^\beta \right)}{\left(c_1^\alpha - c_1^\beta \right)} \right] \tag{418}$$

for the *i*th component. We have defined the *surface excess* in the adsorption process as, $\Gamma_i \equiv n_i^S / A^S$, by Equation (223) in Section 3.3. In order to insert the surface excess terms in the above equation, we divide both sides of Equation (418) with the constant surface area, A^S, giving

$$\Gamma_i - \Gamma_1 \frac{\left(c_i^\alpha - c_i^\beta \right)}{\left(c_1^\alpha - c_1^\beta \right)} = \frac{n_i}{A^S} - c_i^\alpha \frac{V}{A^S} - \left(\frac{n_1 - c_1^\alpha V}{A^S} \right) \left[\frac{\left(c_i^\alpha - c_i^\beta \right)}{\left(c_1^\alpha - c_1^\beta \right)} \right] \tag{419}$$

Then, a new term called *relative adsorption* is defined so that

$$\Gamma_i^{(1)} \equiv \Gamma_i - \Gamma_1 \frac{\left(c_i^\alpha - c_i^\beta \right)}{\left(c_1^\alpha - c_1^\beta \right)} \tag{420}$$

By combining Equations (419) and (420) we obtain

$$\Gamma_i^{(1)} = \frac{n_i}{A^S} - c_i^\alpha \frac{V}{A^S} - \left(\frac{n_1 - c_1^\alpha V}{A^S} \right) \left[\frac{\left(c_i^\alpha - c_i^\beta \right)}{\left(c_1^\alpha - c_1^\beta \right)} \right] \tag{421}$$

where $\Gamma_i^{(1)}$ is the relative adsorption of component *i* with respect to component (1) and this can be experimentally determined by using Equation (421). For dilute solutions, component (1) is generally taken to be the solvent, and when the usual Gibbs convention is applied by choosing the location of the Gibbs dividing plane arbitrarily to a place where $\Gamma_1 = 0$ for the solvent (1), then we have $\Gamma_i^{(1)} \equiv \Gamma_i$, from Equation (420). When binary solutions are considered, containing a solute (2) in solvent (1), we can write for the solute surface excess from Equation (421)

$$\Gamma_2^{(1)} = \frac{n_2}{A^S} - c_2^\alpha \frac{V}{A^S} - \left(\frac{n_1 - c_1^\alpha V}{A^S} \right) \left[\frac{\left(c_2^\alpha - c_2^\beta \right)}{\left(c_1^\alpha - c_1^\beta \right)} \right] \tag{422}$$

where $\Gamma_2^{(1)}$ is the difference between the amount of solute contained in a given volume (or mass) containing unit area of surface, and a similar volume (or mass) in the interior. It must be noted that $\Gamma_2^{(1)} = 0$ for a pure solute (2), since a layer of similar thickness in the interior would consist of almost pure (2), and then the surface layer could not contain more of this component.

For a solution having an interface with air or its vapor under ambient conditions, we usually take the gas phase as β, and then $[c_1^\beta = c_2^\beta = 0]$; because of the negligible concentrations of the solute and the solvent in the gas phase under 1 atm pressure, we then have

$$\Gamma_2^{(1)} = \frac{n_2}{A^S} - c_2^\alpha \frac{V}{A^S} - \left(\frac{n_1 - c_1^\alpha V}{A^S} \right) \left(\frac{c_2^\alpha}{c_1^\alpha} \right) = \frac{n_2}{A^S} - \frac{n_1}{A^S} \frac{c_2^\alpha}{c_1^\alpha} \tag{423}$$

The concentration profile of a solute (2) dissolved in solvent (1) is shown in Figure 5.2. In this figure, the Gibbs dividing plane is located at a place where $\Gamma_1 = 0$ according to the convention. The dividing line is drawn so that the two areas, shaded by the horizontal lines, are equal to each other, on each side of the solvent curve. Thus the surface excess of the solvent is zero. The surface excess of the solute, $\Gamma_2^{(1)}$, is shown by the difference in the dark shaded areas inside the solute curve. It is seen that the dark shaded area of the solute on the right-hand side of the Gibbs dividing plane is larger than the area on the left, and the difference between these dark shaded areas gives the surface excess of the solute, $\Gamma_2^{(1)}$ (it is positive in this case). It should be noted that the unit of $\Gamma_i^{(j)}$ is mol m^{-2} in the SI system, and can be converted to the number of molecules per unit area, $\Gamma_{i(\text{molecule})}^{(j)}$, (molecule nm^{-2}) by multiplying $\Gamma_i^{(j)}$ by Avogadro's number, N_A; and if we need the average surface area available for one molecule (nm^2 molecule^{-1}), we must take the reciprocal of this result. We must note that this is not the cross-sectional area per molecule, and if we want to calculate this, we need to know the area of the close-packed, monomolecular layer where all the molecules are in contact with each other. The experimental determination of this area is possible by using a film balance (Langmuir trough), as we will see in Section 5.6.2.

For a two-phase, ternary solution, where a solute is distributed within two immiscible liquids, Equation (421) can also be used to calculate the surface excesses of solute (3), as $\Gamma_3^{(1)}$ and $\Gamma_3^{(2)}$ are the relative adsorptions of solute (3) with respect to solvents (1) and (2) (see also Section 5.6).

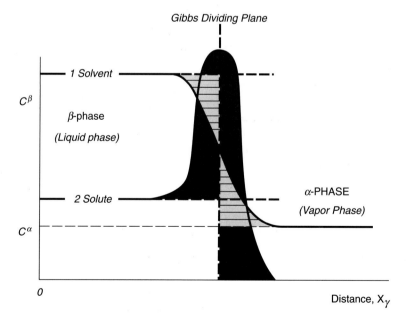

Figure 5.2 Concentration profile of a solute (2) dissolved in solvent (1) with the interfacial distance: α is a vapor phase and β is a liquid phase.

When the surface tension and concentration data are available for a solution, the most common method is to plot the surface tension against the logarithm of the solute activity and to obtain $[d\gamma/d\ln a_2]$ graphically. Nevertheless, very precise experimental results are needed to calculate the surface excess accurately from such a plot. The Gibbs equations give the area per molecule if γ is determined over a range of solution concentrations. We may compare the surface excess results found from Equations (422)–(423) with the results found from Equations (227)–(229), after we plot $[d\gamma/d\ln c_2]$ or $[d\gamma/d\ln a_2]$, and obtain their slopes, but we must remember that we located the Gibbs dividing plane arbitrarily at a place where $\Gamma_1 = 0$ for the solvent (1), in the Gibbs convention while we were deriving Equations (227)–(230), and this is equivalent to $\Gamma_i^{(1)} \equiv \Gamma_i$ in Equation (420).

5.5.2 Spreading pressure

When surfactant is added within a loose circle of thread floating on the water surface, it stretches the thread to a well-defined circular shape by pushing from inside. This shows the presence of a lateral pressure from the surfactant monolayer to the clean water surface. This lateral pressure of a monolayer to a barrier floating on the same sub-phase can be measured in a PLAWM (Pockels, Langmuir, Adam, Wilson and McBain) trough, which is shown in Figure 5.3 *b*. Since most surfactants are partially soluble in water, we need to use a flexible membrane, which is fixed to the barrier to separate the surfactant solution and pure water departments; otherwise the dissolved surfactant molecules will pass into the pure water department beneath the barrier. The force on this movable and friction-free barrier can be measured using any suitable physical method, and it shows the lateral pressure along the length of the barrier. If the solute is practically insoluble in water, then a Langmuir trough (or film balance) is used instead of the PLAWM trough, as shown in Figure 5.3 *a* (see Section 5.6.2 on the experimental determination of spreading pressure in monolayers).

When positive adsorption takes place at the solution surface, it lowers the surface tension of the pure solvent, γ_o, and the surface tension of the solution, γ, can be determined experimentally. If the solution is dilute, then γ can also be calculated from the Gibbs adsorption equation (Equation 224). The *spreading pressure* (or *surface pressure*), π, is defined as the decrease in the surface tension with the presence of an adsorbed monolayer

$$\pi = \gamma_o - \gamma \tag{424}$$

(π was also given in Section 3.3.2). The spreading pressure, π, can be regarded as a two-dimensional lateral pressure exerted by the adsorbed molecules in the plane of the surface. It is easy to understand the reason for the presence of this lateral pressure, if we consider the contractile forces exerted on the barrier from the solvent and the solution sides. Since the molecules in the pure solvent are attracted downwards (towards the bulk solvent) more strongly than the solute molecules, then the γ_o of the solvent is higher than the γ of the solution; consequently it is as if the monolayer is exerting a lateral force on the barrier from the solution side, along the length of the barrier (force per unit length, $N\,m^{-1}$) to the pure solvent side. This may also be shown by a simple units analysis, so that the units of surface tension ($N\,m^{-1}$), are the two-dimensional analogue of the bulk pressure units in three

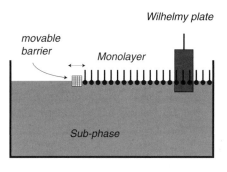

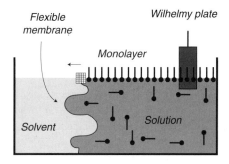

a. Langmuir trough
(Film Balance)

b. PLAWM trough

Figure 5.3 Film balances: **a.** Langmuir trough having a movable barrier and a Wilhelmy tensiometer to measure the spreading pressure, π, for water insoluble monolayers. **b.** PLAWM (Pockels, Langmuir, Adam, Wilson and McBain) trough used for partially water-soluble monolayers, where a flexible membrane, which is fixed to the barrier, separates the surfactant solution and pure water departments to prevent the passage of dissolved surfactant molecules into the pure water department beneath the barrier.

dimensions ($N\,m^{-2}$). In fact, surface tension may be regarded as a two-dimensional negative pressure. In theory, it is possible to convert a two-dimensional pressure, π, into a three-dimensional pressure, P, by simply dividing it by the thickness of the monolayer ($P = \pi/h_{\text{monolayer}}$). We may test this with standard experimental values: $\pi = 5\text{--}15\,mN\,m^{-1}$ and $h_{\text{monolayer}} = 1\,nm$, giving $P = 0.5\text{--}1.5 \times 10^{7}\,Pa \cong 50\text{--}150\,atm$, showing the large magnitude of the spreading pressure in three-dimensional terms.

On the other hand, the magnitude of π depends on both the amount of solute adsorbed and the area over which it is distributed. Thus, we must control both the quantity of the solute added onto the sub-phase and the exact area over which the monolayer spreads. The monolayer area control can be accomplished by compressing the monolayer of a practically nonvolatile solute by using frictionless movable barriers made of low-energy, inert materials such as polytetrafluoroethylene (PTFE) (the use of both single and double barriers is also possible, firstly to sweep the sub-phase surface to clean the insoluble impurities). We can measure both γ_o and γ in the same experiment to determine π. In general, the Wilhelmy plate method is used to measure the surface tension of solutions in film balances, but any suitable surface tension measurement method may also be used for this purpose (see Chapter 6). During the barrier compression process, we simultaneously record the decrease in surface area, A^S and the increase in π, which are somewhat inversely proportional to each other. Nevertheless, we must remember that the molecules present in a monolayer may freely move in the two-dimensional interfacial area, if their interaction with the sub-phase molecules is weak. As a result, the initially measured monolayer area in a film balance experiment is sometimes much larger than their close-packed area, where all the molecules are touching each other. From another point of view, π can also be defined as the pressure required to confine the monolayer to a given area.

The Helmholtz surface free-energy change due to the formation of a monolayer is the difference between the Helmholtz free energy of the pure solvent surface and that of the solution surface ($F^{\text{monolayer}} = F^S - F_o^S$). The Helmholtz free energy at the surface was defined by Equation (211) in Section 3.2.5. We assume the Gibbs dividing plane approach, where the volume of the interface, $V^S = 0$, to give

$$F^S = \gamma A^S + \sum_i \mu_i^S n_i^S \tag{425}$$

from Equation (211). When two insoluble components are present in a binary solution, say water, W, and oil, O, where O represents all the water insoluble materials, the excess Helmholtz free energy at the interface may be expressed from Equation (425) as

$$F^S = \gamma A^S + \mu_O^S n_O^S + \mu_W^S n_W^S \tag{426}$$

If the Gibbs convention is also adopted so that the Gibbs dividing plane is located such that the surface excess of the water sub-phase is zero ($n_W^S = 0$), we may then write

$$F^S = \gamma A^S + \mu_O^S n_O^S \tag{427}$$

Equation (427) shows that when no monolayer is present over the water surface ($n_O^S = 0$), there is only pure water and we may write, $F_o^S = \gamma_o A^S$. Then, the Helmholtz surface free energy due to the presence of a monolayer may be given as

$$F^{\text{monolayer}} = F^S - F_o^S = \gamma A^S + \mu_O^S n_O^S - \gamma_o A^S = \mu_O^S n_O^S + A^S(\gamma - \gamma_o) \tag{428}$$

When Equation (428) is combined with Equation (424) we have

$$F^{\text{monolayer}} = \mu_O^S n_O^S - \pi A^S \tag{429}$$

In a film balance experiment, the rate of change of $F^{\text{monolayer}}$ with the surface area for a closed system, where $n_O^S = 0$ can be found as

$$\frac{\partial F^{\text{monolayer}}}{\partial A^S} = -\pi \tag{430}$$

5.5.3 Gaseous monolayers: two-dimensional perfect gas

It should be clear that monolayer molecules may either merely rest on the surface without appreciably penetrating it or be almost wholly immersed in the top surface molecules of the underlying material. Adsorption on solids belongs to the second type where localized monolayers are formed (see Section 8.4).

   For the first type, the Gibbs monolayers are often described as *two-dimensional gaseous* (or *gas-like*) monolayers formed on the surface of dilute solutions, having a low surface excess, and the surface pressure, π, is regarded as the *two-dimensional osmotic pressure* of this solution. When a two-dimensional gaseous monolayer is present on a dilute solution, the surface tension decreases linearly with the increase in concentration of the added surfactant, at constant temperature

$$\gamma = \gamma_o - mc \tag{431}$$

where m is a constant depending on the solvent and the solute type. The derivative, $(\partial \gamma / \partial c)_T = -m$ can be found from the slope of a γ versus c plot. We can rearrange Equation (431) so that

$$mc = \gamma_o - \gamma = \pi \tag{432}$$

If we combine the $(\partial\gamma/\partial c)_T = -m$ expression and Equation (432) with the Gibbs adsorption isotherm (Equation (229)), we have

$$\Gamma_2 = \frac{mc_2}{RT} = \frac{\pi}{RT} \tag{433}$$

where R is the gas constant and T is the absolute temperature. Equation (433) relates π to the surface excess $[\pi = \Gamma_2 RT]$ and it can be called a *surface equation of state* (see the similar derivation of Equation (433) for gas adsorption on a solid surface in Section 8.3.5). This equation is a two-dimensional analogue of the three-dimensional gas law:

$$\pi = RTf(\Gamma) \tag{434}$$

However, there are conceptual differences between the surface equation of state and the adsorption isotherm, so that the surface equation of state is only concerned with the lateral motions of the monolayer molecules and their lateral cohesive and adhesive interactions with the solvent molecules present in the monolayer, whereas an adsorption isotherm is also concerned with the interactions normal to the surface, between the monolayer molecules (as adsorbate) and solvent molecules (as adsorbent).

Since the surface excess is the inverse of the area available per mole of solute, A^S_{mole}, in the monolayer, $[\Gamma = 1/A^S_{mole}]$, then from Equation (433) we have

$$\pi A^S_{mole} = RT \tag{435}$$

If we divide both sides of Equation (435) by Avogadro's number we have

$$\pi A^S_{molecule} = kT \tag{436}$$

where $A^S_{molecule}$ is the area available per solute molecule in the monolayer and k is the Boltzmann constant $(R = kN_A)$. However, we must distinguish the term, $A^S_{molecule}$, from the term, $\sigma_2^{molecule}$, which is the cross-sectional area of the solute molecule related to its geometric size.

The derivation of Equations (435) and (436) from dilute solutions is only approximate. It is also possible to thermodynamically derive more fundamental types of these equations by using the activity concept, but initially we need to define non-localized, ideal and non-ideal monolayers. If all the solute molecules are mobile in a monolayer, this is called a *non-localized* monolayer. We may consider three types of molecules in non-localized monolayers: ideal point molecules having no mass and volume where no lateral interactions are present between these point molecules; non-ideal molecules having their mass and volume but no lateral interactions taking place between them, as above; and non-ideal molecules having their mass and volume, and in addition appreciable lateral interactions taking place between them.

Ideal monolayers

For ideal point molecules, the surface equation of state is as given in Equation (435) and can also be written as

$$\pi A^S = n_2^S RT \tag{437}$$

where A^S is the measured area of the monolayer [$A^S = n_i^S A_{mole}^S$], and n_2^S is the number of excess moles contained in it. Since $n_2^S = N_2^S / N_A$, where N_2^S is the number of adsorbed molecules (2) contained in the monolayer, and N_A is Avogadro's number. Then the above equation can be written as

$$\pi A^S = N_2^S kT \tag{438}$$

where k is the Boltzmann constant. (This equation can also be derived directly from statistical thermodynamics.) The rate of spreading pressure change by the number of molecules can be found from Equation (438) as

$$d\pi = \frac{kT}{A^S} dN_2^S \tag{439}$$

When the Gibbs adsorption approach is applied to any monolayer in which the Gibbs dividing plane is located, where $n_1^S = 0$ for the solvent excess, then we have, from Equation (225)

$$-[d\gamma]_T = \Gamma_2 d\mu_2 \tag{440}$$

By combining Equation (440) with [$d\pi = -d\gamma$], from Equation (424), this results in

$$[d\pi]_T = \Gamma_2 d\mu_2 \tag{441}$$

Since [$d\mu_2 = RTd\ln a_2$] is also valid thermodynamically, then we have

$$[d\pi]_T = \Gamma_2 RTd\ln a_2 \tag{442}$$

Now, since the surface excess, [$\Gamma_i \equiv N_i^S / A^S N_A$] by definition from Equation (223), then Equation (442) becomes

$$[d\pi]_T = \frac{N_2^S}{A^S N_A} RTd\ln a_2 = \frac{N_2^S kT}{A^S} d\ln a_2 \tag{443}$$

When Equations (439) and (443) are combined we obtain

$$\frac{N_2^S kT}{A^S} d\ln a_2 = \frac{kT}{A^S} dN_2^S \tag{444}$$

After simplification, we have

$$d\ln a_2 = \frac{dN_2^S}{N_2^S} \tag{445}$$

Integration of Equation (445) gives

$$a_2 = K N_2^S \tag{446}$$

where K is the integration constant. Equation (446) is the basic equation of a non-localized and ideal two-dimensional monolayer. If pressures instead of activities are used in the above derivation [$d\mu_2 = RTd\ln P_2$], then we have

$$P_2 = K N_2^S \tag{447}$$

for the gas monolayers on a liquid sub-phase, which is analogous to Henry's law in two-dimensions and indicates that for the adsorption from a perfect gas the amount adsorbed is directly proportional to the gas pressure. It should be noted that Equations (446) and (447) are valid for ideal and non-localized monolayers where no lateral attractions are present, and can be applied approximately for only very dilute monolayers.

Non-ideal, monolayers without molecular interaction

For non-ideal molecules, the size of the molecules takes part in the derivation of the equation of state expressions. This case is valid for a more concentrated, non-ideal and non-localized monolayer where the lateral molecular interactions are also absent. In the above derivation for ideal, non-localized monolayers, the fluid molecules were assumed to be point molecules, which could move freely in the actual area, A^S, which is the product of solute mole number and area available, per mole of solute $A^S = n_2^S A_{mole}^S$. Since $[n_2^S = N_2^S/N_A]$ we have

$$A^S = \frac{N_2^S}{N_A} A_{mole}^S = \frac{N_2^S}{N_A}(N_A A_{molecule}^S) = N_2^S A_{molecule}^S \qquad (448)$$

where $A_{molecule}^S$ is the area available per solute molecule in the monolayer. However, for non-ideal monolayers, the fluid molecules cannot approach more closely to each other than the sum of their hard sphere radii will permit. Volmer modified Equation (438) in 1925 for non-ideal monolayers so that

$$\pi(A^S - N_2^S A_{molecule}^{SO}) = N_2^S kT \qquad (449)$$

where $A_{molecule}^{SO} = A_2^S/N_2^S$ is the co-area of the molecule where A_2^S is the incompressible solute area and the $(A^S - N_2^S A_{molecule}^{SO})$ term is the free area where the solute molecules present in the two-dimensional non-ideal and non-localized monolayer can move freely. The $A_{molecule}^{SO}$ term gives the area that any molecule occupies when all the molecules are in contact with each other in a two-dimensional gaseous monolayer. For two different molecules touching each other, having their hard sphere radii as r_a and r_b, there is an area of $[\pi(r_a + r_b)^2]$ around the center of each molecule from which the centers of all other molecules are excluded. This excluded area is common for these two molecules, and therefore the excluded area per molecule is only half of this, $[A_{molecule}^{SO} = \pi(r_a + r_b)^2/2]$. If there is only one type of molecule, then this analysis gives $A_{molecule}^{SO} = 2\pi r_i^2$. We must realize that, $A_{molecule}^{SO} \neq \sigma_2^{molecule}$, where $\sigma_2^{molecule}$ shows the cross-sectional area of the solute molecule. However for rigid spherical molecules, $A_{molecule}^{SO} = 2\sigma_2^{molecule}$ is valid from simple geometrical considerations.

If Equations (448) and (449) are combined and divided by (N_2^S), then we have

$$\pi(A_{molecule}^S - A_{molecule}^{SO}) = kT \qquad (450)$$

and when spreading pressure, π, is differentiated with respect to $A_{molecule}^S$ we have

$$d\pi = -\frac{kT}{\left(A_{molecule}^S - A_{molecule}^{SO}\right)^2} dA_{molecule}^S \qquad (451)$$

Since the $[d\pi]_T = \Gamma_2 d\mu_2$, $[d\mu_2 = RTd\ln a_2]$ and $[\Gamma_2 \equiv N_2^S/A^SN_A]$ expressions are also valid, then by combining these equations with Equation (451) we have

$$d\ln a_2 = -\frac{A_{molecule}^S}{\left(A_{molecule}^S - A_{molecule}^{SO}\right)^2} dA_{molecule}^S \tag{452}$$

After integration, Equation (452) yields

$$\ln a_2 = -\left[\ln\left(A_{molecule}^S - A_{molecule}^{SO}\right) - \frac{A_{molecule}^S}{\left(A_{molecule}^S - A_{molecule}^{SO}\right)}\right] + \text{integral constant} \tag{453}$$

If pressures instead of activities are used for the gas monolayers in the above derivation, from the $[d\mu_2 = RT\,d\ln P_2]$ expression, we have

$$\ln P_2 = -\left[\ln\left(A_{molecule}^S - A_{molecule}^{SO}\right) - \frac{A_{molecule}^S}{\left(A_{molecule}^S - A_{molecule}^{SO}\right)}\right] + \text{integral constant} \tag{454}$$

which can be further transformed into *surface coverage*, $[\theta = A_{molecule}^{SO}/A_{molecule}^S]$ terms, which are occasionally used for adsorption on solid surfaces (see Section 8.4).

Non-ideal monolayers with molecular interaction

This case is valid when non-ideal and non-localized monolayers are formed on concentrated solutions. Lateral molecular interactions are present in such monolayers. The equation of state takes the form of the two-dimensional van der Waals equation

$$\left(\pi + \frac{a\left(N_2^S\right)^2}{\left(A^S\right)^2}\right)\left(A^S - N_2^S A_{molecule}^{SO}\right) = N_2^S kT \tag{455}$$

where a is the two-dimensional van der Waals constant which shows the magnitude of the lateral interactions in the monolayer. The behavior represented by Equation (455) has been observed experimentally on both liquid and solid surfaces.

5.5.4 Adsorption on a water surface

Water is especially suitable to be used as the high density, sub-phase solvent, because it is partially and sometimes practically immiscible with most organic chemicals, surfactants, synthetic polymers and biological macromolecules. The phrase *practically immiscible* is used in the above sentence because there are no *completely immiscible* materials in theory, and all chemicals are miscible within each other always from large quantities down to a very small (parts per billion) extent, and this small miscibility must be neglected to assume complete immiscibility. Another important factor towards water being selected as the sub-phase solvent is its high surface tension: for pure water, it is equal to $72.8\,\text{mN}\,\text{m}^{-1}$, which is much higher than most organic liquids (except liquid metals) at room temperature, so that it allows other liquids to spread on the water sub-phase as a monolayer (we will see the reasons for this mechanism in the spreading section given below).

Direct experimental verification of the Gibbs adsorption equation in aqueous solutions is difficult, because physical separation of the monomolecular layer at the water surface is, required to compare the concentration differences between the surface layer and the bulk solution. Several attempts have been made on this subject from 1910 to the present day, and although an exact fit has never been obtained, the results show a good agreement with the theory. McBain and co-workers used a suitable microtome to cut off a thin layer of approximately 50–100 μm from the surface of phenol, *p*-toluidine etc. solutions and verified the Gibbs equation within experimental error in 1932. Later, isotopically labeled solute molecules were employed for this purpose. Beta-emitter molecules, such as ^{3}H, ^{14}C and ^{35}S have also been used and the radioactivity close to the surface measured. Since electrons only travel a short distance, the recorded radioactivity comes from the interface or very near the interface.

Typical plots of the variation of the surface tension with the logarithmic concentration difference of a surfactant, a saturated hydrocarbon alcohol and an inorganic salt in an aqueous solution are shown in Figure 5.4. Surfactant molecules find a lower free-energy environment at the interface of an aqueous solution than they have in the bulk solution and prefer to concentrate at the surface. The adsorption isotherm of most surfactants shows a sharp decrease initially with the increase in the solute concentration until a plateau appears, due to the formation of *micelles* following the *critical micelle concentration (CMC)*, where all the solution surface is covered with surfactant molecules. Any further addition of surfactant does not decrease the surface tension of the solution because it is directly consumed in the micelle formation process (see also Section 5.7). Saturated hydrocarbon alcohols also show a decrease in surface tension with increases in their aqueous solution concentration because they do not like to stay in the bulk solution and prefer to enrich on

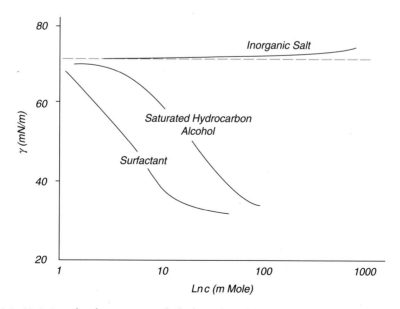

Figure 5.4 Variation of surface tension with the logarithm of the concentration difference of a typical surfactant, saturated hydrocarbon alcohol and an inorganic salt in an aqueous solution.

the surface. However, there is not a CMC point for these alcohols, and after they completely cover the water sub-phase surface, any more alcohol addition results in a two-phase solution, and the surface tension of the upper organic phase becomes equal to the surface tension of the alcohol itself. On the other hand, a hydrophilic inorganic salt, which gives ions in aqueous solution, prefers to stay in the bulk solution rather than to concentrate at the solution surface and negative adsorption occurs. The extent of this negative adsorption shows the decrease in the concentration of this salt in the surface layer rather than in the bulk solution. The presence of very minor amounts of high surface free-energy ions at the solution surface increases the surface tension of the solution over that of the pure water surface. Inorganic salts increase the surface tension almost linearly with rising concentration. The more strongly hydrated ions cause a greater increase in surface tension and the order for the monovalent cations is $Li^+ > Na^+ > K^+$, and for monovalent anions, $Cl^- > Br^- > I^-$. (When non-aqueous sub-phases such as alcohols, are used instead of water, the presence of salts in the solution also increases the surface tension of these solutions.)

The adsorbed films of soluble substances on the water sub-phase are mostly of the *gaseous monolayer* (or gas-like) type, where their lateral interactions are weak so that the molecules in the surface film move somewhat independently on the surface, sharing the translatory motions of the underlying water molecules (see Section 5.6 for the description of gaseous, liquid and condensed monolayers at the surface). Traube showed in 1891 that the solute concentration is directly proportional to the depression of the surface tension for very dilute "*gaseous*" surface solutions ($\pi = Kc_2$), after he examined the surface behavior of many dilute organic compounds in aqueous solutions. He also showed that during the adsorption of hydrocarbon derivatives on the water surface, the longer-chain paraffins were more frequently adsorbed than the shorter, and unsaturated hydrocarbons are more strongly adsorbed than the saturated ones.

We should be careful that the *gaseous* type of adsorption does not necessarily mean the adsorption of only gas or vapor on the water surface; many liquids give *gaseous* monolayers when adsorbed on an aqueous solution surface. In principle, the term *gaseous* describes the behavior and mobility of the adsorbed molecules in the monolayer. On the other hand, organic vapors are also adsorbed on the water surface, and they cause a decrease in the surface tension. When hydrocarbon vapors are adsorbed, they decrease the surface tension a few $mN\,m^{-1}$, also giving *gaseous* surface films. For some hydrocarbons, the decrease in the surface tension is approximately proportional to the partial pressure of the hydrocarbon vapor over the sub-phase surface. Traube's rule also holds for the gas-phase adsorption case where shorter-chain paraffin vapors are adsorbed less than the longer-chain variety.

5.5.5 Adsorption on surfaces other than water

In the 1930s many workers investigated the adsorption of vapors on liquid mercury surfaces, which reduced the surface tension of mercury considerably from $488\,mN\,m^{-1}$. However, it is a very difficult task to prepare pure fresh liquid mercury surfaces for such experiments, and there is great confusion in the data reported. On the other hand, many papers were published on the adsorption of solutes over organic solvent sub-phases. In general, the decrease in surface tension of the organic solvent with the increase in solution concentration was less than the results obtained for water sub-phases.

5.5.6 *Molecular orientation at the interface*

The intermolecular forces described in Sections 2.2–2.7 are operative across interfaces of two immiscible liquids. When molecules are not spherically symmetrical, the molecular orientation at interfaces should also be considered. The molecules at an interface should be oriented so as to provide the most gradual transition possible from one phase to the other. That is, the molecules will be oriented so that their mutual interaction energy will be a maximum. For example, between a liquid *n*-alkanol such as an octyl alcohol–water interface, the polar end of the organic molecule (the hydroxyl group of octyl alcohol) is oriented towards the water and the hydrocarbon chains extended into the alcohol. Langmuir showed that when π – surface area measurements were done in a film balance, the length of the hydrocarbon chains in saturated fatty acids and alcohols made no difference, provided there were more than 14 carbons in the molecules. He varied the length from 14 to 34 carbons, and hardly any change was found in the close-packed areas of the molecules, showing that the molecules are oriented steeply to the surface in all of the films. The head groups of —OH and —COOH attached to alcohols and acids have different roles in short and long chains. In shorter-chain compounds, such as ethyl alcohol and acetic acid, these groups confer solubility on the whole molecule in water, whereas in longer-chain groups they cannot pull the whole molecule into the water due to the resistance of the long hydrocarbon chains to immersion; instead they spread on the water sub-phase as a monomolecular film, as a film of the head group of the molecule in the water, the rest refusing to be dragged in. The lateral adhesion between these long-chain organic solute molecules also affects spreading. They assist in keeping the molecules out of the water by causing them to pack side by side forming an insoluble film on the sub-phase surface. The effect of compression on such films will be examined in Section 5.6.3. The orientation of water molecules at an interface is the subject of some uncertainty; several unsuccessful attempts have been made to show whether the water dipoles have a special orientation at the interface.

Various methods of statistical mechanics are applied to the calculation of surface orientation of asymmetric molecules, by introducing an angular dependence to the intermolecular potential function. The Boltzmann distribution can also be used to estimate the *orientational distribution* of molecules. The pair potential $V(r)$ may be written as $V(r, \theta)$ if it depends on the mutual orientation of two anisotropic molecules, and then we can write for the angular distribution of two molecules at a fixed distance, r, apart

$$\frac{X_{(\theta_1)}}{X_{(\theta_2)}} = \exp\left[-\frac{V(r,\theta_2)-V(r,\theta_1)}{kT}\right] \tag{456}$$

This type of interaction aligns the molecules mutually, such as placing the solvent molecules around a dissolved solute molecule in a solution.

5.5.7 *Marangoni effect*

It is clear that the molecules present in a monolayer interact with the molecules of the underlying liquid phase and cannot be assumed merely to consist of molecules moving freely in two dimensions. Such interactions result in molecular motions energized by surface tension gradients; this is called the *Marangoni effect*. In a glass of wine, ethyl alcohol evaporates more rapidly than water from the meniscus of the wine solution due to its high vapor pressure (and low boiling point), and this leads to a rise in surface tension locally at the wine surface. This evaporation induces a surface flow (Marangoni flow) which is accompanied by a bulk upwards flow from inside the solution due to the hydrodynamic instability, and the accumulating liquid returns in the form of drops (or *tears*) inside the glass.

5.6   Langmuir Surface Layers of Insoluble Materials on Liquids

When a small amount of non-volatile and practically insoluble material is placed on a liquid sub-phase (generally on clean water), it usually forms a *Langmuir insoluble mono-layer* (or duplex film) on the surface of the sub-phase (or between two immiscible phases separated with a recognizable meniscus), if its molecules attract the water molecules more than they attract each other. This means that the cohesion energy between these molecules is less than their adhesion energy with the water molecules. If the film is thicker than a monolayer, then we have a *duplex film*, so that the two interfaces (liquid–film or film–air) are independent of each other and possess their individual surface tensions. Partially immiscible (slightly soluble) materials also form Langmuir monolayers like the practically insoluble materials. Many water-insoluble solid materials can also be spread on the water subphase, by pouring their solution in a highly volatile organic solvent onto the im-miscible water surface (chloroform or hexane is commonly used). Such monolayers are called *spread monolayers*. In principle, the same thermodynamic rules (Gibbs adsorption approach) apply for all the types of insoluble monolayers, but determination of the solute concentration in the solution is neither a matter of interest for this case, nor a convenient quantity to measure experimentally. Instead, the more important matter is the direct deter-mination of interfacial tension and spreading behavior.

There are several examples of the spreading of a monolayer of an insoluble or slightly soluble liquid at a liquid–air interface. Historically, Benjamin Franklin, in 1774, observed that one teaspoon (2 ml) of olive oil sufficed to calm a half-acre surface of a pond on a windy day, giving an oil film on the water surface. It was not until over a hundred years later that Lord Rayleigh suspected that the maximum extension of an oil film on water rep-resents a layer one molecule thick. It was later calculated that olive oil spread on water as a film of one molecule thick (approximately 25 Å) due to the attraction of water for the polar groups orienting the olive oil molecules with polar group down and hydrocarbon "tail" up, creating a low energy surface. Since 25 Å is the length of the hydrocarbon chains present in olive oil as a monomolecular layer, this phenomenon was the first conclusive proof of atomic and molecular theory. Nevertheless, olive oil is a mixture of several organic materials, and it is instructive to repeat this calculation with a pure substance such as palmitic acid, with sixteen carbons in the molecule ($C_{15}H_{31}COOH$) having a molecular mass of $M_W = 256.42 \times 10^{-3}$ kg mol^{-1}, and a density of $\rho = 853$ kg m^{-3}, giving a molar volume of $V_M = 3.0061 \times 10^{-4}$ m^3 mol^{-1}. Therefore a single palmitic acid molecule has a volume of $V_{molecule} = 0.499$ nm^3 molecule^{-1} = 499 (Å)^{3}molecule^{-1}, which is found by dividing the V_M value by Avogadro's number. Its cross-sectional area was measured experimentally as, $A_{molecule} = 20.5$ (Å)^2 molecule^{-1}, so that its length measured perpendicular to the surface must be $l = 24.34$ Å, if the density in the film is the same as in the bulk liquid. The diameter of the cross-sectional area can be found as $\phi = 5.11\,\text{Å}$, showing that palmitic acid is 4.76 times as long as it is thick. These calculations are of course only approximate, and for a more detailed investigation we need to consider the effect of additional —CH_2— groups. It has been determined experimentally that most saturated carboxylic acids of whatever length, having carbon atoms between 14 and 34, showed almost the same molecular close-packed area of, $A_{molecule} = 20.5\ (\text{Å})^2$ molecule^{-1}, indicating that the additional —CH_2— groups only add to the height of the vertically positioned acid molecule. The molar volume of a —CH_2— group is approximately, $V_M = 1.78 \times 10^{-5}\,\text{m}^3\,\text{mol}^{-1}$, and its molecular volume is roughly, $V_{molecule} = 29.5\ (\text{Å})^3$ molecule^{-1}, and by dividing this value by the cross-sectional molecular area we have a vertical length of, $l = 1.44\,\text{Å}$ per —CH_2— group. However, X-ray investigations have shown that the carbon atoms are arranged not in a straight line, but in a zig-zag conformation, as expected from the stereochemistry, and the lines joining successive carbon atoms are inclined at the tetrahedral angle of 109° 28′. The distance between two carbon atoms in the chain was measured at approximately 1.54 Å, giving a height of 1.26 Å of one carbon atom above another along the vertical axis due to the zig-zag conformation. The resultant height difference (between 1.44 and 1.26 Å) is probably due to the density differences in the film and the bulk liquid. In general, the hydrocarbon chain of the substance used for monolayer studies has to be long enough to be able to form an insoluble monolayer. A rule of thumb is that there should be more than 12 hydrocarbons or groups in the chain (($CH_2)_n$, $n > 12$). If the chain is shorter, though still insoluble in water, the amphiphile on the water surface tends to form micelles. These micelles are water-soluble, which prevents the build-up of a monolayer at the interface (see Section 5.7). If the length of the hydrocarbon chain is too long the amphiphile tends to crystallize on the water surface and consequently does not form a monolayer. It is difficult to state an optimal length for the hydrocarbon chain because its film-forming ability also depends on the polar head group of the amphiphile.

We owe our ability to characterize monolayers on an air–water interface to a German woman scientist named Agnes Pockels. She developed a rudimentary surface balance in her kitchen sink, which she used to determine (water) surface contamination as a function of area of the surface for different oils. She sent her experimental findings to Lord Rayleigh in England, who encouraged her to publish her results in *Nature* in 1891. Irving Langmuir, who was the first Nobel laureate in surface science, was the first to perform systematic studies on floating monolayers on water in the late 1910s and early 1920s. The term *Langmuir film* is normally reserved for a floating monolayer. Later, the formation of insoluble monolayers on a liquid became an important branch of surface science. These mono- or multi-layers are very important in nature and also in the chemical industry, especially when water is chosen as the high-density, sub-phase solvent. In biology, the membranes present in cells of living organisms bear a close resemblance in structure to insoluble monolayers. Lipids and proteins form monolayers to construct cell membranes. The lipid components include phospholipids, sterols, glycerides of long-chain fatty acids and others that are known to form stable insoluble monolayers over a water surface, which has been well documented in physical chemistry. Thus the investigation of these monolayers helps in understanding the behavior of cell membranes. The cell membrane is also a highly selective barrier whose permeability characteristics are intimately involved in cell metabolism. Apart from holding the cell together, it also controls the diffusion of nutrients that are taken in and wastes excreted. However, simple diffusion equations derived from basic physics are inadequate to explain the transport rates at which all nutrients, wastes, salts and other chemical substances pass across the cell membrane, so it seems that monolayer chemical structure is the most important parameter in deriving an experimentally acceptable diffusion model in biology. There are other examples of monolayers in biology. One example is *mitochondria*, which are present in both plants and animals in order to oxidize foodstuffs to produce usable energy. Mitochondria contain extensive internal membrane systems. Another example is the billions of tiny membranes which are located in sub-cellular units in green plants to collect and utilize the energy of visible light for photosynthesis.

In the chemical industry, the presence of Langmuir monolayers is also very important. Emulsions and foams are liquid or gas dispersions formed in other liquids, and are stabilized by interfacial films, which consist of mono- or multi-layers. The reduction of evaporation losses of stored water and other liquids is also an important area where monolayers can be successfully applied. On solid surfaces, some technological areas such as adsorption, ore flotation, lubrication, friction and catalysis can be explored by applying mono- and multi-layer techniques, although the processes on solids are much more complicated than on liquid surfaces. The transfer of insoluble mono- or multi-layers onto other solid substrates is also possible using a controlled dipping process, such as the Langmuir–Blodgett (LB) method, which is important in the electronics and nanotechnology industries (see Section 5.9). Recently, such transferred monolayers have become the source of high expectations as being useful components in many practical and commercial applications such as sensors, detectors, displays and electronic circuit components. The possibility of synthesizing organic molecules with the desired structure and functionality, in conjunction with a sophisticated thin-film deposition technology, such as LB, enables the production of electrically, optically and biologically active components on a nanometer scale.

Monomolecular solute layers show ordered phases similar to three-dimensional systems. When Langmuir monolayers form, it does not necessarily mean that the molecules touch each other in the monolayer. Instead, if the anchoring from the sub-phase molecules is light, the molecules move freely in a two-dimensional interfacial area. Thus, the physically measurable monolayer area is sometimes much larger than the close-packed area where all the molecules touch each other. This is in analogy with the free translational motion of a gas molecule in a definite volume. When the surface area of a Langmuir monolayer is decreased by means of suitable barriers, initially there is very little change in the surface tension of the underlying liquid, but after the monolayer is confined to an area corresponding to their per-molecule close-packed area, the surface tension changes very rapidly due to the full coverage of the top molecules. This shows that the top molecules are floating or moving freely at the interphase until they actually touch each other. Such monolayers are called *gaseous monolayers* (gas-like) which occupy a much greater surface area per molecule and are very compressible laterally. *Condensed monolayers* (solid-like) are at the opposite extreme, and this describes monolayers which occupy only small surface areas per molecule and are very incompressible laterally because of the closely packed molecules (see Section 5.6.3). Condensed monolayers are similar to condensed fluid molecules in a limited volume, in three dimensions. *Expanded liquid monolayers* are intermediate between condensed and gaseous monolayers. The molecules are not packed closely in the expanded monolayers. The increase in temperature and decrease in the chain length of the

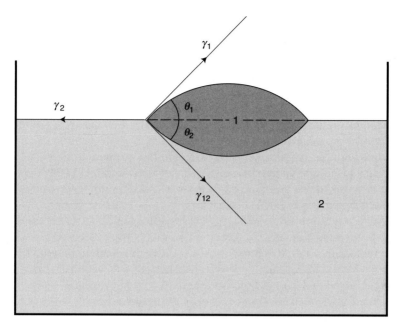

Figure 5.5 Vectorial equilibrium of the lens of liquid (1) on the surface of a sub-phase liquid (2). In equilibrium, (1) and (2) must be mutually saturated in each other and hence the surface and interfacial tensions will not necessarily be those of pure liquids.

monolayer molecules both reduce the cohesion between chains and tend to give expanded monolayers (see Section 5.6.4).

Insoluble monolayers may also exist at water–air as well as water–oil interfaces. In general, a monolayer of the same material tends to be more expanded at the water–oil interface than at the water–air interface, and usually it is recognized that a condensed monolayer forms at the water–air interface whereas it sometimes becomes gaseous at the water–oil interface. In summary, investigation of the basic principles of monolayer formation is a requisite in surface physical chemistry. We will start by defining the *spreading* concept in section 5.6.1, then explaining the experimental methods in section 5.6.2.

5.6.1 Spreading of one liquid on another

When we place a drop of appreciably non-volatile low density liquid (1) on the surface of a high density sub-phase liquid (2) which is practically immiscible with liquid (1), there are three possibilities:

1 Liquid (1) forms a non-spreading liquid lens with a defined edge on the sub-phase liquid (2), as shown in Figure 5.5, leaving the rest of the surface clean. The shape of the lens is constrained by the force of gravity.
2 Liquid (1) spreads as a monolayer on the surface of the sub-phase liquid (2), if space on the surface permits. The thickness of the monolayer is so small that gravitational effects are negligible.

3 If there is not enough space for all of the liquid (1) to spread fully, it spreads as a polylayer or a relatively thick film on the surface of the sub-phase liquid (2), where the corresponding liquid surface tensions, γ_1 and γ_2 retain their bulk values and the interfacial tension of the mutually saturated solutions, γ_{12}, can be measured experimentally.

In some instances, lens, monolayer and thick liquid film formation occurs simultaneously, depending on the strength of molecular interactions between these two immiscible materials and the availability of free surface. The surplus material (1) may remain as a lens in equilibrium with the monolayer. In general, a monomolecular film and some local small drops containing the excess material are seen when a pure substance is used to form the insoluble phase over water. However, much thicker films are formed for complex organic mixtures such as kerosene on water.

It is possible to perform a physical analysis to predict either liquid lens or thick film formation, and the strength of adhesion between the two phases. In order to assess the adhesion strength, initially we need to formulate the work of cohesion and adhesion. In Section 2.1, we defined the term *cohesion* to describe the physical interactions between the same types of molecule, so that it is a measure of how hard it is to pull a liquid (and solid) apart. In Section 3.5.3, we defined, the *work of cohesion*, W_i^c as the reversible work, per unit area, required to break a column of a liquid (or solid) into two parts, creating two new equilibrium surfaces, and separating them to infinite distance. (In practice, a distance of a few micrometers is sufficient.) The work of cohesion required to separate liquid layers into two parts having unit area can obviously be expressed from the definition of surface tension as

$$W_i^c = 2\gamma_i \tag{457}$$

At constant pressure and temperature conditions, $W_o^c = -\Delta G_o^S$ for pure liquids, where the minus sign arises from the thermodynamical inverse directions of the required work outside the system and the surface free energy of the liquid within the system.

The term *adhesion* is used if the interaction occurs between different types of molecule, and the *work of adhesion*, W_{12}^a is defined as the reversible work, per unit area, required to separate a column of two different liquids at the interface (or to separate a liquid from an underlying liquid), creating two new equilibrium surfaces of two pure materials, and separating them to infinite distance. However, the derivation of W_{12}^a is different from that of W_i^c because of the presence of equilibrium interfacial tension of the mutually saturated solutions, γ_{12}. Since two new surfaces (1) and (2) are formed, and the interfacial area (12) disappears during the separation process of two different liquids, then the work of adhesion can be formulated as given by Dupre

$$W_a = \gamma_1 + \gamma_2 - \gamma_{12} \tag{458}$$

Now, in order to set criteria for the requirements of spreading of a liquid on another sub-phase liquid, we need to formulate the vectorial equilibrium of the lens of liquid (1) on the surface of a sub-phase liquid (2), from Figure 5.5. In equilibrium, (1) and (2) must be mutually saturated in each other, and hence the surface and interfacial tensions will not necessarily be those of pure liquids. By applying the vectorial summation rule, we can write at hydrostatic equilibrium

$$\gamma_2 = \gamma_1 \cos \theta_1 + \gamma_{12} \cos \theta_2 \tag{459}$$

where θ_1 and θ_2 are the contact angles of the phase (1) lens with the air and phase (2). Angle θ_1 is not necessarily equal to θ_2, depending on the molecular interactions between (1) and (2) so mostly, $\theta_1 > \theta_2$. This is the general equation of a floating lens under another liquid in air. If the surface of (2) remains planar beneath the drop, then Equation (459) turns into Young's equation for solid surfaces, which we will see in Section 9.1.

When $\theta_1 + \theta_2 = 0$, this implies the complete spreading of the lens of liquid (1) over (2), and there are two possibilities. First, the spreading liquid, if not restrained, may expand to a film of molecular thickness. If so, this film will not completely cover the underlying liquid, and there will be a boundary (and a triple point) between (1), (2) and the air at the edge of the monolayer. Second, the spreading liquid (1) may be restricted by the walls of the vessel so that it forms a thick layer on the sub-phase. In this case, the three-phase boundary (the triple point) at the edge vanishes and the contact angle is non-existent, and sometimes this may end with two immiscible bulk phases with a recognizable meniscus between them. This is a two-phase solution made from two components. In practice, this is a time-dependent process so that first mutually dissolved solutions form in each other; later an equilibrium is reached between these mutually dissolved solutions, and then two phases appear with a meniscus between them. If the liquid viscosities are small, the formation of these two phases does not depend on the history of mixing; for example, we can obtain two immiscible phases having the same properties (in most cases), by placing one liquid carefully on another, or by mixing them mechanically and leaving the phases to separate for a suitable period.

In order to analyze the spreading process, Harkins introduced the concept of S, the *spreading coefficient* in 1919. He realized that spreading occurs if the surface tension of the underlying liquid surpasses the sum of the tensions of the interface and the top liquid. When the converse condition occurs, the drop assumes and retains a lens shape. Harkins defined the *initial spreading coefficient*, $S_{1/2}^i$, for a liquid (1) on another liquid (2),

$$S_{1/2}^i = \gamma_2 - (\gamma_1 + \gamma_{12}) = \gamma_2 - \gamma_1 - \gamma_{12} \tag{460}$$

and for the usual case of an oil (O) spreading on water (W), Equation (460) becomes

$$S_{O/W}^i = \gamma_W - (\gamma_O + \gamma_{OW}) = \gamma_W - \gamma_O - \gamma_{OW} \tag{461}$$

By combining Equations (457), (458) and (460), the initial spreading coefficient, $S_{1/2}^i$ may be related to the work of adhesion, cohesion and the Gibbs free energy of spreading per unit area:

$$S_{1/2}^i = W_a^{12} - W_c^2 = \Delta G_c^2 - \Delta G_a^{12} \tag{462}$$

and for the oil on water case, by combining Equations (461) and (462), we obtain

$$S_{O/W}^i = W_{OW} - W_O = \Delta G_O^S - \Delta G_{OW}^S \tag{463}$$

In summary, an oil spreads spontaneously on water when $S^i \geq 0$, and it forms a lens when $S^i < 0$. Spreading occurs when a liquid of low surface tension is placed on one of high surface tension.

In thermodynamic terms, the spreading coefficient, S^i, gives the free energy change for the spreading of liquid film (1) over liquid (2), varying with the creation and vanishing of the new areas. It is given as

$$S^i_{1/2} = -\left(\frac{dG}{dA_1}\right)_{T,P} \tag{464}$$

It is possible to prove that Equation (464) is valid. The total differential of the Gibbs free energy given below for (1), (2) and (12) shows the small change in the surface free energy at constant temperature and pressure due to the spreading of liquid (1) over liquid (2),

$$(dG)_{T,P} = \left(\frac{\partial G}{\partial A_2^S}\right)_{T,P} dA_2^S + \left(\frac{\partial G}{\partial A_1^S}\right)_{T,P} dA_1^S + \left(\frac{\partial G}{\partial A_{12}^S}\right)_{T,P} dA_{12}^S \tag{465}$$

Since the area of liquid (1) increases at the expense of the area of liquid (2) to form a new interfacial area of (12), we may write

$$dA_1^S = dA_{12}^S = -dA_2^S \tag{466}$$

By combining Equations (465) and (466), we have

$$\left(\frac{dG}{dA_1^S}\right)_{T,P} = -\left(\frac{\partial G}{\partial A_2^S}\right)_{T,P} + \left(\frac{\partial G}{\partial A_1^S}\right)_{T,P} + \left(\frac{\partial G}{\partial A_{12}^S}\right)_{T,P} \tag{467}$$

Since, $\gamma_i = (\partial G/\partial A_i^S)_{T,P}$, by the definition of surface tension, then we have from Equations (464) and (467)

$$\left(\frac{dG}{dA_1^S}\right)_{T,P} = -\gamma_2 + \gamma_1 + \gamma_{12} = -S^i_{1/2} \tag{468}$$

Equation (468) is equivalent to Harkins's equation (Equation (460)). However, we must also consider the effect of the mutual saturation of the liquids on the *equilibrium spreading coefficient*, $S^e_{1/2}$. After the initial contact of oil and water molecules [or liquids (1) and (2)], they will become mutually saturated within each other after a while, so that γ_W will change to $\gamma_{W(O)}$ and γ_O to $\gamma_{O(W)}$. At equilibrium, Equations (460) and (461) turn into

$$S^e_{1/2} = \gamma_{2(1)} - \gamma_{1(2)} - \gamma_{12} = \gamma_{w(O)} - \gamma_{O(w)} - \gamma_{ow} \tag{469}$$

where $S^e_{1/2}$ is the corresponding equilibrium (or final) spreading coefficient. For the case of benzene on water, it has been determined experimentally that $S^i_{B/W} = 72.8 - (28.9 + 35.0) = 8.9\,\mathrm{mN\,m^{-1}}$, and benzene must spread on pure water, whereas after the mutual saturation takes place, $S^e_{B/W} = 62.2 - (28.8 + 35.0) = -1.6\,\mathrm{mN\,m^{-1}}$, and benzene must form a lens on a benzene-contaminated water surface. Thus, when benzene is added to a pure water surface, a rapid initial spreading of benzene occurs over the water surface, and then as mutual saturation takes place, the benzene retracts to a lens in equilibrium with an adsorbed monolayer. All the n-alkanes below nonane spread spontaneously on water, but nonane and upwards are non-spreading liquids at 20°C. Therefore, the critical S^i value for the spreading of alkanes on water lies between the surface tensions of octane and nonane, which are 21.8 and 22.9 mN m^{-1}, respectively. It appears that this critical S^i value is a property of the water sub-phase alone. When unsaturated hydrocarbons are used in the spreading experiments, high critical S^i values are obtained for water, and Zisman suggested in

1967 that only dispersion forces contribute to liquid–liquid adhesion for *n*-alkanes, whereas both dispersion and other (polar, donor–acceptor) forces contribute to liquid–liquid adhesion for unsaturated hydrocarbons, and thus change the critical S^i values.

5.6.2 Experimental determination of spreading pressure in monolayers: Langmuir balance

Historically, Irving Langmuir was, in 1917, the first to apply film balances to examine the behavior of monolayers. He used a large tray, where a float separates the monolayer film from the clean solvent surface, as shown schematically in Figure 5.3 *a*. The direct measurement of the horizontal force on this floating barrier was carried out using a torque balance to give the film pressure, π, directly. Langmuir used a two-barrier trough, a rigid but adjustable barrier on one side and a floating barrier on the other. He realized that it is possible to sweep a film off the surface quite cleanly by moving the sliding barrier, which is in contact with the surface, to give a fresh surface of clean water, which would form behind the barrier as it was moving along. Later, several researchers made improvements on Langmuir's first floating barrier and then it became known as a *PLAWM* (Pockels, Langmuir, Adam, Wilson and McBain) trough, where a flexible membrane separates the solution and pure water departments to prevent any leakage of the monolayer beneath the barrier; otherwise the partially dissolved solute molecules would pass into the pure water department.

When the solute is practically insoluble in the water sub-phase, an insoluble monolayer forms and a film balance (or Langmuir trough) are used instead of a PLAWM trough as shown in Figure 5.3 *a*. In this trough, there is no need to separate the compartments with a flexible membrane because practically all the solute molecules are located on the surface. For monolayer formation, an amphiphile solute is initially dissolved in a water-insoluble and highly volatile spreading solvent (mostly chloroform or *n*-hexane, and rarely cylohexane or benzene are used for this purpose) in 0.1–2.0 mg ml^{-1} concentrations, and is placed on the water sub-phase surface with a micro-syringe; then the organic solution spreads rapidly to cover the available area. As the solvent evaporates, a monolayer is formed. The solution is allowed to drop from the micro-syringe held a few millimeters away from the subphase surface, and many workers prefer to distribute the drops over the surface to be covered. The injection of the spreading organic solution beneath the surface of the sub-phase is not recommended due to the risk of contamination from the needle.

The trough holding the sub-phase is usually made of PTFE polymer to prevent any leakage of the subphase over the edges. PTFE is the most inert material available and very suitable to be used for Langmuir troughs because it is resistant to all strong acids and is extremely hydrophobic. It does not contaminate the subphase with surfactants or mineral ions. In addition, the cleaning of PTFE trays is also very simple; it is generally wiped with a surfactant-free tissue soaked in chloroform and rinsed with detergent and pure water to remove materials not soluble in chloroform, such as proteins. The major disadvantage of PTFE is its lack of dimensional stability. Other plastic materials such as polymethylmethacrylate, nylon, polyethylene and polypropylene have also been used as trough material, although they are not as successful as PTFE. In some film balances, the trough is thermostatted by circulating water in channels placed underneath the trough, in order to experiment under constant temperature conditions. The trough edges are important in preventing film leakage and subphase spills. The depth of the sub-phase confining chamber in the trough top is generally 4 mm to minimize the sub-phase volume, which improves temperature control and reduces consumption of ultra-pure water. The trough can be filled exactly level to the rim, which prevents any errors in area calculation arising from meniscus curvature. (In practice, it is usually overfilled initially and then the surface is cleaned by removal of the top contaminated layer using an aspirator pump.) The purity and the protection of the subphase solvent are of utmost importance because impurities in the ppm range can cause serious errors if they enrich at the surface. An overflow channel is also machined into the trough periphery, which allows controlled overfilling and prevents minor spillages down the side of the trough.

The surface area of the trough can be varied by sweeping movable barriers over the subphase surface in the trough. The movable barriers rest across the edges of the trough. They are made of PTFE, which ensures a good seal between the trough and barrier, and prevents leakages from the monolayer. They have a square or rectangular cross-section, and their edges are carefully machined flat. It is customary to sweep the liquid surfaces with the barriers two or three times immediately before spreading the monolayer film under study, to remove any insoluble greasy material. The barrier and its drive mechanism have been designed to have minimum inertia and friction in order to achieve low friction and slip-free positioning and good dynamic properties. The surface pressure, π, and the mean surface area, A^S, are continuously monitored during the compression. The surface pressure is generally measured by the Wilhelmy plate method (see Section 6.4). In this method a paper or platinium Wilhelmy plate is suspended across the air–water interface, and the force on the plate is measured with a micro-balance, to be used as a surface pressure sensor. The force on the plate can be converted to units of surface tension or surface pressure (mN m^{-1}) by using the measured dimensions of the plate and calibrating the micro-balance with a known weight. When paper Wilhelmy plates are used, the contact angle of water to plate is then guaranteed to be 0°. When, non-absorbent materials like platinum plates are used they need to be flame-cleaned before every measurement. The cleaning of the trough, barriers and other parts, which come into contact with the subphase solvent is very important and the avoidance of air-borne contamination is also necessary. In a normally clean laboratory, an exposed water surface will almost certainly have accumulated a layer of grease within an hour, and thus it is recommended that the film balance experiments be carried out within 10 min after cleaning the water surface with aspiration in open laboratory conditions. This period may be much longer in dust-free clean rooms, depending on the conditions. However, if a film has already been spread, the deposition of contamination is greatly impeded, since this monolayer is no longer so attractive to the contaminant molecules in comparison with the water surface. There are other ways to control the area of the monolayer and to measure the surface pressure, but the constructions above are the most commonly used.

Incomplete spreading sometimes occurs due to poor selection of spreading conditions. This can be detected by simple visual observation using suitable lamps to illuminate the surface, such as dark-field illumination. The best test for completeness of spreading is the reproducibility of the experimental results. When we are satisfied with the homogeneous monolayer formation, we can start to examine its properties. When the available area for the monolayer is large, the distance between adjacent solute molecules in the monolayer is also large, and their intermolecular interactions are weak giving a *gaseous monolayer*. Under these conditions, the monolayer has little effect on the surface tension of subphase water. When the barriers reduce the available surface area of the monolayer in a film balance (see Figure 5.3 *a* and *b*), the molecules start to exert a repulsive effect on each other.

5.6.3 Expanded and condensed Langmuir monolayers

The surface pressure, π, is measured as a function of the area of the water surface available to each molecule at constant temperature, and the resulting plot is known as a *surface pressure–area isotherm*. The barriers compress the surface area of the film preferably at a constant rate. Depending on the film material being studied, repeated compressions and expansions are necessary to achieve reproducible results.

A number of distinct regions are immediately apparent on examining the isotherm. When the monolayer is compressed, it can pass through several different phases, which are identified as discontinuities in the isotherm, as shown in Figure 5.6. The phase behaviour of the monolayer is mainly determined by the physical and chemical properties of the amphiphile, the subphase temperature and the subphase composition. Various monolayer states exist depending on the length of the hydrocarbon chain and the magnitude of cohe-

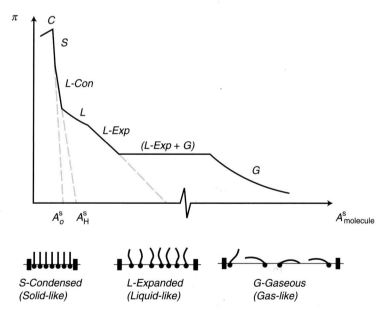

Figure 5.6 *Surface pressure–area isotherm* where a number of distinct regions are present: (*G*) is the *gaseous state* and can undergo a phase transition to the *liquid-expanded state* (*L-Exp.*) on compression. The gaseous and liquid-expanded regions simultaneously exist in the large intermediate region, (*L-Exp.+G*). Upon further compression, the (*L-Exp.*) phase undergoes a transition to initially the *liquid-like state* (*L*) and then to the *liquid-condensed state* (*L-Con.*), and at even higher compressions the monolayer finally reaches the *solid-like state* (*S*). The monolayer will collapse into three-dimensional structures, in the region termed the *collapsed state* (*C*), if it is further compressed after reaching the (*S*) state.

sive and repulsive forces existing between head groups. A simple terminology to classify different monolayer phases of fatty acids was proposed by W.D. Harkins in 1952. At large, the monolayers exist in the *gaseous state* (*G*) and can, on compression, undergo a phase transition to the *liquid-expanded state* (*L-Exp.*), as shown in Figure 5.6. However, there is a large region where gaseous and liquid-expanded regions simultaneously exist (*L-Exp.+G*). Upon further compression, the (*L-Exp.*) phase undergoes a transition to initially the *liquid-like state* (*L*) and then the *liquid-condensed state* (*L-Con.*), and at even higher densities the monolayer finally reaches the *solid-like state* (*S*). If the monolayer is further compressed after reaching the (*S*) state, then it will collapse into three-dimensional structures, the *collapsed-state* (*C*). The collapse is generally seen as a rapid decrease in the surface pressure or as a horizontal break in the isotherm.

During the gaseous state (*G*), very little increase will be observed in the surface pressure, π, as the molecules are randomly distributed on the sub-phase surface with no ordering, and are behaving as would be expected of a two-dimensional gas phase. Further reduction of the surface leads to increases in π, as the intermolecular spacing in the monolayer enters the range of the liquid-expanded state (*L-Exp.*), where intermolecular interactions are operative. The molecules may be touching each other but there is no lateral order, and the hydrophobic chains may lie flat on the subphase. The head-groups are highly hydrated. The hydrophobic tails, which were oriented randomly near the surface of the subphase begin to be lifted away by a further decrease in the surface area. An increase in the chain length of solute molecules increases the attraction between them and helps them to condense. On the other hand, if an ionizable amphiphile is used as solute, then the ionization of the head groups induces repulsive forces tending to oppose phase transitions. When the surface area of the monolayer film is further reduced, the rate of increase of surface pressure is even greater and a nearly vertical linear relationship of π–A^S characterizes the highly ordered and closely packed liquid condensed state (*L-Con.*). In this phase, the solute molecules are almost standing upright with the hydrophobic tail group nearly perpendicular to the subphase. The monolayer film is relatively stiff at this stage but there is still some water present between the head-groups (small hydration). Both (*L-Exp.*) and (*L-Con.*) phases are just two of several liquid crystal phases in which a monolayer can exist. For the condensed (non-penetrating) insoluble monolayers, where the molecules rest on the interface, the surface pressure, π, is not dependent on the interactions with the underlying liquid. In the liquid condensed state (*L-Con.*) of the isotherm, the head-groups are hydrated with water molecules and closely packed; then it is possible to obtain quantitative information on their hydrated dimensions and shapes. The hypothetical molecular area, for the hydrated state, A_H^S, can be obtained by extrapolating the slope of this (*L-Con.*) phase to zero pressure, as shown in Figure 5.6.

If we further compress the surface area, the head-groups become dehydrated and the π–A^S isotherm is also linear with a steeper slope. The close-packed and dehydrated molecular area, A_o^S, can be obtained by extrapolating the slope of this (*S*) phase to zero pressure. However, in reality, it is not possible to obtain both A_o^S and A_H^S values simultaneously for many π–A^S isotherms, because the upper part of the isotherm curve is frequently not so linear. The reason is that the phase transition between *S* and (*L-Con.*) is usually not sharp, but smooth. In these circumstances, only one of these molecular areas can be determined. If we further compress the area very slowly, the surface pressure increase will stop; at a specific surface pressure value, the collapse of the monolayer (*C*) occurs, and π

eventually falls sharply. The monolayer irreversibly loses its mono-molecular form at the collapse pressure. At this point, the forces exerted by barriers upon the monolayer are too strong for confinement in two dimensions, and then molecules are ejected from the mono-layer plane to form layers which are piled on top of one another resulting in disordered multilayers. Some striations can be seen on the subphase surface, which are the areas of collapse.

5.6.4 Monolayers between two immiscible liquids for three-component solutions

What happens if a third solute is dissolved in one of the two immiscible liquids, stirred mechanically with the other liquid, and then the two phases allowed to separate? The answer is that if the third solute is soluble in both solvents, it distributes between them according to the *Nernst distribution law*, due to the equality of the chemical potentials in each phase at equilibrium. The concentration of this solute at the interface is different (generally larger) than that of its bulk concentrations in both solutions. For example, acetic acid or butyric acid forms a gaseous-type monolayer between practically immiscible benzene and water phases. We may use Equation (421) in Section 5.5 to calculate the surface excesses of the solute (acetic or butyric acid) as $\Gamma_{acid}^{(benzene)}$ and $\Gamma_{acid}^{(water)}$, the relative adsorption of the solute with respect to the solvents (benzene) and (water). It has been shown that when the same solution concentrations are applied, acetic acid is more closely packed between the water and benzene interface than between water and air, but the lateral adhesion between acetic acid molecules was found to be nearly the same. The reason for this behavior is the stronger attraction between benzene and the methyl groups in the acetic acid, resulting in the close packing of more acetic acid molecules in the interfacial region for a given surface pressure. In general, the lateral attractions are smaller at the water–air interface than the water–immiscible liquids, for most solutes. However during experimentation with monolayers, one should be careful to avoid the interfacial reactions occurring in a three-component, two-phase solution, which will confuse the experimental findings, especially for varying sub-phase pH ranges.

  On the other hand, if the third solute is not soluble in the sub-phase solvent and if we dissolve the third solute in a solvent and pour it over a high-density sub-phase liquid which is immiscible with the first solvent, then depending on the density differences and molecular interactions, two behaviors can be seen after the evaporation of the upper solvent layer. Either the solute molecules form an insoluble monolayer on the sub-phase solvent where we can use all the insoluble monolayer equations, or the solute material sinks down inside the subphase solvent if the density of the solute is appreciably higher than the subphase density.

5.7 Micelles and Critical Micelle Concentration (CMC)

An amphiphilic molecule (a surfactant) can arrange itself at the surface of the water such that the polar part (head group) interacts with the water, and the hydrocarbon part (tail group) is held above the surface, as shown in Figure 5.3. This kind of organization of surfactant molecules can be described as a head-to-head/tail-to-tail type. If there is an immiscible liquid over the water sub-phase surface, the surfactant molecules take the same positions, because the upper liquid phase is also non-polar, similar to air, in order to be immiscible with the water. As we know from the previous chapters, the surfactant molecules prefer to enrich on the sub-phase surface, and their presence lowers the surface tension of the solution. Meanwhile, few surfactant molecules are also dissolved in the bulk sub-phase simultaneously. Nevertheless, after all the available surface area is covered with these surfactant molecules, the addition of any more surfactant onto the sub-phase surface will force these molecules to enter into the bulk water phase. Then they associate to give a new molecular arrangement in the water phase in order to minimize their surface free energy, so that each hydrophilic head-group and hydrophobic hydrocarbon tail-group of the surfactant can interact with its favored environment. This association process is thermodynamically driven and is spontaneous similar to the formation of the Langmuir monolayers. As a result, aggregates called *micelles* (or *associated colloids*) are formed in the bulk water sub-phase, where the hydrophobic portions are oriented within the cluster, and the hydrophilic portions are exposed to the water phase. These surfactant molecules are also organized as head-to-head/tail-to-tail type in these micelles. The interactions among the surfactant molecules are usually physical in nature, rather than by covalent bonding. The greater potential energy of water molecules in the vicinity of a hydrocarbon chain results in the withdrawal of the hydrocarbon from the surrounding water molecules, and this process reduces the total free energy. The withdrawn tail groups are forced to aggregate with each other. In general, *spherical micelles* form to achieve the lowest interfacial area, as shown in Figure 5.7 *a* in cross section, and 5.7 *b* in three dimensions. In water, each spherical micelle consists typically of 30–100 surfactant molecules, having an oily (hydrocarbon) phase inside. The typical outer diameters of spherical micelles are around 3–6 nm. However, other types of geometric shapes can also be formed by the association of surfactant molecules, such as cylindrical (rod-like) micelles as shown in Figure 5.7 *c*, or bilayers (Figure 5.10 *a*), vesicles and liposomes (Figure 5.10 *b*), and inverted micelles (Figure 5.10 *c*) (see also Section 5.8 for the reasons for taking these different micelle shapes).

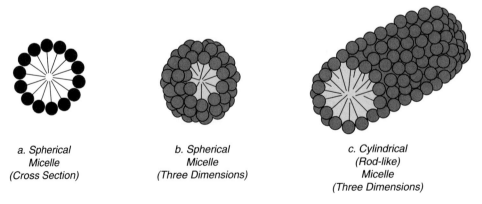

a. Spherical
Micelle
(Cross Section)

b. Spherical
Micelle
(Three Dimensions)

c. Cylindrical
(Rod-like)
Micelle
(Three Dimensions)

Figure 5.7   Schematic representation of spherical and cylindrical micelles.

The proportion of surfactant molecules present on the solution surface or as micelles within the bulk phase depends on the concentration of the added surfactant. As the surface becomes crowded with surfactant, more molecules will arrange into micelles. At some concentration, the surface becomes completely loaded with surfactant molecules and any further additions must arrange only as micelles. Thus, no further decrease in surface tension takes place after this concentration. This critical concentration is called the *critical micelle concentration* (CMC). The CMC may be determined experimentally by the measurement of surface tension of the solution. A plot of surface tension versus concentration of surfactant added, *c* (or preferably a $\gamma - \ln c$ plot) is used to find the CMC, as shown in Figure 5.8. The location of the CMC is found as the point at which two lines intersect; the baseline of minimal surface tension and the slope where surface tension shows a linear decline. For CMC determination, the solution surface tension may be measured by ring or Wilhelmy plate methods (see Section 6.4). There are two ways to perform a CMC measurement either by adding a concentrated solution of surfactants to an initially surfactant-free solution, or by diluting a concentrated surfactant solution with pure solvent. Generally the first method is used and the surface tension of the subphase is initially measured prior to any surfactant addition. Next, a predetermined amount of surfactant solution is added to bring the system closer to the CMC value. Then the solution is mechanically stirred, and after a stabilization period, the values of surface tension of the solution will be measured. (In general, several measurements are made in order to gain experience and to determine the CMC precisely for a specific surfactant in a solution.) The CMCs of most commercial surfactants are reported in tables in order to optimize their

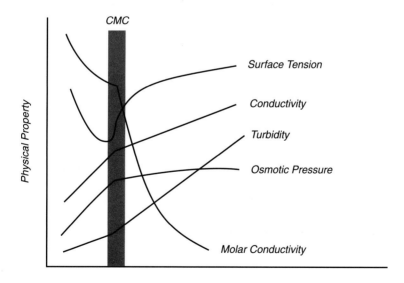

Figure 5.8 CMC determination from a plot of a solution physical property such as surface tension, conductivity, turbidity, osmotic pressure, molar conductivity etc. versus concentration of surfactant added, *c* (preferably, a *property* – ln *c* plot can be used).

detergency power to a specific application, and also to minimize waste to restrict environmental pollution.

The CMC is also well defined experimentally by a number of other physical properties besides the variation of the surface tension. The variation of solution properties such as osmotic pressure, electrical conductance, molar conductivity, refractive index, intensity of scattered light, turbidity and the capacity to solubilize hydrocarbons with the increase of surfactant concentration will change sharply at the CMC as shown in Figure 5.8. The variation in these properties with the formation of micelles can be explained as follows. When surfactant molecules associate in solution to form micelles, the concentration of osmotic units loses its proportionality to the total solute concentration. The intensity of scattered light increases sharply at the CMC because the micelles scatter more light than the medium. The turbidity increases with micelle formation, because the solution is transparent at low surfactant concentrations, but it turns opaque after the CMC. Hydrophobic substances are poorly dissolved in aqueous solutions at concentrations below the CMC, but they start to be highly dissolved in the centers of the newly formed micelles, after the CMC.

Micelles are not static: they are dynamic structures, so that the surfactant molecules leave the micelle and go into the solution while other surfactant molecules enter the micelle from the solution. The timescale for this dynamic rearrangement depends mostly on the length of the hydrocarbon chain: the longer the chain the longer the residence time in the micelle (but the dependence is not linear). Micelles are also continuously moving in the solution; their mobility rates are dependent on the type of surfactant and the temperature of the solution. Micelles in solutions can be examined by light scattering, small-angle X-ray scattering (SAXS), and small-angle neutron scattering (SANS) analytical techniques. It has been found that micelle interiors behave like bulk liquids, as determined by nuclear magnetic resonance (NMR).

The effect of cooling on surfactant solutions is important because, after cooling down to a specific temperature, known as the *Kraft temperature*, the surfactant molecules present in the micelle precipitate as hydrated crystals. Below the Kraft temperature, the solubility of the surfactant in water is so low that the solution contains no micelles. The solution temperature has little influence on the micelle structure, and when the micelles are made of surfactants having ionic head-groups, the increase in temperature generally increases the surfactant solubility slightly, so micelles lose some of their surfactant molecules to the water. However, when micelles are made of non-ionic surfactants, an opposite temperature effect is seen. The non-ionic surfactants become less soluble at elevated temperatures due to breakage of the hydrogen bonds with the water molecules. As the temperature is raised, a point is reached, which is called the *cloud point*, at which large aggregates precipitate out into a separate phase. Experiments to determine the *cloud point* temperatures for non-ionic surfactants usually give less sharp plots than the experiments to determine the *Kraft temperature*.

Both ionic and non-ionic surfactants associate to form micelles. However, there are several property differences between the micelles formed. The CMC value depends on the length of the hydrophobic groups for both types, but it also depends on the state of charge for ionic surfactants and on the length of the hydrophilic part for non-ionic surfactants. The core of a micelle is assumed to be filled with liquid hydrocarbon, and the maximum size of the micelle radius is equal to the length of the fully extended hydrocarbon tail. In any tail length calculation, we need the projections of the covalent bond lengths along the direction of the hydrocarbon chain, and thus we need to know the zig-zag angle, which is 109.5° for many saturated hydrocarbon chains. The tail chain itself occupies a certain volume called the *tail volume*, V_t, and the tail chains may either overlap in the micelle center or part from each other upon penetration of water molecules to a degree (the inner core of the micelle is assumed to be water-free). However, in some cases some surfactant molecules protrude from the surface of the core further than others, allowing crowding at the center of the micelle.

In micelles made of ionic surfactants, variation of the nature of the ionic group is not significant; only the state of charge is important. There is a shell around the hydrocarbon core consisting of ionic heads, and the counter-ions in the solution are located around these ionic heads in a region called the *Stern layer*. Free water molecules and water of hydration are also present in this region. Both the ionic head-group area and the bound counter-ion volume are affected by dissolved electrolytes in the solution. Extension of the ion atmosphere in the medium decreases as the electrolyte content of the solution increases. Ionic micelles will migrate in an electric field in the solution, and the ion atmosphere of the micelle is dragged along with it. Conductivity measurements show the presence of micellar mobility.

The CMC value of non-ionic surfactants is much lower than for surfactants having ionic head-groups, being of the order of $10^{-4}\,mol\,l^{-1}$. They form micelles at concentrations about one-hundredth that of anionic surfactants possessing comparable hydrophobic groups. This shows that non-ionic surfactants form micelles easily at low solute concentrations, and low CMC values correspond to high micellar unit weights. The ether oxygens present in polyethylene oxide chains are heavily hydrated, and these are tangled into coils to an extent which their chain length and hydration allow. When the length of the polyethylene oxide chain is increased to give a more hydrophilic surfactant, their CMC increases too (while the concentrations are expressed as weight per volume). This is expected because, with the increase of its hydrophilicity, a surfactant will prefer to be dissolved in the water phase rather than forming aggregates. (However, when the concentrations are expressed in molar units, several non-ionic surfactants show a downward trend due to the long chain and high molecular mass per molecule.) If we examine the solution behavior of non-ionic surfactants by reducing the length of their hydrophilic groups, we see a similar trend as we reduce the ionic repulsion of the head-groups in ionic surfactants, thus showing that hydration and electrostatic effects are parallel.

The shapes and structures of micelles are dependent on the molecular architecture of the surfactant they are made of. The shapes of micelles can be explained through packing considerations based on simple geometrical features of the surfactant, as shown in Figure 5.9. Such geometric arguments can also be successful in predicting changes in the micelle structure when the pH, charge, ionic strength, temperature and hydrocarbon tail length are varied. There are three effective geometric parameters of a surfactant:

1 The optimal head-group area, A_h, depending mainly on the hydrophilic, steric and ionic repulsions between adjacent head-groups trying to enlarge the head-group area; also, to a lesser extent, depending on the opposite hydrophobic attraction of hydrocarbon tail groups trying to shrink the head-group area. The value of A_h is usually much greater than the cross-sectional area of the present head-group. It depends on both surfactant

and electrolyte concentrations for ionic surfactants, whereas non-ionic surfactants are relatively insensitive to external conditions.

2 The tail volume, V_t, which is the volume of hydrocarbon liquid per hydrocarbon molecule. It is assumed that there is an incompressible hydrocarbon liquid in the core of the micelle due to the presence of hydrocarbon tail molecules. If we cut a uniform conic volume per surfactant from a spherical micelle, as shown in Figure 5.9, we can find the value of V_t, from simple geometry. However, if the micelle is not spherical, V_t does not represent a uniform cone; instead it shows the form of a truncated cone to produce a *cylindrical micelle, vesicle* or *liposome*; a cylinder to form a *bilayer micelle*, or an inverted cone to form an *inverted micelle*.

3 The height of the cone, L_c, is slightly less than the radius of the micelle and depends on the effective chain length of the hydrocarbon tail. On the other hand, the radius of the micelle is assumed to be less than the length of the fully extended hydrocarbon tail. We need projections of the covalent bond lengths along the direction of the chain for maximum tail length calculations, and thus we need to know the zig-zag angle. In general, L_c is defined as a semi-empirical parameter, of the same order as the maximum extended chain length of the hydrocarbon molecule. Several semi-empirical equations have been derived to calculate L_c from the number of ethylene units in a hydrocarbon chain.

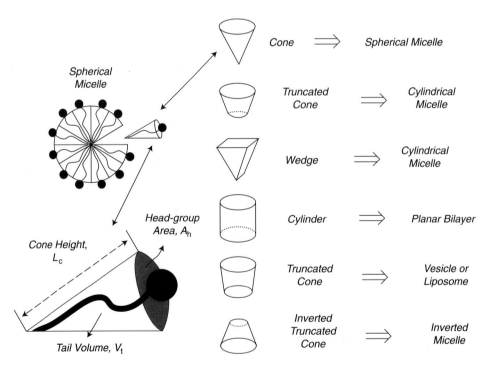

Figure 5.9 Dependence of the shape and structure of micelles on the molecular architecture of the basic surfactant units. The packing possibilities, based on simple geometrical features of the surfactants, build the final shape of the micelles.

If we have the estimated values of A_h, V_t and L_c, then it is possible to define a *surfactant packing parameter*

$$\lambda_p = \frac{V_t}{L_c A_h} \tag{470}$$

Equation (470) can be used to predict the shape of the formed micelles. The surfactant packing parameter, λ_p, relates the molecular geometry of the surfactant to the preferred curvature of the micelle formed. By geometry, we can calculate that spherical micelles can be formed if $\lambda_p < 0.33$. The increase in the λ_p value decreases the curvature of the micelles. We may test this with simple geometric shapes. For a cone, Equation (470) gives $\lambda_p = L_c A_h/3(L_c A_h) = 0.33$; for a wedge having a square or rectangular base, we have $\lambda_p = L_c A_h/2(L_c A_h) = 0.5$; for a cylinder, $\lambda_p = L_c A_h/(L_c A_h) = 1.0$, and for a truncated cone $\lambda_p > 0.33$. In practice, we mostly encounter spherical micelles. All the surfactants having single chains with large head-groups form spherical micelles. Most ionic surfactants also form spherical micelles at low salt concentrations, because the electrostatic repulsion in these conditions leads to large head-group areas, but if the salt concentration in the solution increases, the value of A_h decreases, and correspondingly λ_p increases to give micelles having different geometrical shapes. Cylindrical micelles, shown in Figure 5.7 c, are formed when the surfactant has a packing parameter of $\lambda_p \approx 0.50$, to give a truncated cone or a wedge per surfactant (see Figure 5.9 for these shapes). Once formed, the cylindrical micelles can grow to varying lengths by adding more surfactant molecules and thus they are generally polydisperse. Surfactants having single chains with small head-groups form cylindrical micelles. Many ionic surfactants form cylindrical micelles at high salt concentrations.

Micelles are used in many applications. Their largest industrial use is in emulsion polymerization, as detailed in Section 5.9 below. On the other hand, micelles made of ionic surfactants can trap hydrocarbon wastes in polluted water, since these hydrocarbon molecules prefer to be in the hydrocarbon interior of the micelle in an aqueous environment. In addition, ionic wastes dissolved in water adsorb onto the polar heads of these micelles. The resulting waste-filled micelles may be removed by simple ultrafiltration. As an example of another application, micelles can affect the rate of several chemical reactions and are used in micellar catalysis, similar to enzyme catalysis, in biochemistry. The rate of the chemical reaction increases with increasing micelle concentration, eventually leveling off. Nevertheless, micellar catalysts are less specific than enzymes.

5.8 Bilayers, Vesicles, Liposomes, Biological Cell Membranes and Inverted Micelles

5.8.1 Bilayers and vesicles

Bilayers and vesicles are non-spherical micelles. If the head-group and tail cross-sectional areas of a surfactant are nearly equal, giving an exactly cylindrical shape, and thus packing parameter $\lambda_p = 1.0$, then a perfect planar *bilayer micelle* is formed, having a tail-to-tail configuration as shown in Figure 5.10 a. If the surfactant concentration in the solution is high, a *lamellar phase* forms, consisting of stacks of roughly parallel planar bilayers. If the packing parameter is not exactly equal to 1.0 but $\lambda_p \approx 0.50-1.0$, then bilayer micelles are again

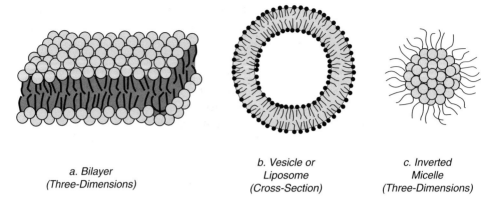

| a. Bilayer (Three-Dimensions) | b. Vesicle or Liposome (Cross-Section) | c. Inverted Micelle (Three-Dimensions) |

Figure 5.10   Schematic representation of bilayers, vesicles and inverted micelles.

formed, but this time they are not perfectly planar; the bilayer micelle has a curvature, the extent of which depends on the packing parameter. Bilayer micelles are generally 3–10 nm in thickness. Bilayer micelles are used in the design of supported membrane biosensors for medical and pharmaceutical applications.

Vesicles are a different kind of bilayer micelle. They are spherical or ellipsoidal micellar structures formed by enclosing a volume of the aqueous solution in a bilayer micelle, as seen in Figure 5.10 *b*. If the head–tail shape of a surfactant is a truncated cone, and thus the packing parameter is $\lambda_p > 0.33$ but less than 1.0, then there is an appreciable curvature on the micellar structure and a vesicle is formed from surfactants having a tail-to-tail configuration. The association of both synthetic and natural surfactants may create vesicles. Both double-tail and single-tail surfactants such as single-chain fatty acids can be used for their formation. When bilayer micelles are fragmented in water by any means, the parts wrap themselves into closed vesicles encapsulating some aqueous solution within during the process. Ultrasonic agitation is the most widely used technique to convert a large bilayer micelle into single-compartment vesicles of a small size. Vesicle diameters may range from 100 nm to 10 μm depending on the conditions of preparation. Electrostatic interactions between the ionic head-groups and the van de Waals interactions between the adjacent layers contribute to vesicle curvature and stability. Vesicles may contain more than one enclosed compartment with concentric bilayer surfaces enclosing smaller vesicles in larger aqueous compartments. Synthetic vesicles find use in the cosmetic, pharmaceutical, agrochemical etc. industries and are a subject of great research interest in controlled release and delivery fields.

5.8.2 Liposomes

When natural phospholipids are the surfactant, the formed vesicles are termed *liposomes*. They are made of fragmented phospholipid bilayers in aqueous solution, and closed liposome structures encapsulate some aqueous solution within. Lipids are natural surfactants having two hydrocarbon tails per molecule and they behave similarly to synthetic surfactants. The alkyl groups in lipid molecules are usually in the C_{16}–C_{24} size range and may be either saturated or unsaturated. Their double alkyl tails are too bulky to fit in a spherical micelle. The most important lipids in biology are the phospholipids having various polar head substituents, such as ethanolamine, glycerol etc. Typical examples are phosphatidyl ethanolamine and phosphatidyl choline. Similarly to synthetic surfactants, a phospholipid molecule consists of a polar, hydrophilic head that is connected to two hydrophobic, hydrocarbon tails. The phospholipid molecules can self-assemble in aqueous solution at appropriate concentrations to form bilayers. The thickness of the phospholipid bilayer can have a profound effect on its permeability. These phospholipid bilayers give closed liposome structures when fragmented. Liposomes shrink and swell osmotically as additives change the activity of water in the surrounding aqueous phase. Phospholipid molecules can flip-flop from one surface to another in the bilayer, determined experimentally by the use of isotopic labels. The presence of embedded proteins in the bilayer facilitates the flip-flop of the phospholipid. Liposomes carry many cargo molecules from one part to another in a living structure. Hydrophobic cargo molecules are carried inside the hydrophobic part of the bilayer, whereas the hydrophilic molecules are carried in the aqueous part of the interior. Some molecules are wholly or partly embedded in the bilayer structure, and some other molecules can be chemically bound to the exterior, or sometimes the interior, of the liposome surface. Liposomes are used in pharmaceuticals, medical technology and genetic engineering. Some drugs can be encapsulated in liposomes and can be delivered much more efficiently and specifically to the affected organs. This will diminish the side effects of these drugs because of the reduction in toxicity to the disease-free parts of the body.

5.8.3 *Biological cell membranes*

In general, synthetic surfactant micelles resemble many biological structures and are used as model systems to mimic living cell membranes and enzymes. For example, phospholipid bilayers form the membranes of biological cells and separate the interior of the living cell from the rest of the environment. Thus phospholipid bilayers are of fundamental importance in biology. The cell membrane consists of a number of functional units, which include many different protein molecules embedded in the membrane. The function of phospholipid bilayers is to avoid the diffusion of some ions (Na^+, K^+, Ca^{2+}, Cl^- etc.) and some polar, hydrogen-bonding molecules (nucleotides, sugars etc.) from one compartment to another. Since the interior of the bilayer membrane is hydrophobic, they do this task well. However, there is also a need for diffusion of nutrients and for some specific ions to be transported into the cell. These functions are performed by *channel protein* macromolecules. Proteins such as hemoglobin, serum albumin, gliadin, glycoproteins, lipoproteins and synthetic polypeptide polymers themselves contain complex hydrophobic and hydrophilic components, structured as helices, and provide pathways for these life-sustaining ions and polar molecules to move across the cell membrane. The inner surfaces of these channel protein helices are hydrophilic, and allow the transfer of some selected ions and polar molecules from one side of the bilayer to the other. (However, these pathways remain closed until an appropriate internal or external stimulus forces them open.) It has been determined that proteins are crystallized in two dimensions within phospolipid monolayers in film balance experiments. In order to understand the lipid–protein interactions, mixed monolayers of phospholipids with helical polypeptides have been studied via rheological and electrical measurements of these monolayers in a film balance. It was suggested that the location of the protein macromolecule in a phospholipid bilayer and its biological activity are very important for the functioning of a living cell. Proteins spread at the water–air interface and provide important enzymatic recognition, and cell membrane behavior. Nevertheless, phospholipid bilayers are mechanically weak, flexible and deformable, so that nature builds an outer cell wall over some cells to resist external forces, in addition to the phospholipid bilayer membrane.

5.8.4 Inverted micelles

We have previously focused on aqueous solution micelles, where hydrophilic parts are oriented to the water phase and hydrocarbon parts associated within the micelle. However, in a non-aqueous medium, surfactant molecules cluster with their polar heads together in the micellar core, and their hydrocarbon tails in the organic medium. Thus, the orientation of the surfactant is reversed and the associated cluster is known as *inverted micelles* (or *reversed micelles*). Water is stabilized in the core of these inverted micelles. Most inverted micelles are spherical, as shown in Figure 5.10 *c*. However, not all surfactants can form inverted micelles, and there are only a few surfactants, which are more oil soluble, that can be used for this purpose. Surfactants with very small head-groups, for example, cholesterol tend to form inverted micelles. This is in conjunction with the shape of an inverted truncated cone, shown in Figure 5.9, so that the head-groups are collected in the core of the micelle while the hydrophobic tails form the outer region. The size of the inverted micelles is much smaller than the micelles formed in aqueous medium. Approximately 10 surfactant molecules may give an inverted micelle, whereas a minimum of 30–50 surfactant molecules are required for the latter. The determination of the CMC for inverted micelles is not easy since the CMC point is blurred when these small micelles are present. Inverted liposomes are also formed in solvents such as cyclohexane, toluene and benzene.

   Alternatively, inverted micelles can also be formed in water-free medium. If no water is present in a two-component system, the difference between the solubility parameters of the hydrocarbon tail of the surfactant and the organic solvent contributes to inverted micelle formation. A large negative enthalpy change is the driving force to form spontaneous inverted micellization, in contrast with aqueous systems.

5.9   Use of Micelles in Emulsion Polymerization

Emulsion polymerization is used in the chemical industry to produce a milky fluid called *latex* which is used as the synthetic rubber raw material in paints, surface coatings, adhesives, paper, textile and leather treatment chemicals and in the manufacture of various other products. In-situ formed micelles are used in order to perform an emulsion polymerization process, carried out in an aqueous medium. There are four basic ingredients required for emulsion polymerization. They are (a) the *monomer*, a polymerizable organic material, (b) the water dispersion medium, (c) the emulsifier (surfactant) and (d) the *initiator*. When correct amounts of the ingredients are mixed together properly in a suitable container, using a mechanical stirrer, and within a certain temperature range, an emulsion of monomer droplets is initially formed in the continuous dispersion medium, which then gives a polymeric emulsion at the end of the process.

Organic monomers commonly used in the emulsion polymerization industry are styrene, acrylonitrile, butadiene, vinyl acetate, acrylic and methacrylic acids, and especially their organic esters, such as butyl and ethyl acrylates and methacrylates. It is quite common to produce copolymers using more than one monomeric substance. Most of these monomers are only slightly soluble in water. The surfactant (emulsifying agent) has a multiple role in emulsion polymerization. First of all, the emulsifier must form the micelles, which will solubilize the monomer; the initiation and early propagation of the polymer particles occurs inside these micelles, as shown in Figure 5.11. The surfactant must also stabilize the relatively immiscible monomer droplets which are initially formed in the water phase, and also the produced polymer particles as they grow and in the final product. Anionic surfactants, having long-chain carboxylates and alkylbenzene sulfonates, are generally used in conjunction with ethoxylate based non-ionic emulsifiers. One special advantage of non-ionics is that they are insensitive over a wide range to the pH of the aqueous solution. The added anionic and non-ionic surfactant molecules form micelles with a roughly spherical shape and consisting of 50–100 surfactant molecules clumped together. An average micelle will have a diameter of approximately 4–5 nm. At the surfactant concentrations normally employed in emulsion polymerization, the solution will contain

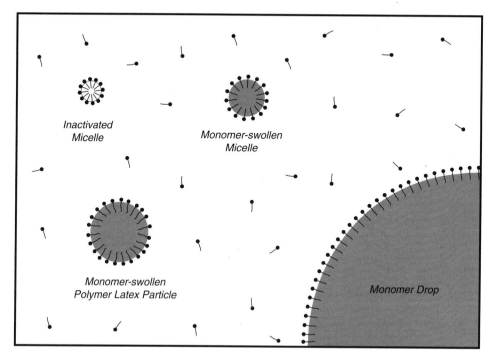

Figure 5.11 Schematic representation of the drops and aggregates present in the water phase in the emulsion polymerization of water-insoluble (approximately) monomers.

approximately 10^{18} micelles per milliliter. Mechanical agitation breaks the immiscible hydrophobic monomer phase into droplets formed in water and these droplets are held in suspension by applying both the agitation action and the surfactant stabilization. The hydrophobic portions of the surfactant molecules are adsorbed onto these monomer droplets, which are about 1μm in diameter. (This means that an average droplet is about 200–250 times as large as an average micelle in the same medium, as shown in Figure 5.11.) The hydrophilic portion of the emulsifier molecules on these monomer droplets remains in the water phase. After the in-situ micelle formation is established, a water-soluble initiator (usually a persulphate) is added while the system is mechanically stirred. The initiator causes the monomer molecules to polymerize by breaking down into charged radicals at a specific temperature (50–70°C for ammonium and potassium persulphates). These initiator radicals in turn react with the monomer molecules to form new monomeric radicals. These newly formed monomeric radicals combine with the monomer molecules present in the solution to start the addition polymerization process. The hydrophobic region of the micelles attracts the organic monomer molecules and the majority of the monomer molecules dissolved in the water phase are located within the micelles. As more monomer molecules dissolve in water, they continue to diffuse into the micellar cores. Most of the polymerization is initiated within the micelles so that a micelle acts as a meeting place for the oil-soluble monomer and the water-soluble initiator. Approximately 10^{13} free radicals per milliliter per second may be produced from a typical persulphate initiator at 50°C, and within a very short time, these radicals will meet the *monomer-swollen micelles* in which the polymerization of a single macromolecule then starts. The next step of polymerization is *propagation* which is the growing of the polymeric chains by the addition of monomeric units present in the micelle. The polymer chains grow in micelles until the process is terminated by reaction with another radical. Thus a polymer chain is either propagating or terminating in micelles at any time; therefore statistically half the micelles contain growing chains under stationary state conditions. As polymerization proceeds, the micelles grow in size through the addition of monomer from the aqueous solution phase via the monomer droplets. As the monomer molecules in the micelles combine with one another to build polymer chains, more will migrate (or diffuse) from the monomer drop through the water to the micelles, as shown in Figure 5.11. After initiation, polymerization inside the micelles proceeds very rapidly and the micelles grow from a tiny group of emulsifier and monomer molecules to larger groups of polymer molecules, held in emulsion by the action of the emulsifier molecules located on the exterior surface of the particle. The active micelles have now grown much larger than the original micelles, depending on the particular polymerization system, 2–15% conversion. They are no longer considered to be *micelles* but *monomer-swollen polymer latex particles*, as shown in Figure 5.11.

The behavior of the surfactant molecules in an emulsion polymerization is complex. The adsorption of the surfactant on the rapidly and continually growing surface of the monomer-swollen latex particles reduces their concentration in the aqueous phase, and also upsets the balance in equilibrium between the dissolved surfactant and the surfactant present in the *inactivated micelles* (those in which polymerization is not occurring), as shown in Figure 5.11. The point is quickly reached at which the surfactant concentration in the solution falls below its critical micelle concentration, CMC. When this occurs, the inactive micelles become unstable and disintegrate to restore the balance. In time all of the micelles disappear and the monomer droplets shrink in size. After a conversion of 10–20%

of monomer to polymer has been reached, no micelles are left and essentially all of the surfactant molecules in the system have been adsorbed by the polymer particles. It is significant that at this point the surface tension of the aqueous phase increases, because surfactant-free water has a high surface tension of $72.8\,\text{mN m}^{-1}$. The monomer droplets are no longer stable from now on and will coalesce if the mechanical agitation is stopped. Later, the monomeric droplets gradually decrease in quantity as polymerization proceeds, and the size of the polymeric particles increases. Finally monomeric droplets disappear completely at 50–80% conversion and the polymer particles contain all of the unreacted monomer, and essentially all of the surfactant molecules are also attached onto the surface of these polymer particles. After this stage no fresh latex particles can be formed since initiation of the polymerization reaction can take place only in a monomer-swollen micelle. The number of latex particles is thus fixed at this point (approximately 10^{15} particles ml^{-1}) and further polymerization occurs only inside these latex particles. Then, the monomer in the monomer-swollen latex particles is gradually used up and the polymerization rate will gradually decrease. Polymerization will cease completely when all the monomer in the particles is consumed, at nearly 100% conversion. The final polymer particles have diameters of the order of 50–200 nm and are intermediate in size between the initial micelles and monomer droplets.

There are several reasons for using emulsion polymerization extensively in the chemical industry. In this process, both the rate of polymerization and the average molecular weight of the polymer depend on the surfactant concentration, via the concentration of the micelles formed, and the production of a high-molecular-weight polymer at a high rate of polymerization is possible. The molecular weight and degree of polymerization can easily be controlled in emulsion polymerization, so that a product having specific and reproducible properties can be obtained. This is important in the chemical industry because there is an inverse relationship between the polymerization rate and the polymer molecular weight in other polymerization processes (bulk, solution, suspension polymerization), so that large increases in molecular weight can only be produced by decreasing the polymerization rate by lowering the initiator concentration or lowering the reaction temperature. Thus, emulsion polymerization is a unique process that affords a means of increasing the polymer molecular weight without decreasing the polymerization rate. In addition, emulsion polymerization offers several other advantages such as the possibility of interrupting the polymerization at any stage for the addition of other materials needed to modify the properties of the finished polymer; better temperature control during polymerization due to more rapid heat transfer in the low viscosity emulsion; better control of the range and distribution of particle size; the versatility of carrying out many copolymerizations, which are difficult to control in bulk or any other polymerization method; and the tendency to restrict the coalescence of polymer particles due to the presence of the adsorbed surfactant, which acts as a protective colloid and thereby prevents the formation of sticky and rubbery products.

5.10 Coating Mono- and Multilayers on Solid Substrates: Langmuir–Blodgett Method

A Langmuir film balance can also be used to build up highly organized multilayers of surfactant films on solid substrates. This is accomplished by successively dipping a substrate vertically up and down through the monolayer formed on the subphase, while simultaneously keeping the surface pressure constant using a computer-controlled feedback system between the Wilhelmy electrobalance measuring the surface pressure, and the mechanical barrier moving mechanism. As a result, the floating monolayer is adsorbed onto the solid substrate. In this way, high-quality and ordered multilayer structures of several (sometimes hundreds of) layers can be produced on solid substrates with high dielectric strengths.

Historically, Irving Langmuir was the first to report the transfer of a fatty acid monolayer from a water surface onto a solid support, in 1920. However, his assistant Katherine Blodgett gave the first detailed description of sequential monolayer transfer several years later, in 1934. These built-up mono- and multilayer assemblies are therefore commonly called *Langmuir–Blodgett* or simply *LB* coatings. After the pioneering work done by Langmuir and Blodgett, it took almost half a century before scientists all around the world started to realize the opportunities presented by this unique technique. The first international conference on LB films was held in 1979, and since then the use of the LB technique has been increasing widely among scientists working in different fields of research. The applications of LB films extend from electronic devices where these thin organic films are used as insulators, to solar energy production systems, and also to biomembrane research in biology.

In practice, an organic thin film can be deposited on a solid substrate by various other techniques such as vapor deposition, sputtering, electrodeposition, molecular beam epitaxy, adsorption from solution, solvent casting, etc. Nevertheless, the LB technique is one of the most promising methods for preparing such thin films as it enables (a) the precise control of the monolayer thickness, (b) homogeneous deposition of the monolayer over large areas and (c) the possibility to make multilayer structures with varying layer composition. The parameters that affect the properties of an LB film are the type and nature of the solid substrate, the spread monolayer, the subphase composition and applied temperature, the surface pressure and the vertical dipping or withdrawal speed during film deposition, and the time the solid substrate is stored in air or in the subphase between deposition cycles. The surfactants that are used in the LB technique should have been carefully selected to have balanced hydrophilic and hydrophobic parts. Metallic salts of these film materials can also be formed on the substrates by previously introducing metallic ions into the water subphase.

5.10.1 Monolayer film transfer to solids

There are several possibilities for coating monolayers onto solids depending on the hydrophilic or hydrophobic nature of the substrate, the type of surfactant and the subphase (which is water in most LB experiments), the initial location of the substrate (inside or outside the subphase) and the direction of the dipper (downstroke or upstroke). Most of the possible LB coating procedures are outlined schematically in Figures 5.12–5.17. In Langmuir's original experiment, a very clean *hydrophilic* substrate is dipped into the water before the monolayer film is spread. Next, the monolayer material is dissolved in a suitable volatile and non-reactive solvent (such as chloroform, toluene, cyclohexane, *n*-hexane, petroleum ether) in concentrations that fit $0.1–1.0\,kg\,m^{-2}$ of solvent on the trough surface and spread over the water subphase. After complete evaporation of the volatile solvent, the barriers are operated to compress the monolayer. Next, the water-wettable, hydrophilic substrate is slowly withdrawn upwards across the water surface and the monolayer is transferred onto it while a constant surface pressure is maintained on the monolayer. This deposition procedure is shown schematically in Figure 5.12 *a*. The monolayer can be transferred quantitatively onto a water-wettable substrate when sufficient lateral pressure is applied on the monolayer, within which the molecules are close to each other. As can be seen in Figure 5.12 *a*, the hydrophilic head-groups (filled dots) attach to the hydrophilic solid and the meniscus is curved upwards, in the same direction as the motion of the substrate. This is important, because the rule for deposition to take place is that the meniscus must curve in the same direction as the motion of the substrate. This means that no deposition will take place if a hydrophilic substrate such as SiO_2, Al_2O_3, MgO or a hydrophilic polymer is immersed into a monolayer-coated water phases, as can be seen in Figure 5.12 *b*. This is because the meniscus is curved upwards due to the capillary forces arising from the strong interaction of water molecules with the hydrophilic solid. After an LB monolayer is coated onto a hydrophilic substrate such as silicon, with a high surface energy of about $50\,mN\,m^{-1}$, this then renders it as a hydrophobic surface with a relatively low surface energy, in the range of 20–$30\,mN\,m^{-1}$. However, a very uniform LB coating is difficult to obtain and many defects may also be present, such as pinholes and island regions due to the collapsed forms on the coated LB monolayer.

Instead of a hydrophilic solid, if we immerse a *hydrophobic* substrate, such as some pure metals, Au, Ag, Ge, or some plastics, polypropylene, polytetrafluoroethylene, silanized SiO_2 etc. into a monolayer-coated water subphase, as shown schematically in Figure 5.13 *a*, then the LB coating process is different. Since both meniscus curve and substrate movement are in the same direction (downwards) during this process, a monolayer can be quantitatively

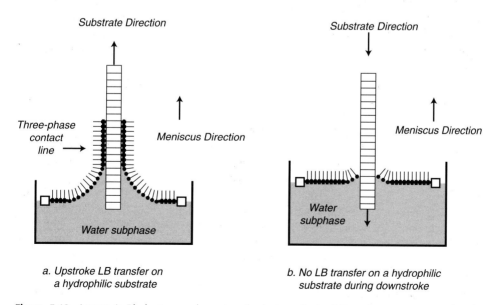

a. Upstroke LB transfer on
a hydrophilic substrate

b. No LB transfer on a hydrophilic
substrate during downstroke

Figure 5.12 Langmuir–Blodgett monolayer transfer onto a hydrophilic substrate: *a.* Upstroke. *b.* Downstroke.

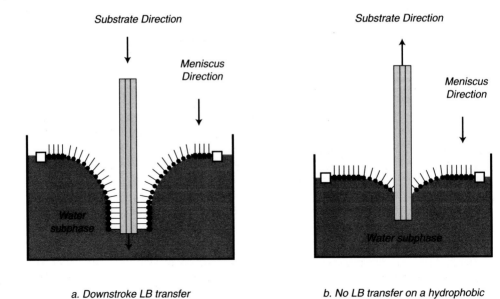

a. Downstroke LB transfer
on a hydrophobic substrate

b. No LB transfer on a hydrophobic
substrate during upstroke

Figure 5.13 Langmuir–Blodgett monolayer transfer onto a hydrophobic substrate: *a.* Downstroke. *b.* Upstroke.

transferred onto the substrate where hydrophobic tail-groups (lines) attach onto the hydrophobic solid. In this case, the meniscus is downwards because the capillary forces due to the water–hydrophobic surface interactions are weak. The cohesion forces between water molecules are stronger than the adhesion forces between the water and the hydrophobic substrate, and water drops have a contact angle of greater than 90° on such non-wettable and flat hydrophobic surfaces. On the other hand, no LB deposition will take place if a hydrophobic substrate is previously located in the water subphase and withdrawn upwards after the water surface is coated with a monolayer, as shown in Figure 5.13 *b*, since the meniscus and substrate directions are reversed.

In the practice of LB coating, the monolayer should be continuously compressed as it is transferred onto the substrate, to maintain the original coating uniformity. The barrier compression must be carried out very sensitively to avoid the film collapsing on the subphase, and the whole system should be vibration free during this process. The LB deposition is generally carried out in the *solid, (S)* or *liquid-condensed state (L-Con.)* phases of insoluble films on the subphase, as shown in Figure 5.6 and explained in Section 5.6.3. In the *S* or *L-Con.* phases, the surface pressure should be high enough to ensure sufficient cohesion in the monolayer, e.g. the attraction between the molecules in the monolayer must be sufficiently high so that the monolayer does not fall apart during transfer onto the solid substrate. This also ensures the build up of the homogeneous multilayers. The surface pressure value that gives the best coating results depends on the nature of the monolayer and is usually established empirically; generally a surface pressure of 20–40 mN m^{-1} is applied. However, surfactants can seldom be successfully deposited when π is lower than

$5\,\mathrm{mN\,m^{-1}}$, or at surface pressures above $40\,\mathrm{mN\,m^{-1}}$, where the rigidity and collapse of the film often creates problems. When monolayers are deposited onto the substrate at low surface pressure $\pi = 5\text{--}10\,\mathrm{mN\,m^{-1}}$, the amount of unwanted water lifted with the LB film is increased. In general, it is often observed that the substrate is visibly wet immediately after the transfer. The incorporation of water from the subphase into the coated organic LB film is an important problem with the LB method. It generally gives a pick-up of water between 5 and 10% of the total weight of the coated LB film. This water layer may be removed by drainage or evaporation in the open air or in an oven. However, in some instances monolayers may appear dry depending on the surfactant and substrate types. In usual LB practice, wet immersion of the coated substrate is preferred, but sometimes drying of the coated substrate is required after forming every single monolayer.

5.10.2 Multilayer film transfer to solids

Different kinds of LB multilayer can be produced by successive deposition of monolayers on the same substrate. For this purpose, the dipper must move up and down many times through the monolayer formed on the subphase in a Langmuir trough, while the surface pressure is kept constant throughout, by applying a controlled barrier compression. There are generally three types of built-up multilayer films, named X, Y and Z-types. The most common is the Y-type multilayer, which is produced when the monolayer deposits onto the solid substrate in successive upstroke and downstroke movements of the dipper holding the substrate. The Y-type multilayers can be formed on both hydrophilic (as seen in Figure 5.14) and hydrophobic (as seen in Figure 5.15) substrates, and the monolayers have both

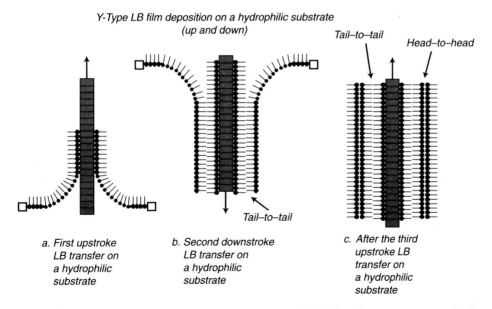

Figure 5.14 Y-type Langmuir–Blodgett film deposition onto a hydrophilic substrate: **a.** First upstroke. **b.** Second downstroke. **c.** After the third upstroke.

tail-to-tail and *head-to-head* configurations of the contained surfactant molecules, but no *head-to-tail* or *tail-to-head* configurations.

In Figure 5.14 *a*, we see that a monolayer is coated on a hydrophilic substrate by withdrawal from the subphase in the vertical direction in the first process, and this monolayer-coated hydrophilic substrate is immersed downwards into the same monolayer on the water phase in the second process, as shown in Figure 5.14 *b*. Since both the meniscus curve and the substrate movement are in the same direction, a second LB coating is then possible, but only in the *tail-to-tail* form, where the alkyl chains are oriented towards the substrate due to the strong interactions between the hydrophobic tail-groups at the outermost surface of the monolayer-coated substrate, and the tail-groups of the surfactant molecules. If we continue this process by withdrawing the bilayer-coated substrate upstroke, a third monolayer can also be coated, as shown in Figure 5.14 *c*, but this time giving a *head-to-head* pattern, due to the strong interactions between the hydrophilic head-groups on the outermost substrate surface, and the hydrophilic head-groups of the monolayer surfactants which exclude the tail-groups so that these hydrocarbon chains are exposed to the air. In summary, a monolayer is deposited on both sides of the substrate on each traverse of the monolayer–air interface. We may continue this process until the desired thickness of multilayer LB film is achieved, having successive tail-to-tail and head-to-head patterns.

In general, the quality of a multilayer LB film, that is, the ordering perpendicular to the substrate surface, depends on the quality of the first monolayer transfer. The subphase also plays a very important role in determining the quality of the deposited film. Ultrapure water is generally used because of its exceptionally high surface tension value. The temperature and pH of the subphase also affects the quality of the LB film, and in most papers researchers have used a water of pH = 7, at room temperature. The adhesion of the first LB layer to the underlying substrate is particularly important, and this may also be called *heterogeneous crystal growth*, since the first monomolecular layer is transferred onto the surface of a different material. However, the deposition will be *homogeneous* to transfer onto the existing first monolayer, for subsequent monolayers.

In Figure 5.15 *a*, we see the reverse process, so that initially a monolayer is coated on a hydrophobic substrate by immersing into the subphase downwards, where the hydrophilic head-groups are now on the uppermost surface of the substrate. If we withdraw this monolayer-coated substrate from the subphase in a second upstroke process, as shown in Figure 5.15 *b*, a second monolayer is coated onto the first coated LB film giving a *head-to-head* pattern, due to the strong interactions between the hydrophilic head-groups, which exclude the hydrophobic tail-groups of the surfactant molecules in the monolayer. Similarly to the above process, if we continue by immersing the bilayer-coated substrate on the downstroke, a third monolayer can be coated, as shown in Figure 5.15 *c*, giving a tail-to-tail pattern due to the strong interactions between the hydrophobic tail-groups present on the outermost surface, which excludes the head-groups. By repeating these transfer cycles, we can achieve the desired thickness of multilayer LB film, having successive head-to-head and tail-to-tail patterns. For fatty acid salts such as calcium stearate deposited onto metallic substrates such as an aluminum plate, there is an ion exchange between the fatty acid salt and the thin aluminum oxide layer forming a new layer of aluminum stearate. The strong chemical bond anchors the polar head of the first LB monolayer to the metallic substrate surface, and in most cases, this monolayer remains on the surface if any mechanical force is applied. However, the outer layers in a multilayer LB film can be wiped off with a tissue (or by any

Y-Type LB film deposition on a hydrophobic substrate
(down and up)

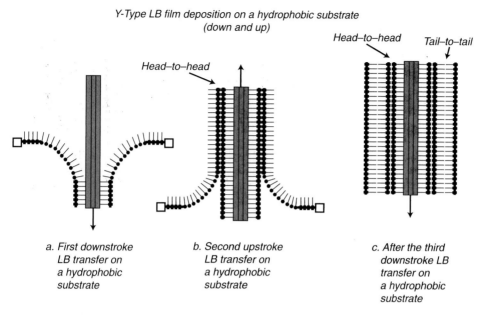

a. First downstroke
LB transfer on
a hydrophobic
substrate

b. Second upstroke
LB transfer on
a hydrophobic
substrate

c. After the third
downstroke LB
transfer on
a hydrophobic
substrate

Figure 5.15 Y-type Langmuir–Blodgett film deposition onto a hydrophobic substrate: *a.* First downstroke. *b.* Second upstroke. *c.* After the third downstroke.

other mechanical means) to remove all the long-chain fatty acid layers, because they are only held by relatively weak van der Waals forces.

In Figure 5.16, we see schematically *X*-type LB multilayer deposition on a hydrophobic substrate by only dipping it downwards into the monolayer coated subphase. Only a *head-to-tail* configuration can be obtained by this process, as seen in Figure 5.16 *c*. If we compare Figures 5.15 *a* and 5.16 *a*, we can see no differences and the question arises: how can we apply a second downstroke immersion such as that in Figure 5.16 *b*? The answer lies in a Langmuir trough containing two wells, where the LB monolayer coated substrate in Figure 5.16 *a* is not removed from its initial well but is removed from the second well, which is free of any monolayer. This can be accomplished by mechanically passing the coated substrate under the same subphase from one well to the other, in this special trough. Then, this initially hydrophobic substrate now having hydrophilic head-groups on its outermost surface can again be immersed into the monolayer floating first well, to coat a second monolayer in a *head-to-tail* configuration. We can continue this process until we obtain the desired thickness of multilayer LB film, having successive head-to-tail patterns. This X-type deposition can also be modified with changes in the dipping conditions and solutions used; for example, we can change the pH of the subphase if desired.

The *Z*-type LB multilayer deposition on a hydrophilic substrate is shown schematically in Figure 5.17. We can only withdraw the substrate upwards from the monolayer-coated subphase. Only a *tail-to-head* configuration can be obtained using this process, as seen in Figure 5.17 *c*. Again we need a two-well Langmuir trough to apply this process, where we immerse the monolayer-coated substrate from a monolayer-free water surface and withdraw it from the neighboring monolayer-coated well. Aromatic-based molecules containing short and

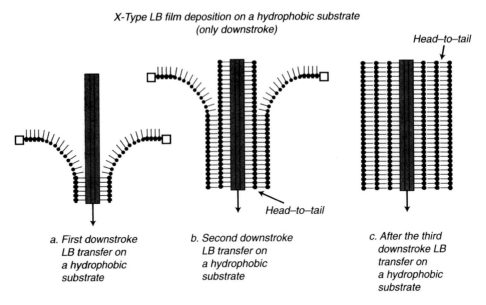

Figure 5.16 X-type Langmuir–Blodgett film deposition onto a hydrophobic substrate: *a.* First downstroke. *b.* Second downstroke. *c.* After the third downstroke.

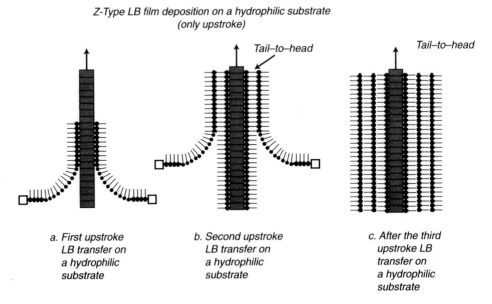

Figure 5.17 Z-type Langmuir–Blodgett film deposition on a hydrophilic substrate: *a.* First upstroke. *b.* Second upstroke. *c.* After the third upstroke.

sometimes no hydrocarbon chains, such as phthalocyanines and fullerenes give Z-type LB coatings; they are different from conventional fatty-acid-based LB films.

It is also possible to coat LB multilayers containing more than one type of monomolecular layer. So-called *alternating layers* may be produced, which consist of two different kinds of surfactant, by using highly sophisticated instrumentation. In such an instrument there is a trough with two separate compartments both possessing a well and a floating monolayer of a different surfactant. These monolayers can then be deposited on one solid substrate in an alternating mode.

5.10.3 Properties of LB films

The quantity and the quality of the deposited monolayer on a solid substrate is measured by a so-called *transfer ratio* or *deposition ratio*, τ,

$$\tau = \frac{A_L}{A_S} \tag{471}$$

where A_L is the area decrease in the monolayer, and A_S is the coated area of the solid substrate during a deposition stroke. For an ideal transfer, τ is equal to 1. This suggests that the LB film bridges over the surface roughness of the substrate at the moment of deposition. However, some monolayers may also be deposited on grooved surfaces. Transfer ratios appreciably outside the range 0.95–1.05 show that coating homogeneity is not good. If the monolayer is not coherent and the surface pressure is low, the transfer ratio is less than one. If the chain–chain attractions are strong and the monolayer is continuous, the film is transferred like a carpet to the substrate. However if the monolayer is extremely rigid it is very difficult to transfer by the LB method, such as Cu^{2+}-stearate monolayers. Coating homogeneity cannot be inspected easily using the naked eye because of the coloring of most LB films. The thickness of LB films can be of the order of the wavelength of visible light (420–700 nm) and LB coatings may appear colored or dark due to the constructive and destructive interference of light waves, which are reflected from the film–air and film–substrate surfaces.

After deposition of an LB film on a substrate, the characterization of mono- and multilayers can be carried out using various methods, such as ellipsometry, electron and Brewster angle microscopy, FTIR, fluorescence spectroscopy, X-ray diffraction, X-ray reflectometry, XPS etc. The thickness of the LB multilayers can be measured by ellipsometry: the behavior of light reflected by an interface allows determination of the thickness of the present inhomogeneous region via its index of refraction and its absorption coefficient. A monochromatic light source (e.g. a laser) sends a beam to the LB-coated multilayer and the reflected beam will be elliptically polarized, the angle of polarization is then determined. This leads to computation of the thickness of the LB film. The reproducibility of LB films can be monitored by measuring a suitable physical characteristic of the organic film, such as film thickness and optical density as a function of time. The *quartz crystal microbalance* (QCM) technique, which is sensitive to nanogram weight increases on a quartz substrate, by checking the frequency shift, is also used to determine actual transfer ratios in the LB method. Alternatively, an LB multilayer-coated substrate can be sandwiched between two metal electrodes and the capacitance measured as a function of the number of dipping cycles. The capacitance varies with the reciprocal of the thickness. Unfortunately, many results for LB deposition are not reproducible because of the factors that cannot be held constant during the coating process.

LB coatings may be modified, for example, when an LB-coated substrate is soaked in a suitable solvent such as acetone, alcohol or benzene, and the multilayer is *skeletonized* due to the dissolving out of the free fatty acid, reducing its actual thickness slightly, but decreasing the refractive index appreciably. This property is used to control the refractive index to produce antireflection coatings for glass. The holes in these skeletonized films can also be filled with other materials in vapor or liquid form.

An alternative way to deposit the monolayer is by the Langmuir–Schaeffer (LS) technique. This technique differs from the vertical technique described above only in the sense that the solid substrate is lowered in a nearly horizontally position, until it is in contact with the monolayer on the water surface. After the substrate has made contact with the monolayer, the rest of the monolayer is cleaned away and the substrate plate is lifted up with a monolayer which is coated on only one side of the substrate. This method is useful for the transfer of highly rigid monolayers. On the other hand, Sagiv and Netzer, in 1983, invented a method of preparing chemically bound multilayers, which is based on the successive adsorption and reaction of suitable organic molecules. Initially these molecules adsorb onto the substrate to form a monolayer as for the normal LB technique. Then, the head-group reacts with the topmost groups on the coating to give a permanent chemical attachment, and each subsequent layer is chemically attached to the previous monolayer. This is a promising method to create versatile structures in nanotechnology.

References

1. Adam, N.K. (1968) *The Physics and Chemistry of Surfaces*. Dover, New York.
2. Aveyard, R. and Haydon, D.A. (1973) *An Introduction to the Principles of Surface Chemistry*. Cambridge University Press, Cambridge.
3. Lyklema, L. (1991) *Fundamentals of Interface and Colloid Science* (vols. I and II). Academic Press, London.
4. Adamson, A.W. and Gast, A.P. (1997) *Physical Chemistry of Surfaces* (6th edn). Wiley, New York.
5. Scatchard, G. (1976) *Equilibrium in Solutions & Surface and Colloid Chemistry*. Harvard University Press, Cambridge.
6. Erbil, H.Y. (1997) Interfacial Interactions of Liquids. In Birdi, K.S. (ed.). *Handbook of Surface and Colloid Chemistry*. CRC Press, Boca Raton.
7. Erbil, H.Y. (2000) *Vinyl Acetate Emulsion Polymerization and Copolymerization with Acrylic Monomers*. CRC Press, Boca Raton.
8. Hiemenz, P.C. and Rajagopalan, R. (1997) *Principles of Colloid and Surface Chemistry* (3rd edn). Marcel Dekker, New York.
9. Butt, H.J., Graf, K. and Kappl, M. (2003) *Physics and Chemistry of Interfaces*. Wiley-VCH, Weinheim.
10. Murrell, J.N. and Jenkins, A.D. (1994) *Properties of Liquids and Solutions* (2nd edn). Wiley, Chichester.
11. Petty, M.C. (1996) *Langmuir–Blodgett Films*. Cambridge University Press, Cambridge.
12. George, J. (1992) *Preparation of Thin Films*. Marcel Dekker, New York.

Chapter 6
Experimental Determination of Surface Tension at Pure Liquid and Solution Surfaces/Interfaces

It is difficult to classify liquid surface tension measurement methods because there are static, dynamic, rapid, slow, old and new methods. It is best to describe these methods in separate sections with their relation to surface tension theory (especially with the Young–Laplace equation given in Section 4.3), and to discuss their use in special conditions. Equipment based on the *ring* and *Wilhelmy plate* detachment methods is commonly used in most modern laboratories working in the surface science field, since these can easily be applied to surface tension measurement of pure liquids and also solutions at the liquid–air surface, and to the interfacial tension between two immiscible liquids at the interface. *Capillary rise, drop volume* and *maximum bubble pressure* are also conventional surface tension measurement methods, and several sophisticated instruments have been developed for their application. The *optical drop shape determination and video image digitization* method compares the optically determined contour of a pendant or sessile drop image with the theoretical contour, which can be predicted by the Young–Laplace equation. When both profiles fit each other by applying a numerical method using computers, then surface tension can be obtained from the classical Young–Laplace equation. On the other hand, *dynamic methods* such as the *oscillating jet* and *capillary wave* methods are applied to systems that are not in equilibrium with the intention to study the changes in surface tension and relaxation effects on a very small time scale.

When compared with the other methods, the *capillary rise* method is the ultimate standard method in terms of the degree of theoretical exactitude, and, although it is the oldest method, it still gives the most precise liquid surface tension results if carefully applied, and when the time of measurement is allowed to be sufficiently long. However, with the improvement in computer-controlled electronic equipment, other methods now also have a very high precision. Some of the surface tension results are summarized in Table 6.1, and the interfacial tension between pure liquids in Table 6.2.

6.1 Liquid Surface Tension from the Capillary Rise Method

As discussed in Section 4.4, when a solid capillary tube is inserted into a liquid, the liquid is generally raised (or rarely depressed) in this tube. In the capillary rise method, the height of a liquid column in a capillary tube above the level of the reference liquid contained in a large dish is measured. The container must be sufficiently large so that the reference liquid

Table 6.1 Surface tension values at the liquid–air (vapor) interface (Values compiled from standard references especially from David R. Lide (ed.) (2003) *CRC Handbook of Chemistry and Physics* (83rd ed.) CRC Press, Boca Raton; Jasper, J. J. (1972) *J. Phys. Chem. Ref. Data*, **1**, 841; Korosi, G. and Kovats, E. J. (1981) *Chem. Eng. Data*, **26**, 323

Liquid	Surface tension $(mN\,m^{-1})$	Temperature (°C)
n-Hexane	18.40	20
Cyclohexane	25.50	20
n-Heptane	20.14	20
n-Octane	21.00	20
n-Nonane	22.85	20
n-Decane	23.83	20
Dodecane	25.44	20
Benzene	28.88	20
Toluene	28.52	20
n-Propylbenzene	28.98	20
Bromobenzene	35.75	25
Nitrobenzene	41.71	40
Aniline	42.67	20
Dimethyl aniline	36.56	20
Chloroform	27.14	25
Carbon tetrachloride	26.90	25
Diiodomethane	50.80	20
Perfluoropentane	9.89	20
Perfluoroheptane	13.19	20
Perfluoromethycyclohexane	15.75	20
Ethyl ether	17.01	25
Butyl acetate	25.09	20
Ethyl acetate	23.39	25
Acetic acid	27.10	25
Propionic acid	26.69	20
Butyric acid	26.51	20
Methanol	22.51	25
Ethanol	21.82	25
n-Propanol	23.58	25
Iso-Propanol	21.22	25
n-Butanol	24.93	20
Sec-Butanol	22.54	20
n-Pentanol	25.36	25
n-Hexanol	25.81	25
Cyclohexanol	32.92	25
n-Octanol	27.10	25
Ethylene glycol	47.99	25
Diethylene glycol	30.90	20
Glycerine	63.40	20
Water	72.80	20
Formamide	57.02	25
Dimethyl sulfoxide	43.54	20
Polydimethylsiloxane (Mw = 3900)	20.47	20
Carbon disulfide	32.32	20
Mercury	486.50	20

Table 6.2 Interfacial tension values at the $liquid_1$–$liquid_2$ interface (Values compiled from Davies, J. T. and Rideal, E. K. (1963) *Interfacial Phenomena* (2nd ed.). Academic Press, New York; Donahue, D. J. and Bartell, F. E. (1952) *J. Phys. Chem.*, **56**, 480; Girifalco, L. A. and Good, R. J. (1957) *J. Phys. Chem.*, **61**, 904; Ivosevic, N, Zutic, V, and Tomaic, J. (1999) *Langmuir* **15**, 7063

$Liquid_2$	Interfacial tension $(mN\,m^{-1})$	Temperature $(°C)$
$Liquid_1$: Water		
n-Hexane	51.0	20
Cyclohexane	51.0	20
n-Heptane	50.2	20
n-Octane	50.8	20
n-Decane	51.2	20
Hexadecane	53.3	20
Benzene	35.0	20
Toluene	36.1	25
n-Propylbenzene	39.1	25
n-Butylbenzene	40.6	25
Bromobenzene	38.1	25
Nitrobenzene	26.0	20
Aniline	5.85	20
Chloroform	28.0	25
Carbon tetrachloride	45.0	20
n-Butanol	1.8	20
Iso-Butanol	2.1	20
n-Pentanol	4.4	25
n-Hexanol	6.8	25
Cyclohexanol	4.0	20
n-Octanol	8.5	20
Ethyl acetate	6.8	20
Heptanoic acid	7.0	20
Ethyl ether	10.7	20
Carbon disulfide	48.0	20
Methylene iodide	45.9	20
$Liquid_1$: Diethylene glycol		
n-Heptane	10.6	20
n-Decane	11.6	20
$Liquid_1$: Mercury		
Water	426	20
Ethanol	389	20
n-Hexane	378	20
n-Heptane	379	20
n-Octane	375	20
n-Decane	372	20
Hexadecane	365	20
Benzene	363	20
Toluene	359	20

has a well-defined horizontal surface. A glass capillary is most commonly used for this purpose because it is completely wettable by most liquids as a result of the high surface free energy of glass and, in addition, it is transparent so that the height of the rising liquid can easily be determined. In physical terms, the capillary rise of a liquid in the tube is a consequence of the adhesion interactions between the liquid and the capillary wall, which are stronger than the cohesion interactions within the liquid. The rise in the liquid is restricted by the opposing hydrostatic (gravity) forces, approaching equilibrium between the capillary and hydrostatic pressures within the column at a definite height of the liquid. Thus, the upward flow of liquid continues until the hydrostatic pressure just balances the Young–Laplace pressure difference, ΔP. The attractive forces between the wall of the capillary tube and the liquid are exerted only along the edge at which the upper surface of the liquid meets the tube, so that only the radius at the upper surface is important and the capillary radius can be wider or narrower lower down.

It has been observed experimentally that there is an inverse proportionality between the height of the liquid present in the capillary tube and the tube radius. As given in Equation (326), the height of the rising liquid, h, is directly proportional to the surface tension of the liquid, γ, but inversely proportional to the radius of the capillary tube, r, and also to the density of the liquid, ρ. So we can write, $h \propto 2\gamma/r\rho$. On the other hand, the liquid surface in the capillary tube usually takes the form of a concave spherical cap, as shown in Figure 4.7. Indeed, the rise of a liquid in a capillary tube is simply the automatic recording of the Young–Laplace pressure difference, ΔP, across the meniscus of the liquid in the tube. The general Young–Laplace equation [$\Delta P = \gamma(1/R_1 + 1/R_2)$] (Equation (325)), which is given in Section 4.3.3, relates the pressure difference to the liquid surface tension and the mean radius of curvature. For the simplest ideal case, if the capillary tube is circular in cross section and not too large in radius ($r < 0.1$ mm), where the contact angle of the liquid on the capillary wall, $\theta = 0°$, so that the liquid completely wets the capillary wall, then the meniscus will be completely hemispherical. For this case, $r = R_1 = R_2$ from simple geometry, as shown in Figure 4.7, and the Young–Laplace equation gives $\Delta P = 2\gamma/r = \Delta \rho g h$ (Equations (327) and (328)) where $\Delta \rho$ denotes the difference in density between the liquid and gas (or vapor) phase, and g is the acceleration of gravity. Then, the *capillary constant* (or *specific cohesion constant*), a_o^2, is defined for $\theta = 0°$ conditions, as the product of the capillary radius and the liquid height [$a_o^2 = rh = 2\gamma/\Delta \rho g$] (Equation (329)) with the units of area (m^2).

Nevertheless, Equation (329) cannot be used for some cases where the liquid meets the circularly cylindrical capillary wall at a stable contact angle, θ, between the liquid and the wall, even for capillary tubes having a small radius, as shown in Figure 4.7. The value of the contact angle depends on the relative magnitudes of the adhesive forces between the liquid and the capillary wall and the internal cohesive forces in the liquid. When the adhesive forces exceed the cohesive forces, θ lies in the range $0° \leq \theta \leq 90°$; and when the cohesive forces exceed the adhesive forces $90° \leq \theta \leq 180°$. When $\theta > 90°$, the ($\cos \theta$) term is negative, resulting in a convex meniscus towards the vapor phase, and the liquid level in the capillary falls below the liquid level in the container (capillary depression occurs for liquid mercury in a glass capillary). In practice, during a capillary rise experiment, θ normally lies in the range $0° \leq \theta \leq 40°$, and the curvature of the meniscus determines the height to which the liquid will rise in the capillary tube. If the radius of the tube is sufficiently small but $\theta \neq 0°$, then we obtain the expression, $R_1 = R_2 = r/\cos \theta$, from the mathematical relation of $\cos\theta = r/R$, as shown in Figure 4.7, and by neglecting the gravitational distortion on the curvature. Then, the *capillary constant with a finite contact angle*, a_θ^2, can be given as $[a_\theta^2 = rh = (2\gamma\cos\theta)/(\Delta\rho g)]$ (Equation (330)) for this case.

In order to determine the capillary height, we need to measure the distance between the bottom of the meniscus (the lowest point in the concave meniscus) and the plane level of liquid in a large container that is in communication with the capillary tube. However, by doing so, we neglect the weight of the liquid in the *crown* of the concave meniscus (the liquid between the topmost and the lowest ends of the meniscus), as given in Figure 4.7. But the weight of this small quantity of liquid also contributes to the hydrostatic pressure, which is in equilibrium with the capillary pressure, ΔP. Thus, several correction factors have been suggested to be included in the capillary constant equation, such as $[a_o^2 = r(h + r/3)]$ (Equation (331)), and for the case of $h \gg r$, Equation (332) is applied.

If the radius of the capillary tube is particularly large, or when the capillary rise method is used to measure the interfacial tension between two liquids (see Section 6.6), the contact angle is larger than zero, $\theta > 0°$, and the meniscus is no longer spherical. Then, Equation (329) must be corrected to allow exact treatment of the capillary rise due to the curved meniscus. In Figure 6.1, we see an interfacial meniscus (curve), which divides two fluids of unequal density. This curve is concave upwards and point A is on this curve at a level (z) above the minimum level point, O. The pressure on the concave side of point O is P_1, and P_2 is that on the convex side, $\Delta P_O = P_1 - P_2$ at point O. The density on the upper side (concave side) of point O is ρ_1, and ρ_2 is the density on the lower side (convex side) of point O. If we assume that both media are continuous, the hydrostatic pressure at point A can be calculated as $P_{A1} = P_1 - zg\rho_1$ and $P_{A2} = P_2 - zg\rho_2$, due to the fact that hydrostatic pressure decreases under gravity with elevation. Then the pressure difference at point A across the curved surface would be

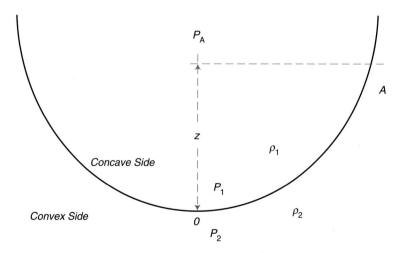

Figure 6.1 Interfacial meniscus, which divides two fluids of unequal density. Point A is on the concave upwards curve at a level (z) above the minimum point, O. P_1 is the pressure on the concave side of point O, and P_2 that on the convex side; ρ_1 is the density on the upper side (concave side) of point O and ρ_2 is the density on the lower side (convex side) of point O.

$$\Delta P_A = P_{A1} - P_{A2} = P_1 - P_2 + zg(\rho_2 - \rho_1) \tag{472}$$

The ΔP_A term can also be calculated by using the Young–Laplace equation (Equation (325) in Section 4.3.3) at the point A. If we combine Equations (472) and (325), then we have

$$\gamma\left(\frac{1}{R_1} + \frac{1}{R_2}\right) = \Delta P_O + zg\Delta\rho \tag{473}$$

where $(\Delta P_O = P_1 - P_2)$ at point O. Equation (473) is called the *fundamental capillarity equation*, and is the general equation of a curved surface, such as a meniscus, under surface tension and gravity effects, and is the basis of calculating interfacial and surface tension of pure liquids from many different measurement methods such as capillary rise, drop profile (Section 6.2), maximum bubble pressure (Section 6.3), and the height of a meniscus at a solid wall (Section 6.4). If the surface tension of a liquid is measured at the liquid–air interface, since $\rho_2 \gg \rho_1$, then the density of the air, ρ_1, may often be neglected without any serious loss of accuracy; however if the interfacial tension between two immiscible liquids is measured at a liquid$_1$–liquid$_2$ interface, then both densities must be used in Equation (473).

As we see in Section 4.3.2, for any curve that is written in the form of a $y = f(x)$ function, where f has continuous first and second derivatives in rectangular Cartesian coordinates, the curvature in two-dimensions can be calculated from Equation (296), so that

$$\frac{1}{R_1} = \frac{\left|\dfrac{d^2 y}{dx^2}\right|}{\left[1 + \left(\dfrac{dy}{dx}\right)^2\right]^{3/2}} \tag{474}$$

By definition of three-dimensional curvature, the other radius of curvature, R_2, must be in the plane perpendicular to that of the plane of R_1. We can write from three-dimensional geometry, as shown in Figure 4.7, that

$$\cos\theta = \frac{x}{R_2} \tag{475}$$

and R_2 can be given as

$$\frac{1}{R_2} = \frac{\cos\theta}{x} = \frac{\sin\varphi}{x} \tag{476}$$

where the *inclination angle*, φ, of the $y = f(x)$ function is $(\varphi = 90° - \theta)$, from plane geometry, as shown in Figure 4.7. Since, $\tan\varphi = (dy/dx)$ for the $y = f(x)$ function, by definition in the x–z coordinate system, we have

$$\frac{1}{R_2} = \frac{\left(\dfrac{dy}{dx}\right)}{x\left[1 + \left(\dfrac{dy}{dx}\right)^2\right]^{1/2}} \tag{477}$$

If we combine Equations (474) and (477) with the Young–Laplace equation (Equation (325) in Section 4.3.3), we obtain

$$\Delta\rho gy = \gamma \left\{ \frac{\left|\dfrac{d^2y}{dx^2}\right|}{\left[1+\left(\dfrac{dy}{dx}\right)^2\right]^{3/2}} + \frac{\left(\dfrac{dy}{dx}\right)}{x\left[1+\left(\dfrac{dy}{dx}\right)^2\right]^{1/2}} \right\} \qquad (478)$$

Equation (478) is the exact analytical geometry expression of capillary rise in a cylindrical tube having a circular cross section, which considers the deviation of the meniscus from sphericity, so that the curvature corresponds to $(\Delta P = \Delta\rho gy)$ at each point on the meniscus, where y is the elevation of that point above the flat liquid level $(y = z + h)$. Unfortunately, this relation cannot be solved analytically. Numerous approximate solutions have been offered, such as application of the Bashforth and Adams tables in 1883 (see Equation (476)); derivation of Equation (332) by Lord Rayleigh in 1915; a polynomial fit by Lane in 1973 (see Equation (482)) and other numerical methods using computers in modern times.

In the case of revolution of half profiles around the axis, the radii of curvature are equal to each other, $R_1 = R_2$, at the apex (at point O in Figures 4.7 and 6.1), because the capillary tube is cylindrical and the point O is on the axis of revolution. Thus, we need to define a new curvature term for this case. If we denote d as the radius of curvature of a curve of revolution around the vertical axis at the apex O, then by definition, we may write $(R_1 = R_2 = d)$ where $r \neq d$ and $\theta \neq 0°$. Then, the ΔP_O term at point O can be expressed as $\Delta P_O = 2\gamma/d$. By combining this equation with Equation (473), we have the $[\gamma(1/R_1 + 1/R_2) = 2\gamma/d + gz\Delta\rho]$ expression which was given as Equation (335) in Section 4.4, where z is the elevation of any point on the meniscus $(z = y - h)$. This equation is general and can be applied to any capillary rise situation where the gravitational distortion on the curvature must be considered. The new capillary constant can be written as $[a_d^2 = dh = 2\gamma/\Delta\rho g]$ for this case, as given in Equation (334). However, it is a very difficult experimental task to measure the d parameter in a capillary tube, and we need a mathematical relation between d and the experimentally accessible radius of the capillary tube, r. In order to find a relation between d and r, we need to convert Equation (335) into another mathematical form. For this process, initially a dimensionless term, β, which is called the *Bond number*, is defined as

$$\beta = \frac{\Delta\rho g d^2}{\gamma} = \frac{2d^2}{a_d^2} \qquad (479)$$

Now, if we multiply both sides of Equation (335) by d, then by rearrangement we have

$$\frac{1}{(R_1/d)} + \frac{1}{(R_2/d)} = 2 + \frac{gz\Delta\rho d}{\gamma} \qquad (480)$$

If we combine Equations (479) and (480), and replace R_2 by its equivalent $(x/\sin\varphi)$ from Equation (476), then we have

$$\frac{1}{(R_1/d)} + \frac{\sin\varphi}{(x/d)} = 2 + \beta\frac{z}{d} \qquad (481)$$

Equation (481) is known as the *Basforth–Adams equation*. Small Bond numbers indicate a high liquid surface tension. The β parameter is positive for a meniscus in a capillary tube, for a sessile drop on a solid and for a bubble under a plate in a liquid, and it is negative for a pendant drop or a clinging bubble attached onto a solid substrate in a liquid, due to the density differences between the immiscible phases (see also Section 9.1). Equation (481) is a differential equation that can be solved numerically by setting one of the β or φ parameters. Using a numerical integration procedure applied manually in a time-consuming process, Bashforth and Adams solved Equation (481) for a large number of values of β between 0.125 and 100 and, in 1883, reported numerous tables compiling values of (x/d) and (z/d) for $(0 < \varphi < 180°)$. After 90 years, Lane found two accurate polynomial fits to numerical solutions of Equation (481), depending on the size of capillary radius. For the case where $2 \geq r/a$,

$$\frac{d}{r} = 1 + \left[\begin{array}{c} 3327.9\left(\dfrac{r}{a}\right)^2 + 65.263\left(\dfrac{r}{a}\right)^3 - 473.926\left(\dfrac{r}{a}\right)^4 + 663.569\left(\dfrac{r}{a}\right)^5 \\ - 300.032\left(\dfrac{r}{a}\right)^6 + 75.1929\left(\dfrac{r}{a}\right)^7 - 7.3163\left(\dfrac{r}{a}\right)^8 \end{array} \right] 10^{-4} \qquad (482)$$

and for the case where $r/a \geq 2$,

$$\frac{r}{d} = \left(\frac{r}{a}\right)^{3/2} \left\{ \exp\left[-1.41222\left(\frac{r}{a}\right) + 0.66161 + 0.14681\left(\frac{a}{r}\right) + 0.37136\left(\frac{a}{r}\right)^2 \right] \right\} \qquad (483)$$

These polynomials must be used in an iterative manner, so that we should calculate the initial approximate value of the capillary constant from the measured height and radius of the capillary $(a_1^2 = rh)$. Then this value is inserted in terms of $(r/a)_1$ or $(a/r)_1$ in Equations (482) (or (483)) to calculate the $(d/r)_1$ or $(r/d)_1$ terms to find d_1. Now since the $[a_d^2 = dh]$ expression is exact in these circumstances, we need to calculate $[(a_d^2)_1 = d_1 h]$ by using the recently found d_1 value. Afterwards, $(a_d^2)_1$ is used in the $(r/a)_2$ or $(a/r)_2$ terms of Equations (482) or (483) to re-calculate the $(d/r)_2$ or $(r/d)_2$ terms. This iteration procedure is repeated until the value of d_n is constant and gives a constant $(a_d^2)_n$ value. Then, the exact value of the surface tension of the liquid can be found from the expression,

$$\gamma = \frac{(a_d^2)_n \Delta \rho g}{2} \qquad (484)$$

The capillary rise method for determining liquid surface tension has the advantages of being well known, widely used and well understood; however, the cleaning of the glass capillaries, determination of capillary radius and the measurements of liquid height are a little more tedious than in some other methods. This method is mainly used to measure the surface tension of pure liquids, although solutions may also be measured, with a lesser accuracy due to the formation of appreciable contact angles. Since it is very rare to observe the complete wetting of the glass wall during an interfacial tension measurement between two immiscible liquids, the capillary rise method is of little value for this purpose.

The capillary rise method is a static method, which measures the tension of practically stationary surfaces that have been formed and equilibrated for an appreciable amount of time. The use of a cathetometer and suitable illumination of the menisci is required during the measurement. The capillary tubes must be very clean, be placed as accurately vertical, and be circular in cross section with an accurately known and uniform radius. The uniformity in the radius is essential practically because of the movement of the meniscus up and down in the tube (otherwise we cannot be sure of the measured tube perimeter). Thin-walled glass tubes are preferred because their interior diameters are more uniform than thick-walled tubes along the length of the tube. Long ago, the weight of the mercury that fills a capillary tube was used to calculate the radius of a capillary from simple cylinder geometry; at present capillary diameters are measured by optical microscopy. The contact angle between the liquid and the glass must be zero for a precise measurement. This can be achieved for water and most of the organic liquids if the interior of the thin (i.e. $r < 0.1$ mm) glass capillary is properly cleaned. We should note that the presence of a contact angle of $\theta = 2.5°$ introduces only 0.1% error in γ measurement.

While determining the liquid height, it is better to measure with a falling (or receding) meniscus, so that the liquid level is initially raised above its equilibrium value by a slight suction above the capillary tube, and then left to equilibrate. On the other hand, two-armed capillary tubes, connected with a cross tube above the liquid level, are also used to ensure that the pressure in both arms of the glass apparatus is the same. An interesting modification of the capillary rise method is to measure the pressure, ΔP, that is required to force the meniscus down until it is on the same level as the plane surface of liquid outside the capillary tube. This method is useful to compare the surface tension of water and its dilute solutions.

On the other hand, if a liquid of known capillary constant (or known surface tension and density) is used in a capillary tube, it is possible to calculate the radius of this tube by measuring the height of this liquid in the capillary and applying the appropriate expression selected from Equations (329)–(332). Later, the same capillary can be used to measure the surface tension of other unknown liquids. However, care must be taken because surface tensions of liquids may be changed depending on the conditions. For example, ultra-pure water has a capillary constant of $a_o^2 = 1.488 \times 10^{-6}\,\mathrm{m}^2$ at room temperature, but if our water sample is not very pure or if we allow an ultra-pure water sample to age in open laboratory conditions for one or more hours, then the capillary constant may easily decrease down to $a_o^2 = 1.35 \times 10^{-6}\,\mathrm{m}^2$ or more. The same purity requirements also apply for all organic liquids, and extra care must be taken; for example, several distillations must be done for even laboratory-grade organic solvents before measurement of the surface tension, or their ultra-pure grades must be used instead. Glass capillary tubes are not suitable for concentrated alkaline solutions due to the possibility of chemical reactions, and in general, surface tension measurements on basic solutions give incorrectly low values.

6.2 Drop Volume and Drop Shape Methods, Video-image Digitization Techniques

6.2.1 Drop volume or drop weight method

This is one of the oldest detachment methods for measuring liquid surface tension. It was first reported by Tate in 1864 and formerly called the *stalagmometer method*. In this method, a stream of drops of a liquid falls slowly from the tip of a thin glass tube having a radius of approximately 2–3 mm. The liquid stream is nipped off into drops having critical sizes by surface tension. In theory, a single drop enlarges to a size where the gravity and capillarity forces are in equilibrium, and then falls down at this critical size, as shown in Figure 6.2 *a*. Thus, the volume (or weight) of a single drop surrounded by air (or vapor) must be found in order to calculate the surface tension. The simple force balance expression at the point of detachment is given below:

$$W_{drop} = m_{drop}g = V\rho g = 2\pi r\gamma \tag{485}$$

where m_{drop} is the mass of the drop, V is the volume of the drop and r is the interior radius of the tube tip for non-wetting liquids (however r is the external radius for wetting liquids such as water on glass, and in some instances the outer part of the tip is coated with melted paraffin wax to prevent adhesion of the liquid, and then the interior radius of the tube can again be used in calculations). The tip of the small glass tube must be ground smooth at the end and must be free from any nicks. Thin metal tubes are also used on occasion. The measurement of surface tension with the drop weight technique is very simple but, unfortunately, sensitive to vibration on the other side. Vibration of the apparatus can cause premature separation of the drop from the end of the tube before it reaches the critical size. In practice, it is impossible to apply Equation (485) directly because the whole of the drop never falls from the tip. A thin cylindrical liquid neck develops while the drop is being detached from the thin tube during falling, so that only a portion of the drop falls, and some portion of the streaming liquid remains attached to the tip. Consequently, drops having weights less than the ideal value, W_{drop}, will be obtained, and this process causes large deviations in Equation (485). Therefore a correction parameter is added to Equation (485)

$$W_{drop}^{actual} = 2\pi r\gamma f \tag{486}$$

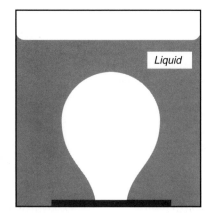

a. Pendant Liquid Drop in Air   *b. Sessile Air Bubble in a Liquid*

Figure 6.2   Liquid surface tension determination by the drop shape method: *a*. A pendant drop is formed by suspending the liquid from the tip of a thin tube. *b*. A sessile air (or vapor) bubble is formed in a liquid by injecting the gas from the tip of a needle connected to a syringe.

where *f* is a correction factor which is related to dimensionless (r/a) or $(r/V^{1/3})$ ratios. The value of term, *f*, can be determined by using liquids whose γ values are known from capillary rise or other precise methods. Approximate correction factors for many liquids are widely available in the literature. Since the weight of a single drop is too low to be determined precisely, a practical solution is to count the number of falling drops (approximately 20–30 drops for each experiment), while collecting them in a container, and to then determine their total weight and divide this value by the number of drops to obtain the average weight of a single drop. Another alternative is to use a volumetric syringe with a motor-driven plunger to obtain an accurate volume for a single drop as shown in Figure 6.2 *a*. Modern video digitization techniques can determine individual drop volumes to $\pm 0.1 \times 10^{-3}$ ml.

The drop weight method eliminates prolonged aging of the liquid surface because a fresh drop forms during the measurement. Usually slow formation of drops is required during experimentation, otherwise the weight of a single drop will be too large. For volatile liquids, a closed system (chambers) is used to restrict the rapid evaporation of the liquid. This method is also suitable to be used for viscous liquids and for solutions. When carefully applied, it is also used as a dynamic method to measure the time-dependent surface tension, since it involves the creation of a new surface by time.

In practice, we need not calculate *r* and *f* for every capillary; we may simply use a liquid whose surface tension is known, and then the surface tension of an unknown liquid can be found from the expression

$$\frac{W_1}{W_2} = \frac{\gamma_1}{\gamma_2} \tag{487}$$

where W_1 and W_2 are the weights of the single drops. However, the volume of the cylindrical liquid neck following every drop is dependent on the nature of the liquid and its interaction with the glass tube, so that neck volumes for different liquids are not equal to each other, and thus, this approach is only approximate.

The drop weight method is preferably used to measure the interfacial tension between two immiscible liquids. For this case, we need to determine the volume or weight of a liquid drop surrounded by a second liquid, as it becomes detached from a tip of known radius. In practice, usually drops of a liquid with a higher density are formed through a thin glass tube, in another immiscible liquid, and the volume of the first liquid is measured after the two phases are separated. The reverse process may also be applied so that drops of the liquid with the lower density are formed through an inverted thin glass tube in an immiscible liquid with a higher density, and the upper phase is separated and weighted for this case. The interfacial tension can be calculated from

$$\gamma_{12} = \frac{V \Delta \rho g}{2 \pi r f} \tag{488}$$

where $\Delta \rho$ is the density difference between the two liquids. The drop weight method has proved convenient for measuring time-dependent interfacial tensions. Although this method is only approximate, it occasionally gives very reproducible results for the surface and interfacial tension of pure components, as well as liquid mixtures and surfactant solutions after the *f* factor is satisfactorily determined. This is because the results of this method are essentially independent of the contact angle of the dropping tip.

6.2.2 *Drop shape method and video-image digitization techniques*

When a liquid is suspended from the tip of a thin tube to form a pendant drop without detachment, as shown in Figure 6.2 *a*, its shape is the result of a balance between the capillary and gravitational forces, and the surface tension of the liquid (or interfacial tension between two immiscible liquids) can be calculated from the shape of this suspended drop. Without gravitational effects, small liquid drops (or sessile air, vapor bubbles formed in a liquid, as shown in Figure 6.2 *b*) will tend to be spherical with a minimum surface area. If the densities of two immiscible liquids are equal, a drop of the first liquid in the second one will also be spherical. If the gravity effect is appreciable, especially for large drops, then Equations (335), (473), (478) and (481) can be applied to calculate the surface tension due to deviations from the spherical shape. Pendant liquid drops in air (or air bubbles formed in a liquid) are photographed, or their video images are grabbed, to measure the dimensions of their profiles. Then, computer-image digitization techniques are applied to define their profiles in two-dimensional Cartesian coordinates, and to calculate their volumes.

The surface or interfacial tension at the liquid interface can be related to the drop shape through the following equation, obtained from Equation (479):

$$\gamma = \frac{\Delta \rho g d^2}{\beta} \tag{489}$$

where β is the shape factor (Bond number). Modern computational methods using iterative approximations allow solution of the Young–Laplace equation for β. This involves computer matching of the entire drop profile to a best-fitting theoretical curve. In general, the better curve fits are obtained with fewer very accurate points on a drop profile, rather than a number of less reliable points. The digital representation of the profile of a sessile drop is necessarily limited in accuracy by the pixel resolution of the captured image. The axisymmetric drop shape analysis is then applied. (Non-axisymmetry is rarely examined for indications of drop shape distortions, and this is essential in the simultaneous determination of γ and contact angle, as we will see in Section 9.1.) The development of computer-image digitization techniques represents a significant improvement, in both ease and accuracy, over traditional methods, and has diminished the importance of the various approximations to the drop shape that have been suggested empirically in the past.

During the drop profile determination, the dimensions of the drop must be measured with a high precision, and there must be no external vibration. The radius of the tip, r, should be small enough so that $(r/a) < 0.05$, to obtain precise results. During interfacial tension measurement a hanging drop is formed through a needle within an immiscible liquid, and its profile is determined. The drop shape methods for measuring surface and interfacial tensions are absolute and dynamic, do not depend on the contact angle between the liquid and the needle, require only small volumes of liquid, are readily amenable to temperature control, and can be used to study viscous liquids, molten materials and also aging effects.

6.3 Maximum Bubble Pressure Method

In this method, the surface tension of a liquid is determined from the value of the maximum pressure needed to push a bubble out of a capillary into a liquid, against the

Young–Laplace pressure difference, ΔP. If a vertical tube of internal radius, r, is inserted into a liquid, and an inert gas (usually air) is pressed from the top of the tube, a bubble is blown at the end of the tube, having the form of a sphere segment initially, as seen in Figure 6.3. If the pressure in the bubble increases, the bubble grows and the radius of curvature diminishes. After a while, it becomes fully hemispherical, as can be seen in the middle shapes of Figure 6.3 *a* (for wetting liquids) and Figure 6.3 *b* (for non-wetting liquids), so that the radius of the bubble is equal to the radius of the capillary tube. The Young–Laplace equation gives the pressure inside the bubble as a maximum at this point. When the bubble is allowed to enlarge further, a larger radius is obtained corresponding to a smaller pressure, and the bubble becomes unstable and can easily be detached from the capillary tube, or alternatively, air rushes in and bursts the bubble. It seems that, if we can measure the maximum pressure when a hemispherical bubble is formed, then we can calculate the liquid surface tension. However, this is not true because we must consider the hydrostatic pressure due to the capillary tube inserted into the liquid. Thus, the maximum pressure measured in this process is the sum of the hydrostatic, P_{hyd}, and capillary, P_{capil}, pressures. When a hemispherical bubble is formed, the pressure balance is given as

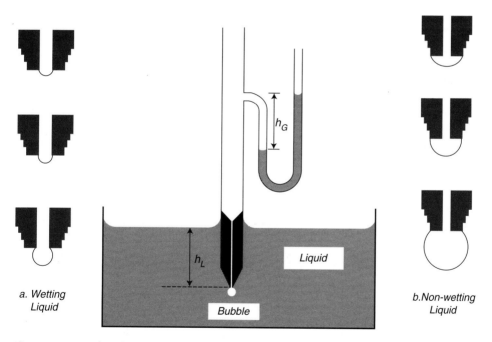

a. Wetting Liquid

b. Non-wetting Liquid

Bubble

Liquid

h_G

h_L

Figure 6.3 Liquid surface tension determination by the maximum bubble pressure method. The maximum pressure, P_{max}, needed to push a bubble out of a capillary into a liquid is determined just prior to the detachment of the bubble; h_L is the distance below the surface of the liquid to the tip of the tube. The value of P_{max} is usually found by measuring the height of a water column, h_G. **a.** If the tube is completely wetted by the liquid, then the radius, r, is its internal radius. **b.** If the liquid is non-wetting towards the tube, then the radius, r, is its external radius. The bubble becomes fully hemispherical, as can be seen in the middle shapes of **a** and **b**.

$$P_{max} = P_{hyd} + P_{capil} = gh_L\Delta\rho + \frac{2\gamma}{r} \qquad (490)$$

where h_L is the distance below the surface of the liquid to the tip of the tube. If the tube is wetted by the liquid, then the radius, r, is its internal radius, since the liquid covers the lower edge of the tube completely. The surface tension value can be calculated by determining the P_{max} value that is attained just prior to the detachment of the bubble. P_{max} is usually found by measuring the height of a water column, h_G, attached to the measuring tube as can be seen in Figure 6.3. In an improved version of this method, double capillary tubes of different radii are used, with a precise differential pressure transducer to sense the pressure difference between them.

If the radius of the capillary is large, so that $(r/a) > 0.05$, then the Basforth–Adams equation (Equation (481)) or the Lane equations (Equations (482) and (483)) can also be used in the surface tension calculation from the maximum bubble pressure method. This method can also be used to determine the surface tension of molten metals. It has been a popular method in the past, but now it is not very common in surface laboratories because of its poor precision.

6.4   Ring, Wilhelmy Plate Detachment and the Height of a Meniscus on a Vertical Plane Methods

The detachment of a ring or a plate (a Wilhelmy plate) from the surface of a liquid or solution is a static surface tension measurement method, which gives the detachment force of a film of the liquid and its extension from the liquid surface. These methods are less accurate than the capillary rise method, but they are normally employed in most surface laboratories because of their ease and rapidity.

6.4.1 du Noüy ring method

This is one of the oldest surface tension measurement methods, first applied in 1878, and later, in 1919, the French scientist du Noüy improved the method by adding a torsion balance. A circular ring (a loop of metal wire), which is usually made of platinum metal or a platinum/iridium alloy, of radius 2–3 cm, is suspended from a balance and is immersed into a liquid surface horizontally. The radius of the wire ranges from 1/30 to 1/60 of that of the ring. The ring is then raised slowly to the liquid–air interface, and the maximum force is measured precisely at the moment of detachment, by the balance. Since metals are high-surface-energy materials, the adhesion of a liquid to a metal ring is greater than the cohesion within the liquid. In addition, the contact angle between the liquid and the ring is generally zero due to complete wetting. Thus, when a ring is detached from the surface of a liquid, the force to be overcome is that of cohesion rather than adhesion. From the definition of surface tension, the force balance at the moment of detachment can be given as

$$F_{max}f_r = 2(2\pi r_{mean})\gamma = 2\pi(r_{ext} + r_{int})\gamma \qquad (491)$$

where F_{max} is the maximum upward pull applied to the ring of mean radius, r_{mean}, $r_{mean} = (r_{ext} + r_{int})/2$, as shown in Figure 6.4, and f_r is the correction factor for the small but significant volume of the liquid that remains on the ring after detachment, and also for the discrepancy between r_{mean} and the actual radius of the meniscus in the plane of rupture. The term F_{max} corresponds to the maximum weight of the meniscus over the liquid surface that can be supported by the ring. The perimeter of the ring is multiplied by 2 because of the presence of two surfaces, created on both sides of the ring. The f_r factor is a function of the mean radius, thickness of the ring and also of meniscus volume, and varies between 0.75 and 1.05 numerically, according to the size and the shape of the ring, and the difference in the fluid density. The f values can be calculated by using the following approximate equation:

$$f = 0.725 + \left(\frac{9.075 \times 10^{-4} F}{\pi^3 \Delta \rho g r^3} - \frac{1.679 r_{wire}}{r} + 0.04534 \right)^{1/2} \tag{492}$$

Equation (492) can be applied in the range $[7.5 \geq \Delta \rho g r^3 / F \geq 0.045]$. In many modern computerized systems, the interfacial tension reading does not require separate calculation of f, since its calculation is incorporated within the software. A ring probe is hung on an electronic balance and, after zeroing the force by subtracting the weight of the ring probe, is brought into contact with the liquid surface (or interface) to be tested. When the ring hits the surface, a slight positive force is recorded due to the adhesive force between the ring and the surface. Then the ring is submerged below the liquid surface and subsequently raised slowly upwards. When lifted through the surface, the measured force starts to increase. While the ring moves upwards, it raises a stretched meniscus of the liquid with it, as shown in Figure 6.4. This meniscus does not have a simple geometrical shape and will become unstable and eventually detach from the ring and return to its original position. During the upward movement, the force keeps increasing until a maximum is reached. It then begins to diminish prior to the actual tearing event. Calculation of surface (or interfacial tension) by this technique is based on the measurement of this maximum force and not the force at the point of detachment. The depth of immersion of the ring and the level to which it is raised when it experiences the maximum pull are irrelevant to this technique. The contact angle must be zero in this process, otherwise results for γ will be low; nevertheless, a zero contact angle is occasionally obtained in practice, because contact angles are always receding angles, which are less than advancing angles, as the ring is initially wetted by the liquid phase (see also Section 9.1).

The system must be vibration-free during the measurement. The ring must be exactly plane and accurately parallel to the surface of the liquid; the presence of a tilt angle of 2.1° introduces an error of 1.6%. If the ring is deformed, it causes major errors. The platinum ring is soft and subject to inadvertent deformation during handling and cleaning. Platinum rings can easily be cleaned by flaming them gently to remove surface contaminants such as grease before every use. Evaporation of the liquid level during measurement should be minimized to obtain better results. The ring method is not suitable for viscous liquids or for some solutions that slowly attain the equilibrium of surface tension.

It is possible to perform interfacial tension measurements between two immiscible liquids by the ring method, just like surface tension measurements, by ensuring that the bulk of the ring probe is submerged in the light phase prior to beginning the experiment.

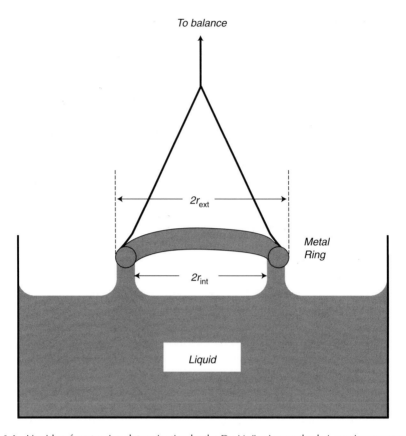

Figure 6.4   Liquid surface tension determination by the Du Noüy ring method: A maximum upward pull is applied to the ring having a mean radius, r_{mean}, $[r_{mean} = (r_{ext} + r_{int})/2]$, where r_{ext} is the external radius, and r_{int} is the internal radius.

The ring is then immersed into the heavy phase, which is under the interface, and lifted up slowly and a stretched meniscus of the heavy liquid forms inside the light liquid. Using the maximum force measured just before the point of detachment, we can calculate the interfacial tension. Poor wetting of the ring by the denser fluid makes the measurement of interfacial tension impossible to carry out. If perfect wetting is not achieved, and the contact angle is above zero in interfacial tension measurements, then additional correction of the instrument reading is needed. For this reason, hydrophobic polytetrafluoroethylene and polyethylene rings are used instead of platinum, in order to diminish the contact angle formed during water–organic liquid interfacial tension measurements.

6.4.2 Wilhelmy plate method

In this widely used method, a thin plate, which is suspended from a balance, is vertically and partially immersed into a liquid whose surface tension is to be determined. A meniscus forms around the perimeter of the suspended plate. If the contact angle between the liquid and the plate is zero, the liquid surface is oriented nearly vertically upwards, as shown in Figure 6.5. Then the surface tension of the liquid present in the meniscus exerts a downward force, which is equal to the weight of the meniscus formed on the perimeter of this thin plate. If the weight of the plate probe is zeroed in the instrument before the liquid surface is raised to the contact position, then the imbalance that occurs on contact is due to the weight of the formed meniscus. Wilhelmy, in 1863, was the first to apply this simple and absolute surface tension method, by using a lever balance. The thin plates are then called *Wilhelmy plates* and are usually made of roughened platinum, platinum/iridium alloy, glass, mica, steel or plastic. Disposable filter or chromatography papers can also be used as Wilhelmy plates, especially in monolayer studies, which always give zero contact angle with the liquid.

There are two modifications to the Wilhelmy plate method. In the first modification, the cup carrying the liquid is mobile and is lowered until the previously immersed plate becomes detached from the liquid surface, and the maximum vertical pull, F_{max} on the balance is noted, similarly to the ring method. Then the capillary force, for the zero contact angle, can be given as

$$F_{capil} = F_{max} - W = P\gamma = 2(w + d)\gamma \qquad (493)$$

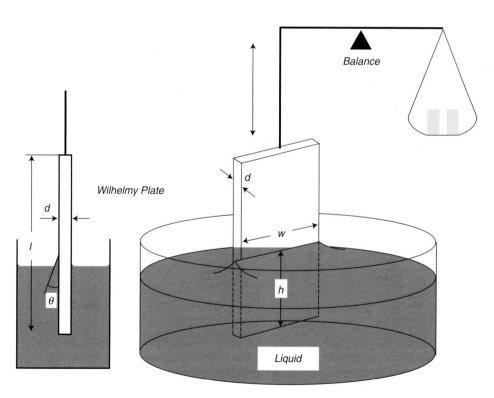

Figure 6.5 Liquid surface tension determination by the Wilhelmy plate method. A rectangular plate of length, *l*, width, *w*, and thickness, *d*, of material having a density of ρ_s, is immersed to a depth of *h* in a liquid of density ρ_L; θ is the contact angle forming between the liquid and the plate.

where W is the weight of the plate probe and $[2(w + d)]$ is the perimeter, P, of the plate. (Modern instruments use plates with standard dimensions, so that measurement of plate size and weight is not required.) The term F_{capil} corresponds to the weight of the meniscus that is formed around the perimeter of the Wilhelmy plate. If a finite contact angle, θ, forms between the liquid and the plate, as shown in Figure 6.5, then the surface tension can be calculated from

$$\gamma = \frac{F_{capil}}{2(w + d)\cos\theta} \tag{494}$$

Equation (494) shows that we are carrying out a single measurement which involves two parameters, γ and θ. If the surface tension of the liquid, γ, is independently known, then we can calculate the contact angle on the Wilhelmy plate, or that of any coated material, such as a polymer (see also Sections 6.4.3 and 9.1). On the other hand, we need to know the value of θ in order to determine γ precisely, and this can be achieved easily if we maintain a zero contact angle with the liquid by choosing a suitable plate material. Equation (494) can also be applied to a fiber or wire instead of the Wilhelmy plate, in the form of $[F_{capil} = P\gamma \cos\theta]$.

In the second modification of the Wilhelmy plate method, which is usually applied to monolayer studies in Langmuir troughs (film balances), it is not desirable to detach the plate completely; instead the force necessary to keep the plate at constant depth of immersion is determined as the surface tension is changed due to the presence of solutes forming the monolayer film. In this case, the forces acting on the plate consist of the gravity and surface tension effects downwards, and the buoyancy effect due to displaced water upwards, and correction due to the buoyancy of the subphase is required. For a rectangular plate of dimensions l, w, and d, of material with a density of ρ_S, immersed to a depth of h in a liquid of density ρ_L, the net downward force is given by the following equation:

$$F_{net} = \rho_S glwd + 2\gamma(w + d)\cos\theta - \rho_L ghwd \tag{495}$$

where g is the gravitational constant. (The density of the air phase is neglected in Equation (495).) The term F_{net} is determined by measuring the changes in the mass of the plate probe, which is directly coupled to a sensitive electrobalance. In general, the force due to the weight of the Wilhelmy plate probe is zeroed in the electrobalance before making any measurements, and thereby eliminating the first term in Equation (495).

In this second modification there are two alternatives:

1 The change in F_{net} is measured when the stationary plate is kept at a constant depth of immersion, h. Since the presence of the solutes in a monolayer film decreases the initial surface tension of the subphase, then F_{net} varies correspondingly. As the plate is always kept at a constant immersion level, then the buoyancy term can also be eliminated from Equation (495), giving

$$F_{net} = 2\gamma(w + d)\cos\theta \tag{496}$$

In addition, if the contact angle is zero, by combining Equations (424) and (496), one obtains the surface (spreading) pressure as

$$\pi = \gamma_o - \gamma = \frac{F_{net}^o - F_{net}}{2(w + d)} \tag{497}$$

As the surface is brought into contact with the probe, modern computerized equipment will notice this event by detecting the change in force. It will register the height at which this occurs as the *zero depth of immersion*. The plate will then be wetted to a set depth to insure that there is indeed complete wetting of the plate (zero contact angle). When the plate is later returned to zero depth of immersion, the force it registers can be used to calculate the surface tension. Maintaining a constant liquid level is important for this modification, especially for volatile solvents. This method is also suitable for studying prolonged aging of the surface, when the plate is kept at a constant height, and an electronic balance gives continuous readings.

The use of the Wilhelmy plate in a fixed position is preferred in monolayer studies because it minimizes the transfer of the monolayer onto the plate. When a freshly cleaned Wilhelmy plate is immersed in the clean liquid surface at the beginning of the experiment, it may be completely wetted; however if the monolayer-forming surfactant solutes are also present, they start to deposit on the plate during its upward movement through the liquid surface. The extent of this deposition depends on the nature of the monolayer, as well as the plate itself, and it may increase the contact angle considerably after deposition. Thus, keeping the plate in a fixed position during the measurement reduces the transfer of monolayer film onto it. Alternatively, a pair of identical Wilhelmy plates are attached to two arms of a balance in some film balance experiments conducted in a Langmuir trough; one plate contacts with the clean subphase surface and the other plate contacts with the monolayer-covered surface on the barrier compressed side. Since the surface tensions are different on each side of the barrier, then the weight of the meniscus formed by the plate immersed in the clean surface is heavier than the other, because of the higher surface tension of the clean surface. If the contact angles are nearly equal on both of these Wilhelmy plates, then the difference between these two meniscus weights measures the difference in the surface tension for these surfaces (thus giving π directly).

2 For the second alternative, the net force, F_{net}, is kept constant in the instrument while changing the immersion depth, h, by the movement of the container. The immersion depth may be measured from the level displacement or by a goniometer. For this modification, we have from Equations (424) and (495)

$$\pi = \gamma_o - \gamma = \frac{\rho_L g w d(h_o - h)}{2(w + d)} \tag{498}$$

and if the thickness of the Wilhelmy plate so thin as to be negligible compared with the width, w, then Equation (498) simplifies to

$$\pi = \gamma_o - \gamma = \frac{1}{2}\rho_L g d(h_o - h) \tag{499}$$

Adsorption of organic compounds from the laboratory environment can be a major source of experimental error when measuring surface tension using the Wilhelmy plate method. Thus, care should be taken in keeping the plates free from organic contaminants during experimentation, and platinum plates can be washed with an organic solvent and water and then flamed before the experiment to remove any wax or greasy contaminants present.

For two immiscible liquids, the Wilhelmy plate may be wetted by either phase. The calculations for this technique are based on the geometry of a fully wetted plate in contact with, but not submerged in, the heavy phase. The maintenance of a constant contact angle is a problem during interfacial tension measurements, and for such systems, the plate should be hydrophobic. Polymer, especially PTFE, plates may be preferable to use, instead of glass or platinum plates, to measure interfacial tension between water and hydrocarbons. Self-assembling of organic amines or similar compounds on the surface of the platinum plate could also be a solution to this problem.

6.4.3 *Height of a meniscus on a vertical plane method*

When a wetting liquid is in contact with a vertical and flat solid plane, the liquid meniscus climbs on the surface, as shown in Figure 6.6. The height of this meniscus can be accu-

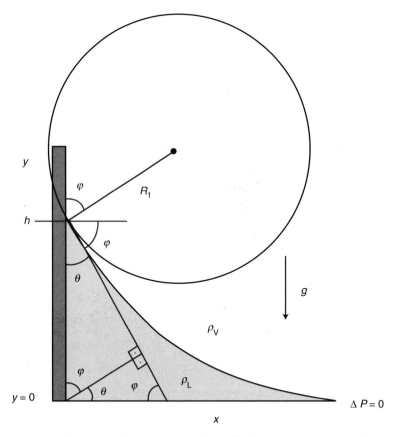

Figure 6.6 Liquid surface tension determination by the height of a meniscus on a vertical plane method: y is the coordinate on the vertical axis and R_1 is the two-dimensional radius of curvature in the plane of this figure. When $y = 0$, the pressure difference, $\Delta P = 0$, and $y = h$ at the triple-point where liquid–solid–air intersect; φ is the *inclination angle* of the $y = f(x)$ function and ($\varphi = 90° - \theta$) at point h. The slope of the $y = f(x)$ curve can be given as $(dy/dx) = -\tan \varphi$. ρ_L is the density of the liquid and ρ_V is the density of the vapor.

rately measured by a cathetometer and, in theory, we may calculate both the surface tension of the liquid and the contact angle simultaneously from this height. This process is a special case of the formation of three-dimensional curvature, where one of the radii of curvature in the Young–Laplace equation (Equation (325)) becomes infinite since the support is planar and hence the $1/R_2$ term disappears; and as ΔP is equal to $(\Delta \rho g y)$, then we obtain

$$\Delta \rho g y = \frac{\gamma}{R_1} \qquad (500)$$

where y is the coordinate on the vertical axis. In Figure 6.6, we see the cross-sectional view of this climbing menicus, where R_1 is the two-dimensional radius of curvature in the plane of this figure and $y = h$ at the triple-point where liquid–solid–air intersect. When, $y = 0$, the pressure difference, $\Delta P = 0$ as well. The angle, φ, is the *inclination angle* of the $y = f(x)$ function and it is made by extension of the normal with the y-axis and also the angle between the tangent and the x-axis, as shown in Figure 6.6. It is clear that $(\varphi = 90° - \theta)$ at point h. The slope of the $y = f(x)$ curve can be given as $(dy/dx) = -\tan \varphi$. Then, the radius of curvature, R_1, can be calculated from Equation (474), giving

$$\frac{1}{R_1} = \frac{\left[\dfrac{d}{dx} \left(-\dfrac{\sin \varphi}{\cos \varphi} \right) \right]}{\left[1 + \left(\dfrac{\sin \varphi}{\cos \varphi} \right)^2 \right]^{3/2}} \qquad (501)$$

After differentiation and simplification, Equation (501) becomes

$$\frac{1}{R_1} = -\cos \varphi \frac{d\varphi}{dx} = -\frac{d\sin \varphi}{dx} \qquad (502)$$

Equation (502) is mathematically equivalent to

$$\frac{1}{R_1} = -\frac{d\cos \varphi}{dy} \qquad (503)$$

If we combine Equations (500) and (503) we have

$$\frac{\Delta \rho g}{\gamma} = -\frac{d\cos \varphi}{y dy} \qquad (504)$$

Equation (504) can be integrated between φ and y

$$-\int_0^\varphi d\cos \varphi = \frac{\Delta \rho g}{\gamma} \int_0^y y dy \qquad (505)$$

$\varphi = 0$ when $y = 0$, showing that the tangent is horizontal at this point, and the integration of Equation (505) gives

$$\cos \varphi = 1 - \frac{\Delta \rho g}{2\gamma} y^2 \qquad (506)$$

since $(\varphi = 90° - \theta)$ at point $y = h$, and $(\cos \varphi = \sin \theta)$ from trigonometry, we can write

$$\sin \theta = 1 - \frac{\Delta \rho g}{2\gamma} h^2 = 1 - \left(\frac{h}{a} \right)^2 \qquad (507)$$

where a^2 is the capillary constant given by Equation (329). Equation (507) shows that if we know the surface tension and the density of the liquid, then we can calculate the contact angle of the liquid with the vertical plate.

The height of a meniscus on a vertical plane wall method can be combined with the first modification of the Wilhelmy plate method, where the previously immersed plate becomes detached from the liquid surface and the maximum vertical pull, F_{max} on the balance is measured. If we square and rearrange Equation (494) so that

$$\cos^2 \theta = \left[\frac{F_{capil}}{2(w+d)\gamma} \right]^2 \qquad (508)$$

since $[\sin^2 \theta + \cos^2 \theta = 1]$ in trigonometry, then we have from Equations (507) and (508)

$$\left[1 - \frac{\Delta \rho g h^2}{2\gamma} \right]^2 + \left[\frac{F_{capil}}{2(w+d)\gamma} \right]^2 = 1 \qquad (509)$$

By rearrangement and simplification of Equation (509) we have

$$\gamma = \frac{\Delta \rho g h^2}{4} + \frac{F_{capil}^2}{4\Delta \rho g h^2 (w+d)^2} \qquad (510)$$

If we combine Equations (494) and (510) we obtain

$$\cos \theta = \frac{2\Delta \rho g h^2 (w+d) F_{capil}}{(\Delta \rho)^2 g^2 h^4 (w+d)^2 + F_{capil}^2} \qquad (511)$$

Equations (510) and (511) can be used successfully to calculate both surface tension and the contact angle in a Wilhelmy plate experiment, if one of these parameters is known.

6.5   Dynamic Surface Tension Measurement Methods

Dynamic surface tension determinations are important for understanding and controlling interfacial processes in multi-phase and multi-component systems. The surface tension of a freshly created surface that is far away from equilibrium can be determined by dynamic methods. Many operations in the chemical industry such as detergency, froth flotation, foam generation etc. involve liquid–fluid interfaces, for which the composition is constantly refreshed and does not reach equilibrium. The *maximum bubble and growing drop pressure* technique and *spinning drop* method are very useful in studying the dynamic interfacial tensions at short intervals. There is also an *oscillating jet* dynamic surface tension measurement method, which has been applied successfully to surfactant solutions. In this method, the unstable oscillations developed in a liquid jet emerging from a noncircular orifice are measured; however it is not suitable to study pure liquid–air interfaces due to rapid equilibration in tenths of a second. Since there are very few publications using this method and no commercial apparatus available, we will not discuss it any further.

6.5.1 Dynamic maximum bubble pressure method

Commercial equipment using the dynamic maximum bubble (or drop) pressure method have been widely used in surface research in recent years. In this method, a gas bubble (or a liquid drop) is formed and released from a capillary tube by using a micro-pump to carefully control its growth rate. A precise pressure transducer continuously measures the pressure inside this bubble (or drop) as it forms and detaches from the end of a capillary. The geometry of the bubble can also be monitored during growth and at the detachment point using a video camera. This ability to simultaneously monitor both pressure and size and shape of bubbles (or drops) as they form allows dynamic interfacial tensions to be evaluated over a range of growth rates, from highly non-equilibrium growth (very rapid bubble growth), to near equilibrium growth (very slow bubble growth).

6.5.2 Spinning drop tensiometer method

This method uses centripetal acceleration to control the shape of a liquid in another liquid. If a liquid drop (*1*) is suspended in an immiscible denser liquid (*2*) in a horizontal transparent tube which can be spun about its longitudinal axis, this drop will go to the center forming a drop astride the axis of revolution, as shown in Figure 6.7. The drop (*1*) elongates from a spherical shape to a prolate ellipsoid upon increasing the speed of

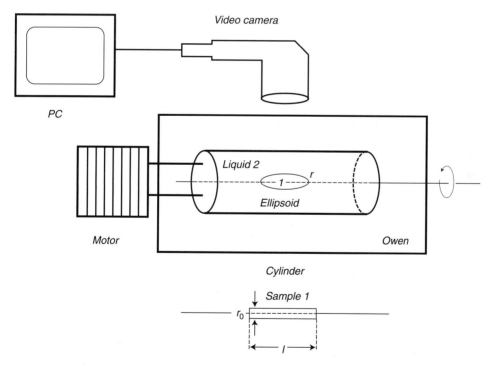

Figure 6.7 Liquid surface tension determination by the spinning drop tensiometer method. A liquid drop (*1*) is suspended in an immiscible denser liquid (*2*) in a horizontal transparent tube which can be spun about its longitudinal axis, and the drop (*1*) elongates from a spherical shape to a prolate ellipsoid with increasing speed of revolution. Later, the drop becomes approximately cylindrical, at very high rotational velocities. A camera with a frame grabber captures the images of the drop inside the transparent tube.

revolution. During centripetal acceleration, the gravitational acceleration has little effect on the shape of the drop. Next, the drop becomes approximately cylindrical at high rotational velocities, because centripetal force increasingly overcomes the surface tension forces. Images of the drop inside the transparent tube can be captured using a camera with a frame grabber (there should be no vibrations at high velocities). If we consider the potential energy of a volume element in the cylinder at distance, r, from the axis of revolution

$$E_{Pv} = \frac{\omega^2 r^2 \Delta\rho}{2} \tag{512}$$

where ω (rads) is the rotational velocity of the spinning horizontal tube and $\Delta\rho$ the density difference between the drop and the surrounding liquid. Since the volume of the cylinder of length, l is $[V = \pi r_o^2 l]$ then the total potential energy of the cylinder can be calculated as

$$E_P = \int_0^V E_{Pv} dV = l \int_0^{r_o} \left(\frac{\omega^2 r^2 \Delta\rho}{2} \right) 2\pi r dr = \frac{1}{4}\pi\omega^2 \Delta\rho l r_o^4 \tag{513}$$

The interfacial free energy between this cylindrical liquid (*1*) drop and the contacting immiscible liquid (*2*) along the surface area can be given from thermodynamics as $[\Delta G = \Delta A \gamma_{12} = 2\pi r_o l \gamma_{12}]$. Then the total energy of the cylinder on revolution can be expressed as

$$E_T = \frac{1}{4}\pi\omega^2 \Delta\rho l r_o^4 + 2\pi r_o l \gamma_{12} = \frac{\omega^2 \Delta\rho V r_o^2}{4} + \frac{2V\gamma}{r_o} \tag{514}$$

Since $dE_T/dr_o = 0$ at equilibrium, then differentiation of Equation (514) and equating it to zero gives

$$\gamma_{12} = \frac{1}{4}r_o^3 \Delta\rho\omega^2 \tag{515}$$

Equation (515) is known as Vonnegut's equation and it is valid on the assumption that the drop is in equilibrium and its length is larger than four times its diameter ($l > 4r_o$). The spinning drop tensiometer method is widely used for measuring liquid–liquid interfacial tension, and is especially successful for examination of ultra-low interfacial tensions down to 10^{-6} mN m^{-1}. In addition, it can also be used to measure interfacial tensions of high viscosity liquids when precise temperature control is maintained.

The spinning drop tensiometer method is particularly suitable for measuring the interfacial tension of melted polymers and is generally used in polymer compatibility, blend and composites research. In this case, a spinning polymer drop (smaller density) is rotated inside another immiscible polymer (higher density). Measurement of drop diameter versus time allows the determination of relaxational and extensional properties of polymeric systems.

6.6 Methods Applicable to Interfaces Between Two Liquids

In theory, every surface tension measurement method can be used to determine the interfacial tension between two liquids. However, the accuracy of these methods is reduced when applied to liquid–liquid interfaces, or when one or both of the liquids is viscous. In practice, the maximum bubble pressure and pendant drop methods are the most suitable, giving consistent and reliable values for interfacial tensions, although there is sometimes the question of whether the liquid wets the inner or the outer radius of a dropping tip. (For such systems, a tapered tip is of value where the drop is actually suspended from a very sharp edge into the heavy immiscible liquid.)

On the other hand, although they are occasionally used, the du Noüy ring, Wilhelmy plate and capillary rise methods are not very suitable for determining interfacial tension because stable contact angles cannot be obtained for liquid–liquid interfaces because of the metastable menisci formed. In both ring and Wilhelmy plate methods the accuracy is reduced because of difficulties in calibrating the weight of the ring immersed in the less dense liquid. Nevertheless, the Wilhelmy plate method is more reliable than the ring method in interfacial tension determination. The drop volume technique may also be used if equilibrium is achieved rapidly (within a few seconds). Capillary rise may be successfully used if the contact angle between the dense liquid and the glass can be determined precisely; otherwise, the capillary rise method should be avoided.

It is important to allow liquid–liquid interfaces to achieve equilibrium before making an interfacial tension measurement. Thus, it is always difficult to measure the interfacial tension of viscous liquids because they are slow in reaching equilibrium. In addition, viscous liquids prevent injection of a liquid sample of the required volume into the instrument. The Wilhelmy plate (at a constant depth, without detachment) and sessile drop methods are generally preferred for viscous liquids, after equilibrating the samples for several hours before measurement.

When the value of the interfacial tension is significantly less than $1\,mN\,m^{-1}$, then we consider the measurement of *ultra-low interfacial tension*, which is common in liquid–liquid emulsification processes when effective surfactant solutions are used. The dynamic spinning drop tensiometer method is especially suitable for this purpose. Ultra-low interfacial tension measurement is important in the chemical industry because the cleaning of solid surfaces of dirt, grease, and oil; the formulation of stable emulsions; the recovery of petroleum, and other applications often rely on lowering the interfacial tension between immiscible liquids to ultra-low values by the use of surfactants.

6.7   Microtensiometry

Microtensiometry is the study of interfacial tension of very small particles in finely dispersed systems. Nanotechnology, biology, pharmaceutical processing and criminology are fields in which materials of interest are in quantities too small to apply conventional surface tension measurement methods. In addition, the interfacial tension of microscopic droplets may differ significantly from that of macroscopic drops of the same surfactant solutions. This is due to dissimilar partitioning of surfactants into two immiscible liquids for microscopic droplets having an enlarged interfacial area on a small length scale. There are two microtensiometry methods to apply: micropipette tensiometry and atomic force microscopy (AFM) tensiometry.

6.7.1 Micropipette tensiometry

The *micropipette* technique was developed in the 1990s to measure interfacial tensions of micrometer-sized droplets such as vesicles. This method is dependent on the pressure differentials across curved interfaces. The droplet is first captured on the tip of the glass micropipette and then sucked into the pipette. The interfacial tension is calculated from the minimum suction pressure at which the droplet extends a hemispherical protrusion into the pipette. (The dimensions of the pipette's internal diameter must be smaller than the diameter of the droplet.) A large pressure difference is required to draw the droplet into the pipette when the droplet wets or adheres to the pipette surfaces. The measured minimum pressure is inserted into the Young–Laplace equation to calculate the interfacial tension. Another more sophisticated technique is the *two-pipette method*, where the separation force between the pipettes is measured, and the interfacial tension is calculated from the force–drop deformation relation.

6.7.2 Atomic force microscopy tensiometry

AFM can directly yield images of nanometer-sized particles and can measure the interactions between substrates and colloidal particles. It is a scanning probe technique based on measuring interaction forces between a cantilever tip and a specimen. Force measurement can be done by measuring the deflection of the cantilever, which has a known spring constant. The cantilever deflection is detected by reflection of a laser beam. A piezoelectric specimen stage controls vertical movement of the specimen very precisely under the cantilever tip, for force measurements and interaction forces as small as $1\,pN$ ($10^{-12}\,N$), measured between the probe tip and the specimen. When a probe tip is inserted into a microscopic liquid drop and detaches from the drop, it is possible to measure the capillarity forces exerted on the tip by the liquid. Then, the surface tension of the liquid can be calculated from the force–distance curves obtained, and the corresponding equations are similar to those for classical macro-detachment (ring or Wilhelmy plate) methods. Fabrication of probe tips with a known geometry, such as spherical or cylindrical, is important for this method. Carbon nanotubes are already being successfully used as AFM tips for this purpose.

6.8 Measurements on Molten Metals

Measurement of the surface tension of liquid metals is difficult due to the relatively high temperatures involved, and also their ease of reactivity with many gases and solids. For example, the presence of oxygen gas reduces the surface tension of many metals, even when present in ppm concentrations. Thus, measurements on liquid metals are carried out in an inert gas environment to avoid chemical reactions. A freshly formed surface is almost a necessity. The maximum bubble pressure, pendant drop, drop weight and sessile drop methods are the preferred methods; the ring, Wilhelmy plate and capillary rise methods cannot be used for this purpose due to the formation of very high contact angles between the molten metal and the substrate, arising from the very high cohesion of liquid metal molecules. The surface tension of molten metals is very high and varies from 400 to $4000\,mN\,m^{-1}$. In general, metals with high boiling points, above 2000°C, have high surface tensions of more than $1000\,mN\,m^{-1}$.

6.9 Surface Tension of Surfactant Solutions

In order to measure the surface tension of solutions containing surfactants, the maximum bubble pressure, pendant drop and Wilhelmy plate (immersed at a constant depth) methods are suitable; capillary rise, ring, mobile Wilhelmy plate, sessile drop and drop weight methods are not very suitable. These methods are not recommended because surfactants preferably adsorb onto the solid surfaces of capillaries, substrates, rings, or plates used during the measurement. In a liquid–liquid system, if an interfacially active surfactant is present, the freshly created interface is not generally in equilibrium with the two immiscible liquids it separates. This interface will achieve its equilibrium state after the redistribution of solute molecules in both phases. Only then can dynamic methods be applied to measure the interfacial tension of these freshly created interfaces.

References

1. Adam, N.K. (1968). *The Physics and Chemistry of Surfaces*. Dover, New York.
2. Aveyard, R. and Haydon, D.A. (1973). *An Introduction to the Principles of Surface Chemistry*. Cambridge University Press, Cambridge.
3. Adamson, A.W. and Gast, A.P. (1997). *Physical Chemistry of Surfaces* (6th edn). Wiley, New York.
4. Davies, J.T. and Rideal, E.K. (1963). *Interfacial Phenomena* (2nd edn). Academic Press, New York.
5. Erbil, H.Y. (1997). Interfacial Interactions of Liquids. In Birdi, K.S. (ed.). *Handbook of Surface and Colloid Chemistry*. CRC Press, Boca Raton.
6. Hiemenz, P.C. and Rajagopalan, R. (1997). *Principles of Colloid and Surface Chemistry* (3rd edn). Marcel Dekker, New York.
7. Butt, H.J., Graf, K. and Kappl, M. (2003). *Physics and Chemistry of Interfaces*. Wiley-VCH, Weinheim.
8. Rusanov, A.I. and Prokhorov, V.A. (1996). *Interfacial Tensiometry*. Elsevier, Amsterdam.
9. Joos, P. (1999). *Dynamic Surface Phenomena*. VSP, Zeist.

Chapter 7
Potential Energy of Interaction Between Particles and Surfaces

In this chapter, we will examine the physical interactions between particles and surfaces, whose sizes are much larger than molecules. In true solutions, the diameter of a solute molecule varies between 0.1 and 10.0 nm (usually less than 1 nm), which is comparable to the size of solvent molecules. On the other hand, colloids and other micro-particles are collectively termed *macrobodies*, and are insoluble and only dispersed in continuous medium. In these dispersions, the particle size varies between 1 and 1000 nm in diameter, which is very much larger than the solvent molecules. These particles may also associate, in some cases, to form an aggregate, which is even larger than the single microparticle. Macrobodies interact with other macrobodies, solvent molecules and also the surfaces of containers. The interactions between particles and surfaces are somewhat different from those between molecules, and they are the main cause of the properties and behavior of the collective dispersion system. Investigation of the potential energy of interactions between particles and surfaces is very important in practice, because colloidal and particulate science is related to a large industrial field consisting of chemical, food, cosmetic and pharmaceutical industries. Such interactions are also operative in nature, such as the behavior of proteins in living systems, etc., as well as in recently emerging nanotechnology applications, so they are the subject of investigation by various scientific disciplines including chemistry, physics, biology and engineering.

7.1 Similarities and Differences Between Intermolecular and Interparticle Forces

As described in Chapter 2, the intermolecular potential energy is defined as *the difference between the total energy of interacting molecules and the sum of their separate molecular energies*. There are two main types of forces between molecules: *short range* and *long range*. These forces together determine the bulk properties of gases, liquids and solids. *Short-range intermolecular forces* are the forces between molecules in close proximity to each other; these forces operate over very short distances of the order of interatomic separations (0.1–1.0 nm) and usually correspond to or are very similar to molecular contact. On the other hand, *long-range intermolecular forces* are mainly van der Waals forces, and their action is more effective between larger molecules within a distance of 100 nm.

In principle, the source of the attraction between macrobodies such as colloids, microparticles and surfaces must be the same as between molecules. The fundamental forces, i.e. Coulombic, van der Waals, repulsive, solvation, hydrogen bonding, hydrophobic and hydrophilic forces, are operative between both molecules and particles. Nevertheless, these forces can manifest themselves in quite different ways and lead to qualitatively new features when acting between large particles or extended surfaces. Apart from various differences, certain semi-quantitative relations describing molecular forces, known as *combining rules*, are applicable to all systems (i.e. to the interactions of molecules, particles, and surfaces) independent of the type of interaction force involved, as will be shown in Sections 7.2 and 7.6. According to the combining rules, when only two components (1 and 2) are present in a system, the associated state of like molecules and particles (1–1) and (2–2) is energetically and thermodynamically preferred over the dispersed state (1–2). However, this is not true for charged particles, where the Coulomb interactions maintain the dispersed state as energetically preferred to the associated state. In addition, hydrogen-bonding interactions do not readily fall into the given behavior format and must be considered separately and cautiously.

So far, we have summarized the similarities between intermolecular and interparticle interactions. However, there are contrasts. As we know from Chapter 2, the molecular properties of pure gases, liquids and solids are determined mainly by short-range forces. This is due to the fact that long-range forces decay very rapidly with increasing separation distance. For example, the van der Waals interaction pair potential, $V(r)$, decreases with the inverse sixth power of the distance between molecules, r^{-6}, so that the van der Waals pair energy of two neighboring molecules is at least 64 times stronger than that between next neighbors, and thus it is effective only for the first contacting molecules. Nevertheless, on the contrary, the Coulomb interactions in a vacuum are effectively long-ranged such that the energy decays slowly as $1/r$; but this is not stable for other media, and when a solvent such as water, with a high dielectric constant, is present the strength of the Coulomb interaction is very much reduced. In summary, long-range interactions play only a minor role in the total intermolecular interactions. However, when interactions between macroscopic particles and surfaces, instead of molecules, are considered, only long-range interactions (especially van der Waals interactions) are operative. As we will see in Section 7.3 the strength of these interactions depends on the geometry and size of macrobodies, and interactions decay much more slowly with the separation distance.

As detailed in Chapter 2, van der Waals interactions consist mainly of three types of long-range interactions, namely Keesom (dipole–dipole angle-averaged orientation, Section 2.4.3), Debye (dipole-induced dipolar, angle-averaged, Section 2.5.7), and London dispersion interactions (Section 2.6.1). However, only orientation-independent London dispersion interactions are important for particle–particle or particle–surface attractions, because Keesom and Debye interactions cancel unless the particle itself has a permanent dipole moment, which can occur only very rarely. Thus, it is important to analyze the London dispersion interactions between macrobodies. Estimation of the value of dispersion attractions has been attempted by two different approaches: one based on an extended molecular model by Hamaker (see Sections 7.3.1–7.3.5) and one based on a model of condensed media by Lifshitz (see Section 7.3.7).

7.2 Combining Rules for Molecular, Particle and Surface Interactions

It is possible to express any molecular interaction semi-quantitatively, independent of the type of interaction involved. All particle–particle, particle–surface, surface–surface and sometimes even complex multi-component interactions obey the *combining rules* of molecular interactions, except hydrogen-bonding interactions.

When two molecules are located at a given separation, their molecular interaction is proportional to the product of their most effective molecular property. For example, if molecule (1) has some effective molecular property such as dispersion forces, then we have $\lambda_1 \propto \alpha_1$, and similarly $\lambda_2 \propto \alpha_2$ for molecule (2), where λ denotes the molecular property and α the polarizability of these molecules. Then the binding potential energy between molecules (1) and (2) in contact may be expressed as

$$V_{12} = -\lambda_1\lambda_2 = K\alpha_1\alpha_2 \tag{516}$$

where K is a constant. Similarly, the binding potential energy between molecules (1) and (1) can be expressed as $[V_{11} = -\lambda_1^2]$, and the binding potential energy between molecules (2) and (2) as $[V_{22} = -\lambda_2^2]$. If the type of interaction is different, for example for charged molecules, then we may use $(\lambda_1 \propto q_1^2)$ and $(\lambda_2 \propto q_2^2)$, or if the molecules are polar, we may use $(\lambda_1 \propto \mu_1^2)$ and $(\lambda_2 \propto \mu_2^2)$, or $(\lambda_1 \propto \mu_1)$ and $(\lambda_2 \propto \mu_2)$, depending on the presence of induction interactions; but Equation (516) remains the same (other than the term on the right-hand side), and the only exception is the reversal of the negative sign for the purely Coulombic charge–charge interactions.

Now, if we assume that a liquid consists of a mixture of molecules (1) and (2) in equal numbers, and there are only six molecules to surround a single molecule to form a two-dimensional cluster, then there are two possibilities for molecular placement, as shown in Figure 7.1. In the *randomly dispersed state*, which is shown in Figure 7.1 *a*, a single (1) molecule will have both molecules (1) and (2) in contact as nearest neighbors so that, on average, three (1) plus three (2) molecules will surround the central (1) molecule, as can also be calculated from probability. This is the same for the two-dimensional cluster where molecule (2) is located at the center. However, in the *associated state*, which is shown in Figure 7.1 *b*, a single (1) molecule will be surrounded by only six molecules of (1) in contact with it to form a two-dimensional cluster; and a single (2) molecule will have only six molecules of (2) in contact as nearest neighbors. Then, let us estimate which state is preferable, from the difference in the interaction potential energy, ΔV. If we count the bonds present between the molecules, there are three (1–1), three (2–2) and 18 (1–2) bonds present in the two small clusters in the dispersed state, as shown in Figure 7.1 *a*. Then we may write

$$V_{Disp} = (-3\lambda_1^2 - 3\lambda_2^2 - 18\lambda_1\lambda_2) = -3(\lambda_1^2 + \lambda_2^2 + 6\lambda_1\lambda_2) \tag{517}$$

On the other hand, there are 12 (1–1) and 12 (2–2) bonds present in the two small clusters in the associated state, as shown in Figure 7.1 *b*. Then we have

$$V_{Assoc} = (-12\lambda_1^2 - 12\lambda_2^2) = -12(\lambda_1^2 + \lambda_2^2) \tag{518}$$

The change in the interaction potential energy, ΔV, on going from the dispersed to associated state can be calculated as

$$\Delta V = V_{\text{Assoc}} - V_{\text{Dis}} = -12(\lambda_1^2 + \lambda_2^2) + 3(\lambda_1^2 + \lambda_2^2 + 6\lambda_1\lambda_2) \tag{519}$$

which can be simplified into

$$\Delta V = (-9\lambda_1^2 - 9\lambda_2^2 + 18\lambda_1\lambda_2) = -9(\lambda_1 - \lambda_2)^2 \tag{520}$$

The number 9 in Equation (520) is in conjunction with the replacement of 18 (1–2) bonds to nine (1–1) and nine (2–2) bonds, on going from the dispersed to associated state, as shown in Figure 7.1. If we apply this procedure for the molecular close-packing conditions in three dimensions, where 12 nearest neighbor molecules surround a central molecule, as shown in Figure 3.8, we would find $\Delta V = -22(\lambda_1 - \lambda_2)^2$, where 44 (1–2) bonds are broken, and 22 (1–1) and 22 (2–2) bonds are formed upon association. Therefore, we may write for any kind of cluster formation between two types of molecules, which are present in equal numbers, irrespective of their sizes and how many molecules are involved

$$\Delta V = -n(\lambda_1 - \lambda_2)^2 \tag{521}$$

where n is the number of like bonds (i.e., 1–1 and 2–2) that are formed in the association process. Since $(\lambda_1 - \lambda_2)^2$ must always be positive in Equation (521), then $\Delta V < 0$, that is $V_{\text{Assoc}} < V_{\text{Dis}}$. From this analysis, it is clear that the associated state of like molecules or particles is energetically preferable to the dispersed state. Consequently, we may conclude that, in a binary liquid mixture, the attraction between like molecules is more efficient and they tend to associate. However, we have not considered the effect of solvent molecules in this analysis, and if a solvent is present as a third component (3), its presence will affect the behavior of solute molecules, as shown in Section 7.6. On the other hand, Equation (521) can also be written as

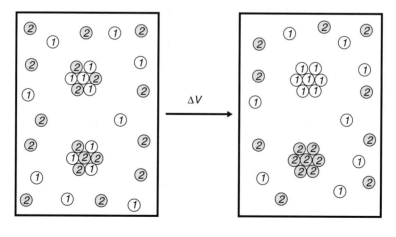

a. Dispersed State   b. Associated State

Figure 7.1 Schematic representation of molecular clusters in two dimensions, where six molecules surround a central single molecule: **a.** The *randomly dispersed state*, in which three (1) plus three (2) molecules surround the central (1) molecule on average, and similarly, three (1) plus three (2) molecules surround the central (2) molecule. **b.** The *associated state*, in which a single (1) molecule will be surrounded by only six molecules of (1) in contact with it, and similarly a single (2) molecule will be surrounded by only six molecules of (2).

$$\Delta V = -n(\lambda_1 - \lambda_2)^2 = -n(\sqrt{-V_{11}} - \sqrt{-V_{22}})^2 \tag{522}$$

which can be converted into

$$\Delta V = -n(\lambda_1^2 - 2\lambda_1\lambda_2 + \lambda_2^2) = n(V_{11} + 2\sqrt{V_{11}V_{22}} + V_{22}) \tag{523}$$

The $(-nV_{11})$ and $(-nV_{22})$ terms are proportional to their respective molar cohesive energies, ΔU_{11}^V and ΔU_{22}^V, as given in Sections 3.5.3 and 5.3.1, which are the molar internal energy of vaporization to the gas phase at zero pressure (i.e. infinite separation of the molecules). Thus, by using Equation (522), the change in the interaction potential energy, ΔV can also be written as

$$\Delta V \propto -\left(\sqrt{\Delta U_{11}^V} - \sqrt{\Delta U_{22}^V}\right)^2 \tag{524}$$

which is the semi-quantitative derivation of Hildebrand's solubility parameter equation (Equation (398)) given in Section 5.3.2. If the molecular interaction potential energies, V_{11} and V_{22} (and thus their respective cohesive energies, ΔU_{11}^V and ΔU_{22}^V,) are largely different from each other, then the value of ΔV will be large from Equations (522) and (524). This is the basis of the *immiscibility* property of two liquids such as water and hydrocarbons within each other, having a large ΔV. Since ΔV is defined as $\Delta V = V_{Assoc} - V_{Dis}$, in Equation (519), thus for a large negative value of ΔV, molecular association and the accompanying *phase separation* is preferred. If however, V_{11} and V_{22} (and thus ΔU_{11}^V and ΔU_{22}^V) are similar to each other, then the difference in interaction potential energy, ΔV, will be small, and molecular dispersion and the accompanying good dissolution of the two components within each other are preferred. This is the basis of the *like dissolves like* rule. In addition, since ΔV is proportional to the number of like bonds, n, then high-molecular-mass organic materials, polymers and large particles with large n values will give large ΔV and will be more easily phase separated in a solvent or in another polymer medium than will small molecules. In practice, this behavioral prediction was found to be valid for the majority of the liquid solutions that were examined experimentally.

However, when hydrogen bonds are present in any interaction, Equations (521)–(524) cannot be used because, for this case, the binding potential energy between molecules (1) and (2) cannot be expressed as in Equation (516). Since the hydrogen-bonding interactions are asymmetrical, $[V_{12} \neq -\lambda_1\lambda_2]$ is valid. As an example, no H-bonding interactions are present between acetone molecules, but they form H-bonds with the OH groups of water molecules via their C=O groups, and we cannot predict acetone–water miscibility by using Equations (521)–(524), depending on the individual properties of acetone and water. An unexpected new interaction (H-bonding) occurs when acetone and water molecules are mixed, so that we can predict only false ΔV values if we use Equations (521)–(524) for this purpose. In practice, acetone is miscible with water due to its H-bonding interactions, so the mixture prefers the dispersion state, $(V_{Assoc} > V_{Dis})$ giving $\Delta V > 0$ from Equation (519).

7.3   van der Waals Interactions Between Macroscopic Bodies

As we saw in Chapter 2, van der Waals forces consist only of long-range forces; the interaction pair potential, $V(r)$, decreases with the inverse sixth power of the distance between molecules, r^{-6}; and the corresponding interaction force, $F(r)$, decreases as r^{-7}. When particle–particle or particle–surface attractions are considered, polar Keesom and Debye terms cancel, unless the particle itself has a permanent dipole, which is rarely encountered. Thus, only orientation-independent London dispersion interactions can be used to derive expressions between macroscopic bodies, by using Hamaker's *microscopic approach*.

7.3.1 Microscopic approach of Hamaker between a molecule and a slab surface

H.C. Hamaker, in 1937, was the first to treat London dispersion interactions between macroscopic objects. He started with the most basic case, to determine the interaction of a single molecule with a planar solid surface. He considered a molecular pair potential and its relation with the molecules present within the solid surface, to derive the total interaction potential by summing the attractive interaction energies between all pairs of molecules, ignoring multibody perturbations. In this way, he built up the whole from the parts. Thus, Hamaker's method is called the *microscopic approach*.

As given in Equation (78) in Section 2.6.1, the dispersion pair interaction potential between hard sphere molecules (for $r > \sigma$, where σ is the diameter of the molecule) is expressed as $[V_d(r) = -C_d/r^6]$, and the dispersion interaction coefficient, C_d, is given as $[C_d = 3\alpha^2 I/4(4\pi\varepsilon_o)^2]$ (Equation (79)), where α is the polarizability, I is the first ionization potential of the molecule, and ε_o is the vacuum permittivity. Hamaker assumed that the interaction between a molecule and a nearby planar surface of a solid is purely attractive. Initially, he calculated the volume of the ring containing the interacting molecules in the solid, as shown in Figure 7.2. This is a usual procedure in physics such that the radius of

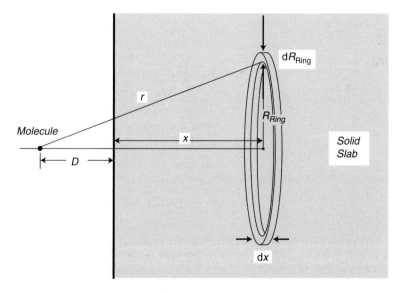

Figure 7.2 Interactions between a molecule and a solid slab surface. D is the distance between the molecule and the surface of the plane solid; R_{Ring} is the radius of the ring; x is the distance of the ring to the side of the solid; r is the distance between the molecule and the ring, which is equal to the hypotenuse of the right-angled triangle formed.

the ring, R_{Ring} and the distance of the ring to the side of the solid, x, may be extended to infinity during the integration process to cover all the molecules in the solid slab. The volume of the circular ring can be given as $[V = 2\pi R_{Ring}dR_{Ring}dx]$, from geometry. If the term, ρ, denotes the number density of the molecules (molecule/m³), then the total number of molecules in the ring will be $= \rho 2\pi R_{Ring}dR_{Ring}dx$. The additivity principle of pair potentials is assumed to sum the interactions of all the molecules present in the solid body with this single molecule, and if we approximate this summation process with an integral, then the net interaction energy for a molecule of the same material, at a distance D away from the surface of the plane solid will therefore be

$$V(D) = \iint [V_d(r)]\rho 2\pi R_{Ring}dR_{Ring}dx = -\pi C_d\rho \int_{x=D}^{x=\infty} \int_{R_R=0}^{R_R=\infty} \frac{2R_{Ring}dR_{Ring}}{r^6} dx \tag{525}$$

If we apply Pythagoras's theorem of plane geometry to the right-angled triangle in Figure 7.2, we have $[r^2 = (D + x)^2 + R_{Ring}^2]$, and inserting this expression into Equation (525) we obtain

$$V(D) = -\pi C_d\rho \int_{x=D}^{x=\infty} \int_{R_R=0}^{R_R=\infty} \frac{2R_{Ring}dR_{Ring}}{\left[(D+x)^2 + R_{Ring}^2\right]^3} dx \tag{526}$$

Since $[2R_{Ring}dR_{Ring} = d(R_{Ring}^2)]$, we have

$$V(D) = -\pi C_d\rho \int_{x=D}^{x=\infty} \int_{R_R=0}^{R_R=\infty} \frac{d(R_{Ring}^2)}{\left[(D+x)^2 + R_{Ring}^2\right]^3} dx$$

$$= -\pi C_d\rho \int_{x=D}^{x=\infty} \left\{ -\frac{1}{2\left[(D+x)^2 + R_{Ring}^2\right]^2} \right\}_0^\infty dx = -\frac{\pi C_d\rho}{2} \int_{x=D}^{x=\infty} \left[\frac{1}{(D+x)^4} \right] dx \tag{527}$$

After integrating the last term in Equation (527), we obtain

$$V(D) = -\frac{\pi C_d\rho}{2} \left[-\frac{1}{3(D+x)^3} \right]_0^\infty = -\frac{\pi C_d\rho}{6D^3} \tag{528}$$

As a result of this analysis, the interaction potential energy of a molecule and a macroscopic surface decreases proportional to D^{-3}, instead of D^{-6} for a molecule–molecule interaction; so the decrease is less steep for the former case than the latter. In order to perform this integration and simplification, Hamaker used various assumptions:

1 Multibody interactions are ignored and the interactions are only pair-wise.
2 The intervening medium is a vacuum.
3 The molecule and the solid body are not distorted by the attractive forces.
4 Interactions due to the Coulomb forces and permanent dipoles are neglected.
5 All the dispersion force attractions are due to a single dominant frequency.
6 The interactions of molecular electron clouds are instantaneous.
7 The solid body is assumed to have uniform density right to the interface.

The corresponding dispersion force, $F(D)$, between a molecule and a solid-plane surface can be calculated from Equation (16) in Section 2.2, by differentiating the $V(D)$ with D as

$$F(D) = -\frac{\partial V(D)}{\partial D} = -\frac{\pi C_d \rho}{2D^4} \qquad (529)$$

Force, $F(D)$ can also be derived in a similar way by integrating all the pair forces resolved along the x-axis. Force, $F(D)$ is usually applied in the force measurement experiments given in Section 7.4, whereas the total interaction energy, $V(D)$ is applied to the free-energy calculations. In thermodynamic terms, $V(D)$ is the work done in bringing the single molecule by attractive forces from infinity to a distance D from the solid slab, at constant temperature, and this work is equal to the Gibbs free energy, due to the dispersion interactions present,

$$\Delta G^d = V(D) = -\frac{\pi C^d \rho}{6D^3} = -\frac{\rho \alpha^2 I}{128 D^3 \pi \varepsilon_o^2} \qquad (530)$$

7.3.2 Microscopic approach of Hamaker between a spherical particle and a slab surface

By applying a similar procedure to that given above, we can now calculate the interaction potential energy between a large spherical particle having a radius of R_{Sph} and a vertical solid slab having a flat surface, as shown in Figure 7.3. The volume of the thin circular slice in the sphere is $\pi R_{Slice}^2 \, dx$. We can relate R_{Slice} with the radius of the sphere, R_{Sph}, by using the *chord theorem* of plane geometry. As shown in Figure 7.4, we can only draw a right-angled triangle ABC in a semi-circle. If we apply Pythagoras's theorem to all such triangles

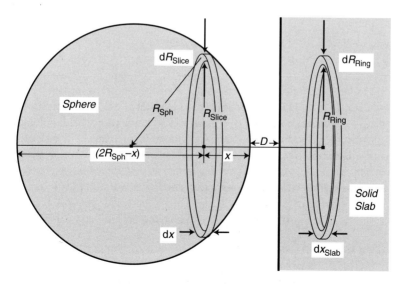

Figure 7.3 Interactions between a spherical particle and a solid slab surface: D is the distance between the particle and the surface of the plane solid; R_{Sph} is the radius of the spherical particle; R_{Ring} is the radius of the ring; R_{Slice} is the radius of the slice in the spherical particle. Other terms are self-descriptive and $[R_{Slice}^2 = x(2R_{Sph} - x)]$ from the chord theorem.

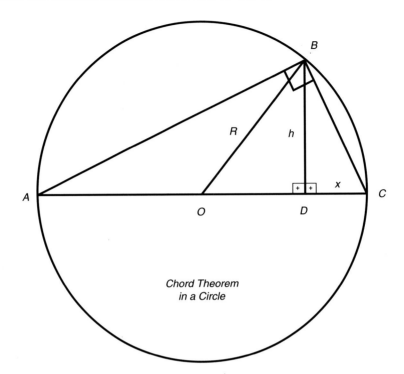

Figure 7.4 Right-angled triangle, *ABC*, captive in a semi-circle. The drawing is used in the proof of the *chord theorem* of plane geometry, resulting in the expression $[h^2 = x(2R - x)]$, which will be used in Hamaker constant calculations.

in Figure 7.4, we have $[(AC)^2 = (AB)^2 + (BC)^2]$ giving $(AC)^2 = [(AD)^2 + (BD)^2] + [(BD)^2 + (CD)^2]$, and after simplification and rearrangement of the terms given in Figure 7.4, we have $[(2R)^2 = (2R - x)^2 + 2h^2 + x^2]$ giving $[h^2 = x(2R - x)]$, which corresponds to $[R_{Slice}^2 = x(2R_{Sph} - x)]$ in Figure 7.3. Then, the total volume of the thin circular slice in the sphere particle can be found from $[V = \pi R_{Slice}^2 \, dx = \pi(2R_{Sph} - x)x \, dx]$, and the total number of molecules in this slice becomes $= \pi\rho(2R_{Sph} - x)x \, dx$. It is obvious that all the molecules in this sphere are at a distance of $(D + x)$ from the plane surface. Since we can use Equation (528) to calculate the dispersion interaction potential of a single molecule with the plane surface, we can then find the total sphere–surface dispersion interaction potential by multiplying the number of molecules within the spherical particle with Equation (528), so that

$$V(D) = \int -\frac{\pi C_d \rho}{6(D+x)^3} \left[\pi\rho(2R_{Sph} - x)x \, dx \right] \tag{531}$$

simplifying, we have

$$V(D) = -\frac{\pi^2 C_d \rho^2}{6} \int_{x=0}^{x=2R_{Sph}} \frac{(2R_{Sph} - x)x}{(D+x)^3} \, dx \tag{532}$$

Equation (532) shows the net dispersion interaction energy for a sphere at a distance D away from the surface of the plane solid, if the additivity principle is assumed. There are two possibilities. When $R_{Sph} \gg D$, only small values of x (for the $x \approx D$ case) can contribute to the integral so that Equation (532) becomes

$$V(D) = -\frac{\pi^2 C_d \rho^2}{6} \int_{x=0}^{x=\infty} \frac{2R_{Sph} x}{(D+x)^3} dx = -\frac{\pi^2 C_d \rho^2 R_{Sph}}{6D} \tag{533}$$

Equation (533) shows that the dispersion interaction potential energy is directly proportional to the radius of sphere, and it does not decay as $1/r^6$, but as $1/D$, which is much slower with the separation distance. On the other hand, if $(D \gg R_{Sph})$, then we can replace $(D + x)$ by D in Equation (532), we then have

$$V(D) = -\frac{\pi^2 C_d \rho^2}{6} \int_{x=0}^{x=2R_{Sph}} \frac{(2R_{Sph} - x)x}{D^3} dx = -\frac{2\pi^2 C_d \rho^2 R_{Sph}^3}{9D^3} \tag{534}$$

We can also derive Equation (534) via another route. Since the total number of molecules in a sphere is $= 4\pi R_{Sph}^3 \rho/3$, then by multiplying the interaction potential energy of one molecule, given by Equation (528), with the total number of molecules present in the sphere, we can obtain the same expression as Equation (534):

$$V(D) = -\frac{\pi C_d \rho}{6D^3} \left(\frac{4\pi R_{Sph}^3 \rho}{3} \right) = -\frac{2\pi^2 C_d \rho^2 R_{Sph}^3}{9D^3} \tag{535}$$

The $(C_d \rho^2)$ term for the same type of molecule–surface and the $(C_d^{12} \rho_1 \rho_2)$ term for different types of molecule–surface are the material-related constants for an interaction. Hamaker collected these material constants in *Hamaker interaction constants*, (A_{11}), (A_{22}) and (A_{12}) so that

$$A_{ii} = \pi^2 C_d^{ii} \rho_i^2 \tag{536}$$

for the interactions containing the same type of materials and

$$A_{12} = \pi^2 C_d^{12} \rho_1 \rho_2 \tag{537}$$

for the interactions containing two different types of materials. We should remember that, for two dissimilar molecules, the London dispersion interaction coefficient C_d, was given in Equation (80) in Section 2.6.1, as $C_d = -3\alpha_1 \alpha_2/2(4\pi\varepsilon_0)^2 (I_1 I_2/I_1 + I_2)$, and should be used when necessary in Equation (537).

The Hamaker constants are usually inserted in expressions for the potential energy of interaction between particles and surfaces. For a spherical particle–planar slab surface interaction of the same material, for the $(R_{Sph} \gg D)$ case, if we combine Equations (533) and (536), we have

$$V(D)_{Sphere-Surface} = -\frac{A_{11} R_{Sph}}{6D} \tag{538}$$

and for the $(D \gg R_{Sph})$ case, by combining Equations (534) and (536)

$$V(D)_{Sphere-Surface} = -\frac{2A_{11} R_{Sph}^3}{9D^3} \tag{539}$$

The A_{11}s will be replaced by A_{12}s in Equations (538) and (539), if two different kinds of interacting materials are present.

7.3.3 *Microscopic approach of Hamaker between spherical particles*

In most colloids, spherical or nearly spherical colloidal particles are present and interact with each other. Then it is important to calculate the sphere–sphere interaction potentials. For the most general case, where two spheres made of different materials with different radii are interacting as seen in Figure 7.5, it has been calculated that, with the same procedure as that given above

$$V(D)_{\text{Sphere–Sphere}} = -\frac{A_{12}}{6}\left\{ \begin{array}{c} \dfrac{2R_1R_2}{2(R_1+R_2)D+D^2} + \dfrac{2R_1R_2}{4R_1R_2+2(R_1+R_2)D+D^2} \\[2mm] +\ln\left[\dfrac{2(R_1+R_2)D+D^2}{4R_1R_2+2(R_1+R_2)D+D^2}\right] \end{array} \right\} \tag{540}$$

However, if we use the separation distance, ξ, between the centers of these spheres $[\xi = D + (R_1 + R_2)]$, instead of the separation distance of their surfaces, D, as shown in Figure 7.5, then Equation (540) can be written as

$$V(D)_{\text{Sphere–Sphere}} = -\frac{A_{12}}{6}\left[\begin{array}{c} \dfrac{2R_1R_2}{\xi^2-(R_1+R_2)^2} + \dfrac{2R_1R_2}{\xi^2-(R_1-R_2)^2} \\[2mm] +\ln\dfrac{\xi^2-(R_1+R_2)^2}{\xi^2-(R_1-R_2)^2} \end{array} \right] \tag{541}$$

For two spheres having the same radii ($R_1 = R_2$), Equation (540) simplifies to

$$V(D)_{\text{Sphere–Sphere}} = -\frac{A}{6}\left\{ \begin{array}{c} \dfrac{2R^2}{4RD+D^2} + \dfrac{2R^2}{4R^2+4RD+D^2} \\[2mm] +\ln\left[\dfrac{4RD+D^2}{4R^2+4RD+D^2}\right] \end{array} \right\} \tag{542}$$

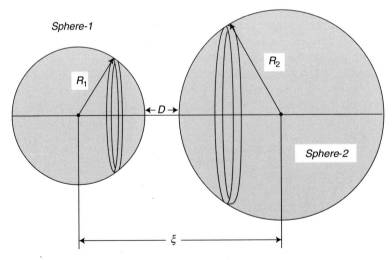

Figure 7.5 Interactions between two spherical particles having different radii, R_1 and R_2. D is the distance between the particles, and ξ is the distance between the centers of particles.

The $(R_{Sph} \gg D)$ case is important for force measurement experiments between spherical particles. If there are two large spheres of different radii R_1 and R_2, and made of different materials, having a small distance D between their surfaces, as shown in Figure 7.5, we can simplify Equation (540) to give

$$V(D)_{\text{Sphere--Sphere}} \cong -\frac{A_{12}}{6D} \left(\frac{R_1 R_2}{R_1 + R_2} \right) \tag{543}$$

If there are two spheres of equal radius, R_{Sph}, then the interaction potential can be calculated as half the value of the sphere–surface interaction potential given by Equation (538)

$$V(D)_{\text{Sphere--Sphere}} \cong -\frac{A_{12} R_{Sph}}{12D} \tag{544}$$

We may also obtain Equation (544) by inserting $(R_1 = R_2)$ into Equation (543) for two spheres having the same radius. On the other hand, for the rarely encountered $(D \gg R_{Sph})$ case, for two spheres having the same radius $(R_1 = R_2)$, which are far apart, it was calculated that

$$V(D)_{\text{Sphere--Sphere}} = -\frac{16 A_{12} R_{Sph}^6}{9D^6} \tag{545}$$

7.3.4 Microscopic approach of Hamaker between parallel slab surfaces

If two parallel planar surfaces of materials (1) and (2) are at a distance D apart, as shown in Figure 7.6, we can calculate their total dispersion interaction energy potential in a similar procedure to that given above. By using Equation (528), which gives the interaction potential energy of a molecule and a macroscopic surface, the total interaction potential energy between two infinitely extended solid slabs (1) and (2) can be calculated by integrating the pair value over all the molecules in the solid slab (1)

$$V(D) = -\frac{\pi C_d^{12} \rho_2}{6} \iiint \frac{\rho_1}{(D+x)^3} \, dV_v = -\frac{\pi C_d^{12} \rho_2}{6} \int_0^\infty \int_{-\infty}^\infty \int_{-\infty}^\infty \frac{\rho_1 dz dy dx}{(D+x)^3} \tag{546}$$

where V_v is the volume of the slab. The integral is infinite because we assume that the solid slabs are infinitely large. Since we need to calculate the dependence of the interaction potential on the separation distance between two slabs, we need to divide the total potential by the area, which is given by the y and z coordinates parallel to the gap:

$$V(D)_{\text{unit--area}} = \frac{V(D)}{A_{\text{Slab--Slab}}} = -\frac{\pi C_d^{12} \rho_2}{6} \int_0^\infty \frac{\rho_1 dx}{(D+x)^3} = -\frac{\pi C_d^{12} \rho_1 \rho_2}{6} \left[-\frac{1}{2(D+x)^2} \right]_0^\infty \tag{547}$$

giving,

$$V(D)_{\text{unit--area}} = -\frac{\pi C_d^{12} \rho_1 \rho_2}{12D^2} \tag{548}$$

Equation (548) can also be re-derived if we consider the unit area of a thin sheet of molecules having a thickness of dx and x away from the surface of a second slab (1), and similar to the derivation of Equation (528), we can write for the same kind of interacting materials

$$V(D)_{\text{unit-area}} = -\frac{\pi C_d \rho^2}{6} \int\limits_{x=D}^{x=\infty} \frac{dx}{x^3} = -\frac{\pi C_d \rho^2}{12D^2} \tag{549}$$

In practice, Equation (549) can be used when D is very small compared with the lateral dimensions of the slab. When Equations (536) and (549) are combined, we have

$$V(D)_{\text{Surface-Surface}} = -\frac{A_{11}}{12\pi D^2} \tag{550}$$

The term A_{12} can be used instead of A_{11} in Equation (550) if the surfaces of two slabs of different materials interact with each other. If the slabs are not very thick, having a thickness of Λ_1 and Λ_2, then the dispersion interaction potential can be given as

$$V(D)_{\text{Surface-Surface}} = -\frac{A_{12}}{12\pi} \begin{bmatrix} \dfrac{1}{D^2} - \dfrac{1}{(D+\Lambda_1+\Lambda_2)^2} \\[2mm] -\dfrac{1}{(D+\Lambda_1)^2} - \dfrac{1}{(D+\Lambda_2)^2} \end{bmatrix} \tag{551}$$

If the thickness of the slabs Λ_1 and Λ_2 is very large, so that $\Lambda_1, \Lambda_2 \to \infty$, then we obtain Equation (550) again.

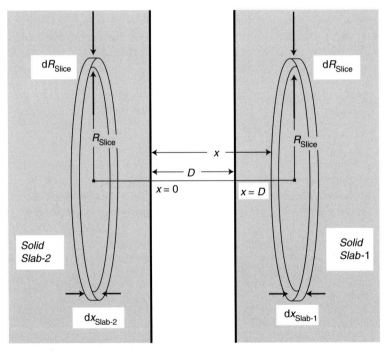

Figure 7.6 Interactions between two solid slab surfaces. D is the distance between the solid slabs, x is the abscissa and R_{Slice} is the radius of the slice in the slabs.

7.3.5 Microscopic approach of Hamaker between cylinder surfaces

If dispersion interactions between two parallel cylinder-shaped materials are considered, as shown in Figure 7.7 *a*, then the final equation can be written as

$$V(D)_{\text{Parallel Cylinders}} = -\frac{\pi^2 C_d^{12} \rho_1 \rho_2 L}{12\sqrt{2}D^{3/2}}\left(\frac{R_1 R_2}{R_1 + R_2}\right) = -\frac{A_{12}L}{12\sqrt{2}D^{3/2}}\left(\frac{R_1 R_2}{R_1 + R_2}\right) \tag{552}$$

where R_1 and R_2 are the radii, and L is the length of the cylinder. If the cylinders are located as crossed with each other, as shown in Figure 7.7 *b*, Equation (552) becomes

$$V(D)_{\text{Crossed Cylinders}} = -\frac{\pi^2 C_d^{12} \rho_1 \rho_2 \sqrt{R_1 R_2}}{6D} = -\frac{A_{12}\sqrt{R_1 R_2}}{6D} \tag{553}$$

Equation (553) shows the important simplification of the dispersion interaction potential between two crossed cylinders, and thus this geometry is preferred for use in the *surface force apparatus*, which we will see in Section 7.4.

7.3.6 Comparison of sphere–surface and sphere–sphere interactions with surface–surface interactions: Langbein approximation

It is clear that the geometry of particles has an important effect on the interaction potential with other particles and surfaces. It is, then, a good idea to investigate the *effective area* of the interaction. For the same dispersion interaction potentials, if we consider sphere–surface and surface–surface interactions, by equalizing Equations (538) and (550), we have

$$2\pi R_{\text{Sph}}D = 1 \tag{554}$$

If we remember that Equation (538) is derived for a unit area = 1 of parallel slabs, it is clear from Equation (554) that the dispersion interaction potentials are the same if the effective interaction area in a sphere–surface interaction is equal to

$$A_{\text{eff}}^{\text{Sph–Surface}} = 2\pi R_{\text{Sph}}D \tag{555}$$

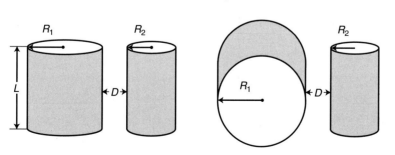

a. Two parallel cylinders   b. Two crossed cylinders

Figure 7.7 Interactions between two cylinder surfaces having the same length, L, but different radii, R_1 and R_2. **a.** Between two parallel cylinders having a distance, D between them. **b.** Between two crossed cylinders.

for the same surface separation of D between the sphere–surface and surface–surface. Since we know that $[R_{Slice}^2 = (2R_{Sph} - x)x]$, from the chord theorem of plane geometry, we can write $[R_{Slice}^2 \approx 2R_{Sph}D]$ for the $(R_{Sph} \gg D$ and $x = D)$ case, so that Equation (555) becomes

$$A_{Eff}^{Sph-Surface} \cong \pi R_{Slice}^2 \qquad (556)$$

which is in conjunction with the geometrical considerations that can be also derived from Figure 7.3. Equation (556) shows that the effective area of interaction of sphere–surface is the circular zone inside the sphere. This conclusion is termed the *Langbein approximation*. It is clear that both $A_{Eff}^{Sph-Surface}$ and R_{Slice} are dependent on the radius of the spherical particle, R_{Sph}, and the separation distance, D, so that the decrease in both R_{Sph} and D lowers the value of R_{Slice}. In practice, during experimental measurement, R_{Slice} values vary between 20 and 60 nm, for small microparticles having a radius of 1–2 μm and for separations of 0.2–1.0 nm from a planar slab (x values are in the range of 1–1.5 nm). However, during actual force measurement, an amount of elastic flattening usually occurs, and this increases the contact area considerably.

If we consider two contacting $(D = 0)$ spheres having the same radius, the effective area of the sphere$_1$–sphere$_1$ interaction for this case can be expressed as $[A_{Eff}^{Sph-Sph} = \pi R_{Slice}^2 = \pi(2R_{Sph} - x)x]$. For the $(R_{Sph} \gg x)$ case, this expression reduces to $[A_{Eff}^{Sph-Sph} \approx 2\pi R_{Sph}x]$. If the two contacting spheres have different radii, then we can write $A_{Eff}^{Sph_1-Sph_2} \approx 4\pi x(R_{Sph_1}R_{Sph_2}/R_{Sph_1} + R_{Sph_2})$ to show the effective area of the sphere$_1$–sphere$_2$ interaction.

7.3.7 Derjaguin approximation

There is a need to convert interaction potential energies into interaction forces in order to apply the Hamaker theory to experimental measurements. The corresponding interaction force can be calculated from Equation (16) in Section 2.2 and Equation (529) in Section 7.3.1, by differentiating $V(D)$ with D for objects having simple geometric shapes. However, when the interacting materials have irregular shapes, it is a very difficult task to calculate such forces by direct differentiation. B. V. Derjaguin, in 1934, proposed an approximation to overcome this problem for the case where the separation distance is small in comparison to the curvature of the interacting surfaces. In this approximation, he generalized the relationship between a sphere–surface interaction with the surface–surface interaction. If we differentiate Equation (538), given for the sphere–surface interaction potential with D, then the force between sphere and surface is

$$F(D)_{Sphere-Surface} = -\frac{\partial V(D)_{Sphere-Surface}}{\partial D} = -\frac{A_{12}R_{Sph}}{6D^2} \qquad (557)$$

If we repeat this procedure for surface–surface interaction, we can write for unit area from Equation (550)

$$F(D)_{Surface-Surface} = -\frac{\partial V(D)_{Surface-Surface}}{\partial D} = -\frac{A_{12}}{6\pi D^3} \qquad (558)$$

For sphere–sphere interaction for spheres having two different radii, from Equation (543) we have

$$F(D)_{\text{Sphere–Sphere}} = -\frac{\partial V(D)_{\text{Sphere–Sphere}}}{\partial D} = -\frac{A_{12}}{6D^2}\left(\frac{R_1 R_2}{R_1 + R_2}\right) \tag{559}$$

and for two crossed cylinders at right angles having different radii, from Equation (553), we have

$$F(D)_{\text{Crossed Cylinders}} = -\frac{\partial V(D)_{\text{Crossed Cylinders}}}{\partial D} = -\frac{A_{12}\sqrt{R_1 R_2}}{6D^2} \tag{560}$$

However, Equation (557) can also be related to the interaction potential energy between surfaces

$$F(D)_{\text{Sphere–Surface}} = -\frac{A_{12} R_{\text{Sph}}}{12\pi D^2}\left(2\pi R_{\text{Sph}}\right) = 2\pi R_{\text{Sph}} V(D)_{\text{Surface–Surface}} \tag{561}$$

and similarly we can write Equation (559) as

$$F(D)_{\text{Sphere–Sphere}} = -\frac{A_{12}}{12\pi D^2}\left[2\pi\left(\frac{R_1 R_2}{R_1 + R_2}\right)\right] = 2\pi\left(\frac{R_1 R_2}{R_1 + R_2}\right) V(D)_{\text{Surface–Surface}} \tag{562}$$

Equations (561) and (562) are examples of the *Derjaguin approximation* which gives the force between two spheres in terms of the energy per unit area of two flat surfaces at the same separation D. If one sphere is very large so that $(R_1 \gg R_2)$, we can obtain Equation (561) from Equation (562), corresponding to the limiting case of a sphere near a flat surface. For two crossed cylinders having different radii, the Derjaguin approximation gives

$$F(D)_{\text{Crossed Cylinders}} = 2\pi\sqrt{R_1 R_2}\left(\frac{V(D)_{\text{Surface–Surface}}}{\sin\phi_{\text{cross}}}\right) \tag{563}$$

where ϕ_{cross} is the cross-angle between the two cylinders. Equation (563) gives Equation (560) if $\phi_{\text{cross}} = 90°$, and in addition, for cylinders having the same radius $(R_1 = R_2)$, we again obtain Equation (561) from Equation (563), corresponding to the case of a sphere near a flat surface.

The Derjaguin approximation can be applied to any type of force law, such as attraction, repulsion, or oscillation, if D is much less than the radii of the spheres. This has been verified experimentally and is very useful for interpreting experimental force data. The Derjaguin approximation shows that, even though the same pair-potential force is operating, the distance dependence of the force between two curved surfaces is quite different from that between two flat surfaces.

7.3.8 Macroscopic approach of Lifshitz

In condensed phases such as liquids, two molecules are not isolated but have many other molecules in their vicinity. As we have seen in Section 2.5, the effective polarizability of a molecule changes when other molecules surround it. In Hamaker's treatment, the additivity of pair-potentials is assumed and the influence of the neighboring molecules is ignored, in contrast with interactions in real condensed systems. In addition, the additivity approach cannot be applied to molecules interacting in a third medium, as we have seen in Sections 2.5.3, 2.6.4 and 2.7.4. Therefore a theory is needed to avoid these weaknesses in the Hamaker approach. E. M. Lifshitz, in 1956, introduced an alternative, molar approach which is called the *macroscopic approach*, to derive the interactions between massive particles. In this theory, Lifshitz treated the objects in continuum, that is, sizes and distances are large compared to atomic dimensions and have bulk properties such as dielectric permittivity and refractive index. In this model, the solvent medium is assumed to be a structureless continuum defined solely in terms of its bulk properties, such as density or dielectric constant etc. These bulk properties hold, right down to molecular dimensions. Lifshitz neglected the discrete atomic structure of the materials. This approach has its origin in Maxwell's equations, where electric and magnetic fields are subjected to fast temporal fluctuations. The van der Waals pressure, according to Lifshitz's theory, can be expressed in terms of continuum properties, i.e. dielectric constant (relative permittivity), ε, of the interacting phases under consideration.

The Lifshitz theory also considered the *retardation effect* originating from quantum theory. Basically, the dispersion interactions arise from mutual interactions between fluctuating dipoles and propagate at the speed of light. However, the question arises: when are they in phase and out of phase? If the separation distance is larger than a specified length, the propagation time for the field becomes comparable to the oscillating dipole period and thus becomes out of phase. The field then interacts with another instantaneous dipole in a different phase, and the attraction force decreases. This retardation effect will diminish the low-frequency parts of the interaction at the longest distances, while the higher-frequency contributions will be suppressed at smaller separations down to 5 nm. Lifshitz applied an analysis of retardation effect through quantum field theory, by applying quantum electrodynamics to calculate the lowering of the zero-point energy of a particle, due to the coordination of its instantaneous electric moments with those of a nearby particle. He finally obtained the same expressions as Hamaker; but in Lifshitz's treatment, the Hamaker constants can only be calculated from different parameters. The polarizability, α, and the first ionization potential, I, in the Hamaker equations are replaced by the static and frequency-dependent dielectric constant, ε, and refractive index, n. There is an important difference though, in Lifshitz's macroscopic approach; the Hamaker constant is not a constant but a variable that depends on the separation distance.

Since Lifshitz's derivation is too difficult and beyond the scope of this book, the Hamaker constant based on the Lifshitz theory is expressed as

$$A \approx \frac{3}{4}kT\left(\frac{\varepsilon_1-\varepsilon_3}{\varepsilon_1+\varepsilon_3}\right)\left(\frac{\varepsilon_2-\varepsilon_3}{\varepsilon_2+\varepsilon_3}\right)+\frac{3h}{4\pi}\int_{v1}^{\infty}\left(\frac{\varepsilon_1(iv)-\varepsilon_3(iv)}{\varepsilon_1(iv)+\varepsilon_3(iv)}\right)\left(\frac{\varepsilon_2(iv)-\varepsilon_3(iv)}{\varepsilon_2(iv)+\varepsilon_3(iv)}\right)dv \qquad (564)$$

where ε_1, ε_2 and ε_3 are the static dielectric constants of the three media, k is the Boltzmann constant, h is Planck's constant, and v is the frequency. As we know from Chapter 2, in the derivation of the London dispersion interaction expressions, the molecules were assumed to have only single ionization potentials (one absorption frequency) in free space; however if many surrounding molecules or a third (solvent) medium are present, then the single absorption frequency cannot be used. It is obvious that the dielectric constants are not constants themselves but dependent on the frequency of the electric field, so $\varepsilon(iv)$s are the values of ε_i at imaginary frequencies. Then, the first term in Equation (564) gives the zero-frequency energy, including the Keesom and Debye interactions, whereas the second term gives the frequency-dependent London dispersion energy contributions; which is the most important part of the equation.

It is essential to know the relation of $\varepsilon(iv)$ with v, in order to calculate the Hamaker constant from the sum over many frequencies. The static dielectric constants, ε_1, ε_2 and ε_3 are the values of this function at zero frequency. The integral in Equation (564) has a lowest value of $v1 = 2\pi kT/h = 3.9 \times 10^{13}\,\mathrm{Hz}\,(s^{-1})$ at 25°C. This corresponds to a wavelength of 760 nm. If we assume that the major contribution to the Hamaker constant comes from the frequencies in the visible light or UV region, the relation of $\varepsilon(iv)$ with v can be given as

$$\varepsilon(iv)=1+\frac{n^2-1}{\left(1+v^2/v_e^2\right)} \tag{565}$$

where n is the refractive index, and v_e is the common *mean ionization frequency* of the material, which is assumed for simplicity. It is now possible to express the Hamaker constant in terms of the McLachlan equation (Equation (92)) given in Section 2.6.4.

$$\begin{aligned}A_{132} \approx &\frac{3}{4}kT\left(\frac{\varepsilon_1-\varepsilon_3}{\varepsilon_1+\varepsilon_3}\right)\left(\frac{\varepsilon_2-\varepsilon_3}{\varepsilon_2+\varepsilon_3}\right)\\ &+\frac{3hv_e}{8\sqrt{2}}\frac{\left(n_1^2-n_3^2\right)\left(n_2^2-n_3^2\right)}{\sqrt{\left(n_1^2+n_3^2\right)}\sqrt{\left(n_2^2+n_3^2\right)}\left[\sqrt{\left(n_1^2+n_3^2\right)}+\sqrt{\left(n_2^2+n_3^2\right)}\right]}\end{aligned} \tag{566}$$

where A_{132} is the Hamaker constant between materials (1) and (2) in a medium of (3), n_1 and n_1 are the refractive indices of molecules (1) and (2), and n_3 is the refractive index of the solvent medium. For two identical objects interacting across medium (3), we can simplify Equation (566) to

$$A_{131} \approx \frac{3}{4}kT\left(\frac{\varepsilon_1-\varepsilon_3}{\varepsilon_1+\varepsilon_3}\right)^2+\frac{3hv_e}{16\sqrt{2}}\frac{\left(n_1^2-n_3^2\right)^2}{\left(n_1^2+n_3^2\right)^{3/2}} \tag{567}$$

When the Hamaker constant is positive, it corresponds to attraction between molecules, and when it is negative, it corresponds to repulsion. By definition, $\varepsilon_3 = 1$ and $n_3 = 1$ for a vacuum. As we know from McLachlan's equation (Equation (92)), the presence of a solvent medium (3) rather than a free space considerably reduces the magnitude of van der Waals interactions. However, the interaction between identical molecules in a solvent is always attractive due to the square factor in Equation (567). On the other hand, the interaction between two dissimilar molecules can be attractive or repulsive depending on dielectric constant and refractive index values. Repulsive van der Waals interactions occur when n_3 is intermediate between n_1 and n_2 in Equation (566). If two bodies interact across a vacuum (or practically in a gas such as air at low pressure), the van der Waals forces are also attractive. When repulsive forces are present within a liquid film on a surface, the thickness of the film increases, thus favoring its spread on the solid. However, if the attractive forces are present within this film, the thickness decreases and favors contraction as a liquid drop on the solid (see Chapter 9).

All the above derivations of Lifshitz continuum approach are valid when the materials are electrically non-conductive (insulating), and the interacting surfaces are farther apart than molecular dimensions ($D \gg \sigma$). However, if we consider conductive materials such as metals, their static $\varepsilon = \infty$ and Equations (566) and (567) are not valid. For this case, it is possible to approximate the metal dielectric constant as

$$\varepsilon(iv) = 1 + \frac{v_e^2}{v^2} \qquad (568)$$

where v_e is the *plasma frequency of the free electron gas* in the range $4-5 \times 10^{15}$ Hz in this case. When Equation (568) is inserted in to Equation (564), we obtain the approximate Hamaker constant for two bodies of a metal interacting across a vacuum:

$$A_{11} \approx \frac{3}{16\sqrt{2}} hv_e \approx 4 \times 10^{-19} \text{ J} \qquad (569)$$

Since metals have high dielectric constants and refractive indices, they have much higher Hamaker constants than those of non-conducting materials.

Values of ε, n and v_e and Hamaker constants for two identical types of a material in a vacuum, which are calculated from Equation (567) by taking $\varepsilon_3 = 1$ and $n_3 = 1$, are given in Table 7.1. Unfortunately, the lack of material constants, such as the dielectric constant, as a function of frequency for most of the substances, and also the complexity of the derived formulae have hampered the general use of the Lifshitz model. However, Lifshitz theory made possible the advent of the first theories on the stability of hydrophobic colloids as a balance between London attraction and electrical double-layer repulsion. Later, these theories were further elaborated by Derjaguin and Landau, and independently by Verwey and Overbeek. The general theory of colloidal stability (which is beyond the scope of this book) is based on Lifshitz theory and has become known as the DLVO theory, by combining the initials of these four authors.

7.4 Experimental Measurement of the Hamaker Constant

With the development of the Hamaker and Lifshitz theories, many researchers have tried to measure forces to verify these approaches. Force measurements between a convex and flat glass and parallel quartz surfaces were carried out in the 1950s. One plate was fixed and the other plate was mounted on a spring where the deflection of the spring was measured by varying the separation distance over the range 100–1000 nm. There were various experimental problems: the surfaces had to be atomically flat, but the surface roughness present in real systems limited the separation distance resolution. For example, if a protrusion was present on a surface, the other surface could not approach any closer. Surfaces also had to be dust-free and charge-free. In order to overcome these experimental difficulties, very flat materials with very small interacting areas were used. Various research experiments carried out between 1954 and 1970 successfully validated the idea of retarded dispersion interactions estimated from the Lifshitz theory.

Later, in 1973, Tabor and Israelachvili developed a *surface force apparatus*, SFA, to measure the interaction force in a vacuum at the 1.5 nm level for the first time. In this equipment, the interaction forces between two crossed cylinders coated with freshly cleaved mica sheets having atomically smooth surfaces were measured. One of the cylinders is mounted on a piezoelectric transducer, and the other cylinder is mounted on a spring of known and adjustable spring constant with a force resolution down to 10^{-8} N. SFA has been further developed for performing measurements in liquids and vapors. In these developed SFA versions, the separation distance between the cylinders can be measured interfero-

Table 7.1   Dielectric constant (ε), refractive index (n), main absorption frequency in the UV region (v_e) and non-retarded Hamaker constants (A_{11}) for two identical liquids, solids and polymers at 20°C interacting across vacuum (or air), calculated using Equation (567), where $\varepsilon_3 = 1$ and $n_3 = 1$. *From Equation (569)

Material	ε	n	v_e $(10^{15}\,Hz)$	A_{11} $(10^{-20}\,J)$
n-Hexane	1.89	1.38	4.2	6.1
n-Octane	1.97	1.41	3.0, 3.9	6.5
n-Hexadecane	2.05	1.43	2.9	5.3
Cyclohexane	2.03	1.426	2.9	5.2
Ethanol	25.3	1.361	3.0	4.2
n-Propanol	20.8	1.385	3.1	4.9
n-Butanol	17.8	1.399	3.1	5.2
n-Octanol	10.3	1.430	3.1	5.8
Acetone	20.7	1.359	2.9	4.1
Benzene	2.28	1.501	2.1	5.0
Toluene	2.38	1.497	2.7	6.3
Chloroform	4.81	1.446	3.0	5.9
Carbon tetrachloride	2.24	1.460	2.7	5.5
Water	78.5	1.333	3.6	4.4
Polyethylene	2.29	1.50	2.6	6.1
Polystyrene	2.55	1.59	2.3	7.2
Polymethyl methacrylate	3.12	1.50	2.7	6.4
Polyethylene oxide	–	1.45	2.8	5.5
Polyvinyl chloride	4.55	1.54	2.9	7.9
Polytetrafluoroethylene	2.1	1.359	2.9	3.8
Polydimethyl siloxane	2.7	1.4	2.8	4.5
Nylon 6	3.8	1.53	2.7	7.1
SiO_2 (quartz)	4.8	1.54	3.2	8.7
SiO_2 (amorphous silica)	3.82	1.46	3.2	6.6
ZnO	11.8	1.91	1.4	8.8
TiO_2 (rutile)	11.4	2.46	1.2	14.6
NaCl	5.9	1.53	2.5	6.6
KCl	4.4	1.48	2.5	5.6
$KAl_2Si_3AlO_{10}(OH)_2$ (mica)	7.0	1.58	3.1	9.5
CaF_2 (fluorite)	6.7	1.43	3.8	7.0
C (diamond)	5.7	2.40	2.7	30.7
Al_2O_3	11.5	1.75	3.2	14.8
Metals (Au, Cu)	∞	–	3–5	25–40*

metrically down to a precision of 0.1 nm, and the apparatus has a sensitivity of better than $10^{-3}\,mJ\,m^{-2}$ for measuring interfacial energies. In addition, mica surfaces have been used as a substrate which is coated with the monolayer or polylayer films of other materials to determine their Hamaker constants.

Atomic force microscopy, AFM, which was developed in the 1980s is also used for interaction force measurement between surfaces. In AFM, the force between the micro-fabricated *tip*, which is placed at the end of an approximately $100\,\mu m$ long and $0.5-5\,\mu m$ thick *cantilever*, with another surface is measured; however this force is usually too small to give reasonable values. Then, a particle with a spherical, conical or any geometrical shape is attached to the cantilever, and good results can then be obtained between like and unlike surfaces. This method is also called the *colloidal probe technique*; the forces can be directly measured in liquids too. The interacting surfaces are much smaller in AFM than in SFA, so surface roughness and contamination problems are reduced, and AFM measurements are more rapid and simple.

7.5 Relation Between Hamaker Constant and Surface Tension

In Section 5.6.1, we defined, the *work of cohesion*, W_i^c, as the reversible work, per unit area, required to break a column of a liquid (or solid) into two parts, creating two new equilibrium surfaces, and separating them to infinite distance $[W_i^c = 2\gamma_i]$, as given by Equation (457). Similarly, the *work of adhesion*, W_{12}^a, is defined as the reversible work, per unit area, required to separate a column of two different liquids (or solids) at the interface, creating two new equilibrium surfaces of two pure materials, and separating them to infinite distance $[W_{12}^a = \gamma_1 + \gamma_2 - \gamma_{12}]$, as given by Equation (458). Under constant pressure and temperature conditions, the free energy of cohesion is the negative of the work of cohesion $[\Delta G_i^c = -W_i^c]$ and adhesion $[\Delta G_{12}^a = -W_{12}^a]$.

With two flat surfaces of material (1), each unit area is placed into a liquid (2), the change in the interaction potential energy, ΔV, on going from the dispersed to associated state is defined as $[\Delta V = V_{Assoc} - V_{Dis}]$ in Equation (519), and ΔV is related to the respective molar cohesive energies of the surfaces and the solvent, ΔU_{11}^V and ΔU_{22}^V, from Equations (523) and (524). In this treatment, $(-nV_{11})$ is proportional to ΔU_{11}^V, and $(-nV_{22})$ to ΔU_{22}^V. Thus, the binding potential energy, V_{11}, between molecules (1) and (1), can be related to the work of cohesion and also to their surface tension as

$$-nV_{11} = W_{11}^c = 2\gamma_1 \tag{570}$$

and the same applies for V_{22} as $[-nV_{22} = W_{22}^c = 2\gamma_2]$. Equation (570) is valid, in a vacuum, and the factor of two in this equation arises from eliminating these two unit areas while bringing the two surfaces into contact. With the same reasoning, the binding potential energy between dissimilar molecules (1) and (2), V_{12}, can be related to the work of adhesion and also to the interfacial tension between them as

$$-nV_{12} = W_{12}^a = 2\gamma_{12} \tag{571}$$

If we combine Equations (523) and (570) we obtain

$$\Delta V = -W_{11}^c + 2\sqrt{W_{11}^c W_{22}^c} - W_{22}^c = -2\left(\gamma_1 - 2\sqrt{\gamma_1\gamma_2} + \gamma_2\right) \tag{572}$$

Now we need to investigate the relation of ΔV with W_{12}^a and γ_{12}. Since ΔV is the change in the interaction potential energy on going from the dispersed to associated state, it is the negative free energy change of bringing two unit surfaces of (1) into adhesive contact with each other in the medium (2). When this happens, two unit areas of the 1–2 interface are eliminated to form a single 1–1 interface. Thus, ΔV can be defined as twice the interfacial energy of the 1–2 interface, γ_{12}, and from Equation (571) we can write

$$\Delta V = -2\gamma_{12} = -W_{12}^a = nV_{12} \tag{573}$$

Thus, it is now possible to rearrange Equation (572) by combining it with Equation (573)

$$W_{12}^a = W_{11}^c - 2\sqrt{W_{11}^c W_{22}^c} + W_{22}^c \tag{574}$$

or

$$\gamma_{12} = \gamma_1 + \gamma_2 - 2\sqrt{\gamma_1\gamma_2} = \left(\sqrt{\gamma_1} - \sqrt{\gamma_2}\right)^2 \tag{575}$$

If the interaction type is only dispersion, then Equation (575) can be used to estimate the interfacial energy, γ_{12}, from the surface energy of the interacting materials in the absence of experimental data on W_{12}^a (see Sections 9.1 and 10.1).

If we want to calculate the surface free energies of liquids and solids from the Hamaker constants, we must seek an equation relating the dispersion interaction potential with the Hamaker constant. Since the work of cohesion, W_i^c, is defined as the reversible work, per unit area, required to break a column of a liquid (or solid) into two parts, creating two new equilibrium surfaces, and separating them to infinite distance, it is clear that we must use a similar form of Equation (550), giving surface–surface interactions between two parallel slab surfaces. We cannot use Equation (550) directly because we only applied pairwise potential energy summation between the molecules in the solid slab with the identical molecules in the other solid slab during its derivation, and we did not apply the potential energy summation to the molecules in the same medium other than these slabs. If we carry out the integration between all the molecules (including the identical molecules in the slabs and all the other identical molecules in the same medium), we should have

$$V(D)_{\text{Total}} = \frac{A_{11}}{12\pi D_o^2} - \text{constant} \tag{576}$$

per unit area, where D_o is the effective molecular size, which is the separation distance between the two imaginary semi-infinite slabs within the material. (That is, these imaginary slabs are separated by just one molecular diameter distance.) In this treatment, the *constant* is the bulk cohesive energy of the molecules with their immediate neighbors at distance $(D = D_o)$ so that (constant = $A_{11}/12\pi D^2$). The first positive term arises from the *unsaturated* (or broken) bonds at the two surfaces upon cleavage of the material. Thus, Equation (576) can be written as

$$V(D)_{\text{Total}} = \frac{A_{11}}{12\pi D_o^2} - \frac{A_{11}}{12\pi D^2} = \frac{A_{11}}{12\pi}\left(\frac{1}{D_o^2} - \frac{1}{D^2}\right) \tag{577}$$

When the two surfaces are in contact, that is $(D = D_o)$, then Equation (577) gives $V(D)_{\text{Total}} = 0$; however when $(D = \infty)$, that is the newly created two new equilibrium surfaces are separated to infinite distance to form two isolated surfaces, we have

$$V(D)_{\text{Total}} = \frac{A_{11}}{12\pi D_o^2} \tag{578}$$

It is obvious that $V(D)_{\text{Total}}$ in Equation (578) corresponds to the free energy of cohesion, so that $[V(D)_{\text{Total}} = 2\gamma_1]$. Equating this to Equation (578) we obtain

$$\gamma_1 = \frac{A_{11}}{24\pi D_o^2} \tag{579}$$

Equation (579) shows that the surface free energy is half the energy needed to separate two flat semi-infinite slab surfaces from contact to infinity, that is, half the cohesion energy. (The Born repulsion interactions are ignored in this treatment.)

However, the use of the molecular diameters, σ (i.e., $D_o = \sigma$), does not give reasonable results in Equation (579). Instead, by considering that there should be nine neighbor molecules in a planar close-packed structure, a semi-empirical equation has been developed, for the cut-off distance for D_o, which is substantially lower than σ,

$$\gamma_1 \approx \frac{A_{11}}{24\pi(\sigma/2.5)^2} \tag{580}$$

This equation is reliable for liquids interacting only with dispersion forces to within ±20%. Unfortunately, this equation greatly underestimates the surface tension of H-bonding and highly polar liquids.

7.6 Solvent Effects on Particle and Surface Interactions

7.6.1 Solvent effects on molecular interactions

The presence of solvent molecules between solute molecules affects all the interaction energies. We have seen that the presence of solvent molecules changes the molecular polarizabilities of solutes in Section 2.5.3 and decreases the strength of van der Waals interactions (Section 2.6.4) and the total intermolecular pair potential energies (Section 2.7.4). When two molecules interact in a condensed liquid medium, there are many solvent molecules interfering in this interaction. Now, this becomes a *many-body* interaction and we have to consider some important new effects:

1 When two solute molecules approach each other, the pair potential includes not only the direct solute–solute interaction potential energy but also any changes in the solute–solvent and solvent–solvent interaction energies. This is because a dissolved solute molecule can approach another only by displacing solvent molecules from its path. The net interaction therefore can be calculated by considering both the solute–solvent and solvent–solvent interactions. As an example, at the same separation distance, the same two solute molecules may attract each other in free space, but they may repel in a solvent medium if the work that must be done to displace the solvent molecules between them exceeds the work that is gained by approaching the solute molecules. The interaction between two dissimilar molecules can be attractive or repulsive depending on refractive index values, whereas the interaction between identical molecules in a solvent is always attractive.

2 When a solute molecule is introduced into a condensed medium, the medium expends *cavity energy*, because it first forms a cavity to accommodate the guest solute molecule by increasing its volume equal to the volume of the solute molecule, as shown in Figure 7.8 a. This cavity energy is an additional energy.

3 Solute–solvent interactions can change the properties of the solute molecules, such as their dipole moment, polarizability and degree of ionization. Consequently, the properties of any solute are different in different condensed media.

4 Solute–solvent interactions perturb the local ordering of solvent molecules as shown in Figure 7.8 *b*. This perturbation effect can produce an additional *solvation* or *structural* force if the free energy of this perturbation varies with the distance between the two solute molecules in this medium. Some associated solvents apply forces on solute molecules that are determined mainly by the orientation of the molecules of (1) and (2) in the medium (3), so that the resultant distribution functions are not only functions of distance, *r*, but also dependent on the orientation angle.

5 If molecules (1) and (2) are in complete contact in liquid (3), the hard-core repulsion interaction must also be considered. This sometimes gives rise to positive values in potential expressions.

These effects are interrelated and can be collectively termed as *solvent effects* or *medium effects*. In addition, the van der Waals forces between (1–1), (2–2) and (1–2) pairs are reduced because of the dielectric screening of the medium (3), which is particularly important for liquid solvents with high dielectric constants. The attraction force is decreased by a factor of the medium's ε_r for Keesom and Debye interactions, and by a factor of ε_r^2 for London dispersion interactions. This strong reduction in the attractive pair potential means that the contributions of molecules further apart tend to be relatively minor, and each interaction is dominated only by its nearest neighbors.

7.6.2 Combining rules for three-component systems: molecules, particles and surfaces in a third medium

In Section 7.2 we described combining rules for molecular, particle and surface interactions in a two-component system of (1) and (2). In this section, we will examine the combining rules when the molecules, particles or surfaces of (1) and (2) are placed in a solvent medium (3).

If we assume that a liquid (3) consists of a mixture of molecules (1) and (2) in equal numbers, and there is only one solvent molecule in contact with one solute molecule, as shown in Figure 7.9 in two dimensions, then there are three possibilities for molecular loca-

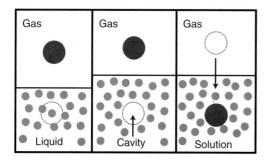

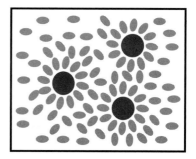

a   *b*

Figure 7.8 a. Condensed medium first forms a cavity to accommodate a guest solute molecule by increasing its volume equal to the volume of the solute molecule. ***b.*** Solute–solvent interactions perturb the local ordering of the solvent molecules, which can produce an additional *solvation* or *structural* force.

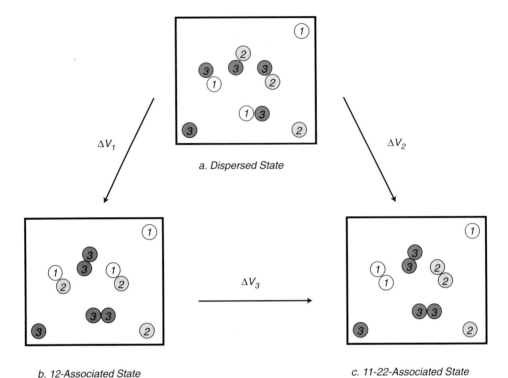

ΔV_1

a. Dispersed State

ΔV_2

ΔV_3

b. 12-Associated State

c. 11-22-Associated State

Figure 7.9 Combining rules for three-component systems in two dimensions: **a.** Schematic representation of two solute (1) molecules in contact with two solvent molecules (3) one by one giving two (1–3) bonds, and the same rule applies to solute molecules (2) giving two (2–3) bonds in the *randomly dispersed state*. **b.** Two (1) molecules are in contact with two (2) molecules giving two (1–2) bonds and the remaining four solvent (3) molecules are in contact as pairs giving two (3–3) bonds in the *12 associated state*. **c.** Two (1) molecules are in contact with each other giving only a single (1–1) bond, two (2) molecules are in contact with each other giving only a single (2–2) bond, and the remaining four solvent (3) molecules are in contact as pairs giving two (3–3) bonds in the *11–22 associated state*.

tions. In the *randomly dispersed state*, which is shown in Figure 7.9 *a*, two solute (1) molecules will have contact with two solvent molecules (3), one by one, giving two (1–3) bonds, and the same applies to solute molecules (2) giving two (2–3) bonds. In the *12 associated state*, as shown in Figure 7.9 *b*, two (1) molecules are in contact with two (2) molecules giving two (1–2) bonds, and the remaining four solvent (3) molecules are in contact as pairs giving two (3–3) bonds. In the (*11–22*) *associated state*, as shown in Figure 7.9 *c*, two (1) molecules are in contact with each other giving only a single (1–1) bond, two (2) molecules are in contact with each other giving only a single (2–2) bond, and the remaining four solvent (3) molecules are in contact as pairs giving two (3–3) bonds. Similarly to Section 7.2, the binding potential energy between molecules (1) and (1) can be expressed as $[V_{11} = -\lambda_1^2]$ and between molecules (2) and (2) as $[V_{22} = -\lambda_2^2]$, between molecules (1) and (3) as $[V_{13} = -\lambda_1\lambda_3]$ and between molecules (2) and (3) as $[V_{23} = -\lambda_2\lambda_3]$, where λ denotes any molecular property. The binding potential energy for the dispersed state can be expressed as shown in Figure 7.9 *a*

$$V_{\text{Dispersed}} = -2\lambda_1\lambda_3 - 2\lambda_2\lambda_3 = -2(\lambda_1\lambda_3 + \lambda_2\lambda_3) \tag{581}$$

The binding potential energy for the *12 associated state* can be expressed as shown in Figure 7.9 *b* so that

$$V_{\text{Assoc-12}} = -2\lambda_1\lambda_2 - 2\lambda_3^2 = -2(\lambda_1\lambda_2 + \lambda_3^2) \tag{582}$$

and the binding potential energy for the *11–22 associated state* can be expressed as shown in Figure 7.9 *c*, so that

$$V_{\text{Assoc-11-22}} = -2\lambda_1^2 - 2\lambda_2^2 - 2\lambda_3^2 = -2(\lambda_1^2 + \lambda_2^2 + \lambda_3^2) \tag{583}$$

Then it is possible to estimate which state is preferable in a three-component system from the difference in interaction potential energy, ΔV, on going from the dispersed to associated states. There are two possibilities: the first one

$$\Delta V_1 = V_{\text{Assoc-12}} - V_{\text{Dis}} = -2\lambda_1\lambda_2 - 2\lambda_3^2 + 2\lambda_1\lambda_3 + 2\lambda_2\lambda_3 \tag{584}$$

After simplification we have

$$\Delta V_1 = V_{\text{Assoc-12}} - V_{\text{Dis}} = -2(\lambda_1 - \lambda_3)(\lambda_2 - \lambda_3) \tag{585}$$

and for the second possibility, we may write

$$\Delta V_2 = V_{\text{Assoc-11-22}} - V_{\text{Dis}} = -2\lambda_1^2 - 2\lambda_2^2 - 2\lambda_3^2 + 2\lambda_1\lambda_3 + 2\lambda_2\lambda_3 \tag{586}$$

which can be simplified to

$$\Delta V_2 = V_{\text{Assoc-11-22}} - V_{\text{Dis}} = -(\lambda_1 - \lambda_3)^2 - (\lambda_2 - \lambda_3)^2 \tag{587}$$

The difference in interaction potential energy, ΔV on going from the *12 associated state*, to the *11–22 associated state*, can also be calculated as

$$\Delta V_3 = V_{\text{Assoc-11-22}} - V_{\text{Assoc-12}} = -2\lambda_1^2 - 2\lambda_2^2 - 2\lambda_3^2 + 2\lambda_1\lambda_2 + 2\lambda_3^2 \tag{588}$$

After simplification we have

$$\Delta V_3 = V_{\text{Assoc-11-22}} - V_{\text{Assoc-12}} = -(\lambda_1 - \lambda_2)^2 \tag{589}$$

Equations (585), (587) and (589) are very important for understanding the behavior of two unlike molecules, particles or surfaces in a third (solvent) medium. If $(\Delta V_1 > 0)$, which means that $(V_{\text{Assoc-12}} > V_{\text{Dis}})$ in Equation (585), then the molecules tend to disperse and a repulsion occurs between the unlike molecules. This condition can be met only if the value of λ_3 is intermediate between λ_1 and λ_2 so that when $[\lambda_1 > \lambda_3 > \lambda_2]$ or $[\lambda_2 > \lambda_3 > \lambda_1]$ then Equation (585) must give a positive value of ΔV_1, and two particles or surfaces in a third medium will repel each other. For the other conditions, we may state that unlike particles may attract or repel each other depending on the values of λ_1, λ_2 and λ_3. The same conclusions apply for the $\Delta V_2 = V_{\text{Assoc-11-22}} - V_{\text{Dis}}$ interaction potential energy difference given by Equation (587). However, when we consider Equation (589) describing the $[\Delta V_3 = V_{\text{Assoc-11-22}} - V_{\text{Assoc-12}}]$ case, it is clear that the most favored final state will be that of particles (1) associated with particles (1) and (2) with (2) and (3) with (3), so that there is always a preferential attraction between like molecules or particles or surfaces in a multi-component mixture, except for the cases where H-bonding and/or Coulombic interactions are operative.

References

1. Israelachvili, J. (1991). *Intermolecular & Surface Forces* (2nd edn). Academic Press, London.
2. Landau, L.D. and Lifshitz, E.M. (1984). *Electrodynamics of Continuous Media* (2nd edn), Vol. 8. Pergamon, Oxford.
3. Erbil, H.Y. (1997). Interfacial Interactions of Liquids. In Birdi, K.S. (ed). *Handbook of Surface and Colloid Chemistry*. CRC Press, Boca Raton.
4. Adamson, A.W. and Gast, A.P. (1997). *Physical Chemistry of Surfaces* (6th edn). Wiley, New York.
5. Hiemenz, P.C. and Rajagopalan, R. (1997). *Principles of Colloid and Surface Chemistry* (3rd edn). Marcel Dekker, New York.
6. Butt, H.J., Graf, K. and Kappl, M. (2003). *Physics and Chemistry of Interfaces*. Wiley-VCH, Weinheim.

PART III

Solids

Chapter 8
Solid Surfaces

8.1 General Properties of Solid Surfaces and Their Experimental Investigation

8.1.1 Properties of solid surfaces

The surface molecules of solids are practically fixed in position, and contrary to the behavior of liquid molecules, they cannot move to any other place. Individual atoms and molecules are only able to vibrate around their mean position. As a result, solid surfaces cannot spontaneously contract to minimize their surface area, and a non-equilibrium surface structure forms. This situation is quite distinct from that of a liquid surface, which attains equilibrium almost as soon as it is formed because of the mobility of the surface molecules. However, this does not mean that surface tension is absent in solids. In principle, surface tension also exists in all solids and the inward pull on the solid surface atoms is always present, owing to cohesion, exactly as in liquids. Nevertheless, the changes of surface shape due to surface tension are very much slower in solids than in liquids; this is not because the cohesion forces are smaller but because the mobility of the surface molecules (or atoms) is very much less. For this reason, measurement of the solid surface tension is a difficult and ambiguous procedure, and indirect methods are mostly applied (see Section 8.2.2 and Chapter 9).

When solids are deformed by external forces at ambient temperature, they generally react elastically. For example, many solid materials recover their original size and shape in a mechanical tensile test, if we remove the applied force. However, if we increase the pulling force until we break the test specimen, then two new solid surfaces form, but the breakage force is not equal to the cohesion forces between the solid molecules because the break occurs at a mechanically weak point (i.e. a previously formed crack on the surface or a crystal defect) present in the specimen, which is not representative of the whole bulk solid material and there is also last-minute viscous flow of the test specimen which alters the interfacial area. Instead, the measured surface tension force depends on the history of formation of the test specimen. On the other hand, the effect of temperature on solids may be decisive regarding their flow behavior. For example, at high temperature some solids, such as sintering powders, become viscous and gain mobility so that they can diffuse laterally under their melting points. They lose their elastic responses and thus capillarity equations can be applied for such cases. The *sintering* process alters the irregularities of the solid surface and viscous sintering materials solidify in a new shape upon cooling. Some polymers also behave similarly; they obey capillary equations above their melting points and show elastic properties below their glass transition temperatures.

A solid surface is mostly contaminated with foreign matter as a thin layer above the topmost surface molecules (or atoms), unless very special precautions are taken. If a solid surface is left unprotected in ordinary air, it becomes coated with a film of greasy material within a short time period (less than an hour). Even if the contamination film is a monolayer, it will alter the surface properties considerably. The cleanest solid surfaces can be obtained by cleaving a single crystal under high vacuum. It is useful to know that freshly split mica surfaces under high vacuum will attract each other, and if connected again, sliding between the crystal planes is very difficult, whereas if the cleavage occurs under ordinary air, the freshly split mica surfaces no longer attract each other and have obviously become coated with greasy materials present in the air.

In comparison with their bulk properties, solid surfaces have two important problems. First, the absolute number of atoms in a solid surface is small. For example, for a typical sample surface area of $1 \, cm^2$, it has been calculated that copper metal (Cu-100) has 1.53×10^{15} atoms at the surface, which corresponds to $2.5 \times 10^{-9} \, mol \, cm^{-2}$. Since the area of the surface probed by an instrument is often nearer to $1 \, mm^2$, detection limits of down to picomoles ($10^{-12} \, M$) may be necessary; this is a real challenge for any analytical instrument. Second, the ratio of the number of surface atoms to the number of bulk atoms is also very small. For the same copper example given above, this ratio has been calculated as $1:5.54 \times 10^7$, and it is generally found that the surface:bulk ratio in a solid is typically $1:10^7$ to $1:10^8$, so that an analytical instrument that is sensitive to $10^{-12} \, M$ is required; otherwise the signal that arises from the bulk solid material will swamp the signal of the surface component in the instrument. Fortunately, the surface analytical instruments discussed below are capable of such sensitivity.

In general, most natural surfaces are amorphous to varying degrees; only a few are completely crystalline. Even many crystalline solids have amorphous surfaces that are different from their bulk crystalline structures. Some solids do not have their original chemical structure at the surface due to the action of gases present in the medium. For example, metals such as aluminum form oxide layers on their surface under ambient conditions. Oxygen is combined by covalent forces with the atoms present in the topmost layers of the metal surface. Thick oxide films can form on some metals. Most metals form these oxide layers over a long period of time, but aluminum is the most rapid, so it is very difficult to wet an aluminum surface with another piece of molten aluminum, and also to solder or weld this metal. In contrast, these metal oxide layers provide protection against metal corrosion. On the other hand, adsorbed monolayers of gases are usually present on the surface of most high-surface-energy solids. Air itself adsorbs on most solids; it alters the settling properties of most powders, including sand. When air is adsorbed onto soil particles, it is difficult to wet the soil, and this is one of the reasons for flooding from summer rain. Once we realize these facts, it seems only academic to investigate crystalline solid surfaces. However, there is a scientific need to compare experiments on well-defined crystalline surfaces in order to understand the adsorption and catalysis processes. In addition, completely crystalline surfaces are now being used in the modern semiconductor industry, and the development of this industry needs tailor-made crystalline surfaces, initially in the laboratory, but then on an industrial scale.

By definition, an ideal solid surface is assumed to be atomically flat and chemically homogenous. However, in reality there is no such ideal solid surface; all real solid surfaces are often extremely uneven at temperatures above 0 K, unless polishing has been employed to smooth out the irregularities present. The number of defects increases with the increase in temperature. Solid surfaces have a *surface roughness* over varying length scales, and also they are chemically heterogeneous to a degree, due to the presence of different chemical groups, impurities or crystals at the surface. Surface roughness is defined as the ratio of real surface area to the plan area, and it is larger than 1 for all practical surfaces. Recently, it has become possible to image surface atoms with the advent of scanning probe microscopy techniques, such as atomic force microscopy (AFM), and scanning tunneling microscopy (STM)..At the atomic level, solid surfaces have been found to consist of a mixture of flat regions, called *terraces*, and defects, *steps*, *kinks* and *point defects*. Thus, each surface site has its own local surface roughness. Since the local distribution of atoms around each of these individual surface sites is different, their electronic properties, physical response and surface chemistry are also different, and this may result in no two adjacent atoms on a solid surface having the same properties. On a larger scale, solid surfaces are also very complex. They have numerous small cracks and often aggregates of small crystals and broken pieces in all possible orientations, with some amorphous materials in the interstices. Even if a solid is wholly crystalline and consists only of a single crystal, then there may be many different types of surface on it; the faces, edges and corners will all be different. The presence of cracks in a surface decreases the strength of the crystals from that deduced from theoretical considerations. Crystal defects affect the crystal's density and heat capacity slightly, but they profoundly alter the mechanical strength, electrical conductivity and catalytic activity.

As stated in Chapter 1, the chemical structure of the top surface layers of a solid determines its surface properties. If these top layers consist of the same chemical groups, then the surface is called *chemically homogeneous*, and if they consist of different chemical groups it is called *chemically heterogeneous*. The presence of two or more chemically different solid substances in a surface layer enormously multiplies the possibilities for variety in the types of surface, such as copolymer surfaces and catalysts having many different atoms at the surface. The chemical heterogeneity of a surface is an important property in industry affecting catalysis, adhesion, adsorption, wettability, biocompatibility, printability and lubrication behavior of a surface, and it must be determined analytically when required.

8.1.2 Experimental investigation of solid surfaces and the requirement for ultra-high vacuum

The experimental and theoretical investigation of rough solid surfaces at the atomic (or molecular) level seems to be an almost intractable problem on a poorly defined surface, since any information obtained contains contributions from a myriad of different combinations of surface sites and chemical compositions. Thus, it is necessary to define precisely the chemical composition and structural, electronic states and bonding properties of molecules of the solid substrate under investigation, in order to ensure we obtain the same reproducible results from experiments. Application of several advanced spectroscopic surface analysis techniques allows a complete surface investigation. X-ray photoelectron spectroscopy (XPS, or the other name ESCA for electron spectroscopy for chemical analysis), ultraviolet photoemission spectroscopy (UPS), extended X-ray absorption fine structure (EXAFS), Auger electron spectroscopy (AES), low-energy electron diffraction (LEED), high-energy electron diffraction (HEED or RHEED), reflection–absorption infrared spectroscopy (RAIRS), high-resolution electron energy loss spectroscopy (HREELS), secondary ion mass spectrometry (SIMS), temperature-programmed desorption (TPD) and temperature-programmed reaction spectroscopy (TPRS) are the main methods used to analyze the surface chemistry of solids. All these methods must be applied under ultrahigh vacuum conditions, which corresponds to the pressure being reduced to below 10^{-9} Torr (1 Torr = 133.3 Pascal, Pa = $N m^{-2}$), otherwise the surfaces of solids are covered with a monolayer of adsorbed molecules of any gas or vapor present in the medium. The application of ultra-high vacuum in a chamber is a solution to preparing a *clean* solid surface and keeping this surface clean and well defined over the duration of a surface experiment up to several hours. For example, it has been calculated from the kinetic theory of gases that it takes 2.6 sec to form a monolayer of carbon monoxide on any solid having an atomic density of 10^{15} atom cm^{-2} (typical for most solids) at 27°C under $P = 10^{-6}$ Torr gas pressure; whereas it takes 7.3 hours under $P = 10^{-10}$ Torr pressure. In general, it is assumed that gas pressures lower than 10^{-7} Pa are required to keep a solid surface clean for a time period of several hours.

There are several methods for obtaining completely crystalline and clean solid surfaces. Many brittle and layered compounds such as mica, highly oriented pyrolytic graphite and some semiconductor crystals are cleaved *in situ*, using a razor blade under ultra-high vacuum to give clean surfaces. The starting material is a pure, three-dimensional single crystal and a slice of the desired orientation is cut from this crystal. Instead of a razor blade, the *double-wedge technique* is also used to cleave some other brittle materials. In this technique, the crystalline material is pre-cut and positioned between very hard wedges to maintain proper cleavage. Depending on the crystallographic orientation, only some definite crystal planes can be obtained. However, the freshly cleaved crystal surfaces may be mechanically stressed, and metastable surface configurations that are different from the equilibrium structure may also be formed.

It is also possible to prepare crystal surfaces from non-brittle materials. Mechanical scraping of the outermost contamination layer will also give clean surfaces. Some hard solids can be ground and polished, and some soft materials are cleaned chemically or electrochemically to give clean solid surfaces. If ultra-high vacuum conditions are not applied during these processes, the crystal surfaces may be contaminated or oxidized, or a monolayer of a gas present in the medium may be formed over them. Polishing a metal surface changes its structure into a completely amorphous assemblage of metal atoms, packed as in liquids. The surface layers of a polished metal are actually liquefied by heat supplied momentarily by the friction action of the moving polisher. The surface temperature can rise rapidly to the melting point of the metal during sliding friction and polishing, but this process can occur only if the melting point of the polisher is higher than that of the substance being polished; hardness alone is of little importance.

Alternatively, samples prepared outside high vacuum conditions can be transferred into an ultra-high vacuum chamber for analysis after formation. This method is important in industry. For example, electrochemical methods in liquids give atomically clean crystalline semiconductor surfaces but these samples should be moved into an ultra-high vacuum chamber for surface analysis. For this purpose, these samples are *capped* with a thin and desorbable protective layer and after transporting through air with this *capping* layer protecting the main surface, the cap may be desorbed subsequently by heating under ultra-high vacuum and the weakly bound substrate removed to leave behind the clean semiconductor surface. In another method, the substrate is heated in an oxygen-rich container to produce gaseous CO and CO_2 to clean the carbonaceous layers on a substrate. If excess oxygen atoms are adsorbed onto a catalysis surface, the substrate is heated gently in a H_2 environment to produce gaseous H_2O to remove O_2.

A general method of cleaning solid surfaces is *sputtering*, where the solid surface is bombarded with noble gas ions. Argon gas is usually preferred for this purpose. The *argon ion bombardment* process can be carried out by inserting argon gas at a pressure of typically 1 Pa into a vacuum chamber and applying a high electric field. The freshly formed high-energy argon ions (100–3000 eV, depending on the applied electric voltage) are used to bombard the sample surface and to remove the monolayers formed at a rate of several monolayers per minute. However, sputtering by argon ion bombardment removes the contaminants, as well as the first few layers of the solid substrate, and gives an undesirable pitted rough solid surface because of the violent nature of this process. Later, adsorbed or embedded argon atoms on the sample surface are removed by annealing which also heals the defects on the crystal surface. Sputtering methods can be applied successfully to clean the surface of pure materials, but they change the surface chemical structure when applied to alloys. Sputtering is also applied for depositing thin metal films onto solid surfaces in a method called *physical vapor deposition*. In this method, a target, which is the source of the metal film, and a solid sample surface are placed in a high-vacuum chamber. The target is bombarded by high-energy argon ions, and the metal atoms that are ejected from the target surface partially condense on the sample surface to form a thin metal film. However, the energy of the argon ions must be strictly controlled because if they have excessive energy, they would penetrate into the target metal without disrupting its surface sufficiently. In a similar method, heated metals are evaporated in a high-vacuum chamber to form a thin metal film on specific solid substrate samples.

On the other hand, optical microscopy, confocal microscopy, ellipsometry, scanning electron microscopy (SEM), scanning tunneling microscopy (STM), atomic force microscopy (AFM) and total internal reflection fluorescence (TIRF) are the main microscopic methods for imaging the surface structure. There are many good books and reviews on spectroscopic and chemical surface analysis methods and microscopy of surfaces; description of the principles and application details of these advanced instrumental methods is beyond the scope of this book.

With the application of these instrumental tools, we can measure the properties of so called *well-defined* surfaces. But, what is a *well-defined* surface? Initially, if we assume that the surface under investigation is *flat* and contains a very high ratio of terraces to defect sites, and it is chemically homogeneous and consists of just one type of atom (or molecule) on the surface, then we can introduce a greater complexity into the system by adding *controlled* amounts of surface defects and chemical heterogeneities. Such surfaces are called *well-defined surfaces*. This is a useful approach, and a large database has been established on the properties of well-defined metal, metal oxide, semiconductor, polymer, catalyst and insulator surfaces, which have been investigated extensively in the past. In a well-defined surface experiment, it is necessary to use a particular crystal plane that can be obtained when cut from a single crystal. Such crystal planes contain atoms in a number of well-defined sites. If required, the geometrical location and the number of adsorption sites on a model solid surface can be varied by simply cutting a single crystal in different directions to expose different crystallographic planes. *Atomically flat* or alternatively *vicinal* surfaces, which consist of short atomically flat terraces separated by atomic steps, can also be obtained by the carefully controlled crystal cleavage method.

8.2   Surface Tension, Surface Free Energy and Surface Stress of Solids

8.2.1   Surface stress and its relation with surface tension and surface free energy of solid surfaces

Since the molecules on a solid surface are generally fixed, a solid cannot spontaneously contract to minimize its surface area. This situation is completely different from that of liquids. As we have seen in Sections 3.2.4, 3.2.5 and 3.3.1, if the Gibbs dividing plane is placed such that $\Gamma_i = 0$, then the surface tension of a pure liquid, γ_o, is equal to the (specific excess) surface free energy per unit area, ΔG_o^S (Equation (222)). However, these two quantities are not in general equal for solid surfaces, because when a fresh surface is formed by cleaving the solid in a vacuum or, if the solid is volatile in its own vapor, then the atoms, molecules or ions present in the freshly generated surface are immobile and will normally be unable to arrange in their equilibrium configuration. Consequently, a non-equilibrium structure forms at the solid surface, and it may take a considerable time (even an infinite amount of time!) for the atoms (or molecules) to come to their equilibrium positions. This situation is quite distinct from that for a liquid surface, which attains equilibrium almost as soon as it is formed because of the ease of mobility of its molecules. Thus, it is convenient to define *surface tension of solids* in terms of the restoring force necessary to bring the freshly exposed surface to its equilibrium state. However, the definition of *surface free energy of solids* is the same as for liquids; it is the work spent in forming a unit area of a solid surface. It is clear from this argument that the surface tension is not equal to the surface free energy for solids. In addition, when the surface area of a liquid increases, the number of surface molecules (or atoms) increases in proportion, according to the Gibbs equation, whereas for solids such a *plastic*, the increase in surface molecules is very limited. Instead, when the surface area of a solid is increased by stretching it mechanically, the distance between neighboring surface atoms changes with the increase in area, while the number of surface atoms remains constant in most cases. For solid surfaces, this *elastic* increase with the increase in surface area is much more important than the negligible *plastic*, the increase. At this point, a new term *surface stress*, ψ, is defined as the external force per unit length that must be applied to retain the atoms or molecules in their initial equilibrium positions (equivalent to the work spent in stretching the solid surface in two directions in a two-dimensional plane). In solids, the surface free energy is not necessarily equal to the surface stress, unlike the situation with liquids, and an expression is required to relate the surface free energy with the surface stress. When a one-component solid is isotropic, then the work done to increase its surface area by an amount dA can be written as $dW = \psi\, dA$. However, this work is also equal to $dW = d(AG^S)$, by the definition of surface free energy. If we equalize both expressions, we obtain

$$\psi\, dA = d\,(AG^S) \tag{590}$$

then, the surface stress can be found as

$$\psi = \frac{d(AG^S)}{dA} = \frac{G^S dA}{dA} + \frac{A dG^S}{dA} = G^S + A\left(\frac{dG^S}{dA}\right) \tag{591}$$

$(dG^S/dA) = 0$ for a one-component pure liquid, and thus we have $\psi = G^S = \gamma_o$. However, this is not true for a freshly formed solid surface, and the magnitudes of G^S and $[A\,(dG^S/dA]$ are comparable. The surface stress may be related to surface free energy via the term *surface strain, λ*. The total surface strain upon extending the surface area of a substance can be expressed as $d\lambda_{tot} = (dA/A)$. If we combine the total strain term with Equation (591) we have

$$\psi = G^S + \left(\frac{dG^S}{d\lambda_{tot}}\right) \tag{592}$$

For solids, the total surface strain has two components, as elastic and plastic strains: $[d\lambda_{tot} = d\lambda_e + d\lambda_p]$. When only elastic strain is present on solid surfaces, we can write $d\lambda_{tot} = d\lambda_e$, so that Equation (592) becomes

$$\psi = G^S + \left(\frac{dG^S}{d\lambda_e}\right) \tag{593}$$

Equation (593) is called Shuttleworth's equation and shows how we can calculate the surface stress. In theory, if we want to calculate the Gibbs free energy per unit area from surface tension and surface stress we may write

$$G_?^S = \left(\frac{d\lambda_p}{d\lambda_{tot}}\right)\gamma + \left(\frac{d\lambda_e}{d\lambda_{tot}}\right)\psi \tag{594}$$

However, $G_?^S$ is not a true thermodynamic quantity, because it does not express the equilibrium; instead it depends on the history of the solid surface formation, as we see above. So far, we have considered only the isotropic solids. If, however, the solid is anisotropic, such crystals can respond differently in different directions when increasing the surface area. This property increases the number of required equations twofold, and the matter is rather complex because it is unclear which shape the anisotropic crystal would prefer for a given volume, if its total surface free energy is minimized; this subject is beyond the scope of this book.

8.2.2 Theoretical estimation of the surface free energy of solids

Some semi-empirical methods are offered to estimate the surface tension of solids which have specific atomic or molecular interactions:

1 The simplest case is a solid bonded with short-range covalent forces, such as a diamond. Harkins assumed that the total surface energy at 0 K is simply half of the energy to break

the total number of covalent bonds passing through a $1\,m^2$ cross-sectional area of the diamond, that is

$$U_i^S = \frac{1}{2}U_i^{cohesion} \qquad (595)$$

If we assume that the Gibbs surface free energy, G_i^S, is nearly equal to the total surface energy, U_i^S, at room temperature, by neglecting entropic effects it is possible to calculate the G_i^S value. As an example, if we split all the C—C bonds in a diamond at the (111) crystal face, there will be 1.83×10^{19} bonds m^{-2}. The energy per C—C bond in diamond has been found experimentally to be $6.2 \times 10^{-19}\,J$ per bond. Then, we can calculate the Gibbs surface free energy value of diamond as, $G_i^S \approx U_i^S = 1/2 \,(1.83 \times 10^{19} \times 6.2 \times 10^{-19}) \times 10^3 = 5670\,mJ\,m^2$. For the (100) crystal face of diamond, a value of $G_i^S = 9820\,mJ\,m^{-2}$ has been calculated. These figures are not absolutely correct but are good estimates and also show the high dependence of the surface free energy of solids on the number of bonds, and the corresponding crystal faces.

2 For solids interacting with only van der Waals pair potentials with no orientation effects, the face-centered cubic crystals of noble gases are good model systems. Using the Lennard-Jones (12–6) pair-potential equation, which considers the van der Waals attraction and Pauli repulsion, and applies computer simulations to both the splitting energy of fixed atoms in the crystal and their rearrangement energies, $G_i^S = 45 \pm 1.8\,mJ\,m^{-2}$ has been calculated for argon crystal, depending on the different crystal orientations. Calculations were also performed for Ne, Kr and Xe crystals, and it was shown that the value of G_i^S increases with the increase in elementary molecular mass.

3 For ionic solids interacting with Coulomb pair potentials, similar calculations can be carried out. However, this is a rather complex matter because Coulomb, van der Waals attraction and Pauli repulsion should all be taken into account. In addition, there are uncertainties in the choice of suitable pair-potential equation (many inter-atomic potential equations, including Lennard-Jones were tried), and the calculated G_i^S results are highly dependent on the particular choice of pair-potential model. As an example, $G_i^S = 212\,mJ\,m^{-2}$ was calculated theoretically for the NaCl (100) crystal, which is near to the experimental value of $G_i^S = 190\,mJ\,m^{-2}$ from extrapolation of the molten salt surface tension values, but far away from $G_i^S = 300\,mJ\,m^{-2}$, which was found from crystal cleavage experiments.

4 There are two main methods for estimating the surface free energy of metallic solids. The pair-potential approach can also be used to predict the surface tension of metal surfaces. Alternatively, a quantum-mechanics-based method has been developed which depends on the behavior of free electrons in a metal box. The wave functions of the electrons have nodes at the walls of the box, whose sides correspond to the surfaces of metals. The wave functions for the standing waves inside the box yield permissible energy states for the electron, which are independent of the lattice type. In a thought experiment, if the metal is broken into two faces through its cross-section, then the electrons are forced to occupy higher energy states because there is no room in their previous locations due to the boundary conditions. The gain in kinetic energy corresponding to the rejected states is therefore assumed to be the surface energy of the

metal; the results obtained are in fair agreement with the experimental surface tension values.

8.2.3   *Experimental determination of surface free energy of solids*

There are many theoretical difficulties in defining the surface free energy of solids, mainly arising from the deviations from ideality due to the immobility of surface atoms and molecules, and also the presence of surface defects. Dislocations, such as a missing atom or molecule and the presence of extra atoms (adatoms) or molecules on the solid surface, and the formation of kinks, edge and screw dislocations, are a few examples of such defects, and thus the surface free energy of solids largely depends on the history of the surface formation. On the other hand, there is a need to pre-determine the average surface tensions of solids in order to predict their usage in many industrial applications. In general, mostly indirect experimental methods are applied to this task:

1 For low-energy solids such as crystalline organic materials and most polymers, surface tension can usually be estimated from the contact angle measurement of liquid drops onto them, using surface tension data of liquids in these calculations (see Section 9.1).

2 If the solid can be melted, its surface tension is measured while it is a liquid, at varyious temperatures, and the derived experimental data are extrapolated to room temperature by means of suitable semi-empirical equations. Since the surface enthalpy of the solid is 10–20% larger than its near-melt temperature values, this fact is also considered in these calculations.

3 The work required to cleave off a unit area of a solid is measured experimentally for some layered solids, which can give flat surfaces. However, the results are reliable only for a very small proportion of solids. This is because mechanical deformations occurring during the cleavage test consume most of the splitting energy, and in addition, surfaces can reconstruct and return to their equilibrium positions after cleaving, a step for which energy cannot be determined during the test. As an example, the cleavage of a mica crystal gives a surface tension of 375 mJ m^{-2} in air and 4500 mJ m^{-2} in a vacuum, showing the effect of water vapor and other greasy material adsorption in air during the measurement.

4 Powder solids generate heat when immersed in liquids due to the solvation action of the solvent molecules, which destroys the solid interface. This exothermic heat can be measured precisely using advanced calorimeters, which are sensitive to just a few Joules per mole of material. If the surface area of this powder has been measured previously using an independent method, such as by BET adsorption, then it is possible to calculate the surface free energy from the heat of immersion.

5 When a solid powder is used as the stationary phase in the inverse gas chromatography method, the interaction of a well-known gas or organic vapor is measured, and the adsorption results for gaseous molecules on the solid powder are used to calculate the difference in surface free energy of bare stationary solid surface and that of the solid–vapor interface ($\gamma_s - \gamma_{sv}$).

6 When compression forces are applied to a small crystal, it is possible to measure the changes in lattice spacing by using X-ray and LEED instruments. Then the

decrease in the lattice constant and solid surface tension can be calculated, but these data are more suitable for calculating non-equilibrium surface stress rather than surface tension.

8.3   Gas Adsorption on Solids

As we saw in Section 3.3, the concentration difference of one constituent of a gas or liquid solution at the surface of another phase is called *adsorption*. In other words, a*dsorption* is the partitioning of a chemical species between a bulk phase and an interface. *Desorption* is the reverse of the adsorption process, showing that the molecules are leaving the interface towards the other phase. Adsorption is different from *absorption* where a species penetrates and is dissolved throughout the bulk phase of a liquid or a solid. For gas or liquid adsorption on solids, the solid material on which adsorption takes place is defined as the *adsorbent*; the material in the adsorbed state (while bound to the solid surface) is called the *adsorbate*, and the gas, vapor or liquid molecule prior to being adsorbed is called the *adsorptive*. In this section, we will investigate only the principles and applications of gas adsorption on solids (see Section 9.4 for liquid adsorption on solids).

Gas or vapor molecules may become bound to a solid surface if they approach close enough to interact. When a gas molecule strikes a solid surface, it may be adsorbed on the solid by interaction with the unsatisfied fields of force of the surface atoms or molecules. The gas molecule loses sufficient energy to the atoms on the solid surface by exciting them, vibrationally or electronically, to become effectively bound to the surface. An ensemble of a monolayer of adsorbed molecules forms with this process, and the average time of stay of a molecule upon the surface is called the *mean stay time*. In this way, the surface tension of the solid surface is diminished, often to a small fraction of its original value. If the strength of the adsorbate–adsorbent interaction is higher than that of the strength of the thermal energy that forces them to mobilize, then the adsorbed molecules spend most of their time on the *adsorption sites* on the surface, so that they are called *localized*. If the reverse is true, the adsorbate molecules can readily move along the surface from one site to another and are called *non-localized*. Whether or not equilibrium has been attained is an important issue in adsorption. Gas adsorption on porous surfaces usually shows *hysteresis*, that is adsorption and desorption isotherms do not coincide. This is not usually an indication of an irreversible process; it may be due to the time differences between the adsorption and desorption processes.

When a solid is in equilibrium with a gas, if the gas is more concentrated in the solid surface than in the bulk of the solid, then the gas is called *positively adsorbed*, whereas if the gas is less concentrated in the surface region, it is called *negatively adsorbed*. The quantity of molecules taken up by a surface depends on several variables including temperature, pressure, surface energy distribution, and the surface area and porosity of the solid. The quantity of molecules adsorbed on the solid adsorbent surface and the pressure of the adsorptive gas at a constant temperature can be plotted to give an *adsorption isotherm*. In an adsorption process, the *fractional coverage* of the adsorbate, θ_f, is defined as

$$\theta_f = \frac{N^S}{N^{tot}} \tag{596}$$

where N^S is the number of adsorbate molecules (or atoms) per unit area on the solid substrate, and N^{tot} is the total number of adsorbate molecules (or atoms) per unit area to produce one complete monomolecular layer (monolayer) on the substrate surface. The value of N^{tot} is not necessarily, but is sometimes, numerically equal to the number of surface molecules (or atoms) of the solid substrate. [If the size of the adsorbed molecules (or atoms) is appreciably larger or smaller than the size of the molecules (or atoms) on the surface of the substrate, then there will be a difference between N^{tot} and the total number of surface sites.] If a molecule can be adsorbed on a surface without fragmentation, then the process is called *associative adsorption*, whereas if fragmentation does occur, it is called *dissociative adsorption*. Time is also an important parameter in the adsorption process, because longer times will increase the amount of adsorption to a final equilibrium value, whereas it may be difficult to reach the equilibrium in a short time for some cases, and for these reasons the *exposure* parameter is defined as the product of the adsorptive gas pressure and the time of exposure during adsorption (the normal unit is a *Langmuir*, where $1\,L = 10^{-6}$ Torr-sec).

8.3.1 Physisorption on a gas–solid interface

Adsorption on solids is usually classified according to the type of force involved in binding the adsorbed molecules (or atoms) to the surface molecules (or atoms) of the solid. When *physical adsorption* or *physisorption* takes place, the only bonding forces are the non-specific and relatively weak van der Waals forces. Since there is no covalent bond formation in this process, there is no electron exchange between the adsorbed molecule and the substrate surface. An *overspill* of electron charge from the solid into the gas molecule results in an imbalance of electron density on either side of the interface. The physisorption process is always exothermic, and the energy given out on adsorption is called the *enthalpy of adsorption*, ΔH_{ads}. Equilibrium is very rapidly established in physisorption, and the value of ΔH_{ads} is low, typically in the region of -10 to $-40\,kJ\,mol^{-1}$, usually not exceeding $-60\,kJ\,mol^{-1}$. The magnitude of ΔH_{ads} is generally of the same order, but slightly larger than the latent heat of condensation, ΔH_{vap}, of the adsorbtive molecules. Physisorbed molecules are fairly free to move and rotate, and are usually not bound to a specific location on the surface, but may diffuse laterally along the surface of the adsorbent. The adsorbate–adsorbate interactions are sometimes greater than the adsorbate–surface interactions. The Born repulsive forces become significant as the molecules (or atoms) closely approach each other. Since the attraction forces arise only from the van der Waals forces, physisorption is non-specific and any molecule (or atom) can adsorb on any solid surface under appropriate conditions. The decrease of temperature lowers the kinetic energy of adsorptive gas molecules and increases the amounts of physisorbed molecules. Thus, physisorption is usually important for gases below their critical temperature, typical examples being the adsorption of nitrogen, methane, or noble gases such as argon and krypton on the surface of any solid adsorbent.

As more adsorptive gas molecules (or atoms) are introduced into the system over time, the adsorbate molecules tend to form a monolayer that covers the entire adsorbent surface. When a monolayer forms, the fractional coverage equals one ($\theta_f = 1$). Being only weakly bound, physical adsorption can easily be reversed, and the physisorbed molecules can easily be desorbed (or removed) from the surface to the gas phase by lowering the pressure of the adsorptive gas. On the other hand, the continued addition of gas molecules beyond monolayer formation leads frequently to the gradual stacking of multiple layers (or multi-layers) on top of each other. These multi-layers also disappear rapidly if the pressure of the adsorptive gas is reduced. When pores are present on the solid surface, the formation of multi-layers occurs simultaneously with capillary condensation. The latter process is described by the Kelvin equation (see Sections 4.7 and 4.8), which quantifies the proportionality between residual (or equilibrium) gas pressure and the size of capillaries capable of condensing gas within them. Next, as the adsorptive gas pressures approach saturation, the pores of the adsorbent become completely filled with adsorbate. Knowing the density of the adsorbate, one can calculate the volume it occupies and, consequently, the total pore volume of the sample. If at this stage, one reverses the adsorption process by withdrawing a known amount of gas from the system in steps, one can also generate the desorption isotherms. For porous adsorbents, adsorption and desorption isotherms rarely overlay because of the *hysteresis* which may be due to the time and energy differences between adsorption and desorption. This hysteresis leads to isotherm shapes that can be mechanistically related to those expected from particular pore shapes.

8.3.2   *Chemisorption on a gas–solid interface*

Adsorption can also result in strong chemical bond formation between the adsorbent surface and the adsorbate molecules (or atoms), by exchanging and sharing electrons, and may be regarded as the formation of a surface compound. The nature of this bond may lie anywhere between the extremes of virtually complete ionic or complete covalent character. Spectroscopic methods can be used to evaluate the nature of the surface bonding involved. Chemisorption may be rapid or slow and may occur above or below the critical temperature of the adsorptive gas. The heats of chemisorption are much larger than for physisorption, and ΔH_{ads} is in the range of -40 to $-1000\,kJ\,mol^{-1}$ (typically in the range of -100 to $-400\,kJ\,mol^{-1}$). As a consequence of the bond strength, chemical adsorption is difficult to reverse and chemisorbed layers are usually very stable to desorption at high temperatures and very low pressures. If desorption does take place, it may be accompanied by chemical changes on the solid adsorbent surface. For example, oxidation can be viewed as the chemisorption of oxygen, and when oxygen gas is adsorbed on carbon, it is held very strongly. During desorption by heating to high temperatures, a mixture of CO_2 and CO gases leaves the surface, thus consuming the carbon adsorbent. Since, in practice, chemisorbed molecules do not desorb at low temperatures, experiments in ultra-high vacuum conditions are possible on these monolayers. Chemisorption is significantly different from physisorption in that it is highly selective and can occur only between certain adsorptive and adsorbent species. Chemisorption typically proceeds only as long as the adsorptive can make direct contact with the surface; it is therefore a monolayer process, as no multi-layers form. Exceptions can exist if the adsorptive is highly polar, NH_3 being an example. However, both physical and chemical adsorption may occur on an adsorbent surface at the same time; a layer of molecules may be physically adsorbed on top of an underlying chemisorbed layer. It is also possible for a gas to be physically adsorbed at first, and then to react chemically more slowly with the surface of the solid. At low temperatures, chemisorption may be so low that only physisorption occurs, but with the increase in temperature and rise in kinetic energy of the adsorbate molecules, physisorption will be very small and only chemisorption may occur. For example, at liquid nitrogen temperature (77 K), the nitrogen gas is adsorbed physically on iron, but at 800 K, where the energy level is too high for physisorption, nitrogen gas is adsorbed chemically to form iron nitride.

In contrast to physisorption, chemisorption involves the formation of strong bonds between adsorbate molecules at specific surface locations known as *active sites*. For a crystalline adsorbent, the strength of adsorption is usually dependent on which face is exposed. Thus, chemisorption can be used primarily to quantitatively evaluate the number of surface active sites, which are likely to promote (catalyze) chemical reactions. Static chemisorption isotherms of hydrogen, carbon monoxide and oxygen give the *active metal area* of heterogeneous catalysts. In an analogous manner, the amount of acidic or basic sites on a substrate is determined from the chemisorption of a basic or acidic gas, such as ammonia or carbon dioxide, respectively. When chemisorption occurs, the molecular structure of the substrate usually changes, and even surface reconstruction may occur on covalent or metallic solids. Upon chemisorption onto a substrate surface, most of the translational, and often the rotational degrees of freedom, are lost, and the order of the system increases, giving a negative entropy change, ΔS_{ads}. Since the constant temperature, T, is always positive, from the thermodynamical expression ($\Delta G_{ads} = \Delta H_{ads} - T \Delta S_{ads}$), it follows that ΔH_{ads} has to be negative (exothermic) for the chemisorption process to proceed spontaneously ($\Delta G_{ads} < 0$).

8.3.3 *Thermodynamics of gas adsorption on solids: relation with the Gibbs adsorption equation*

The amount of material to be adsorbed on a surface is described by the adsorption function, $\Gamma_i = f(P,T)$, where Γ_i is the *excess moles of the component, i*, adsorbed at the interface per unit area of an interphase. The value of Γ_i can be determined experimentally. Since it largely depends on the temperature, keeping the temperature constant and measuring the increase in the adsorbed amount by varying the adsorptive gas pressure gives a graph of Γ versus P, which is called an *adsorption isotherm*. From these graphs, adsorption isotherm equations such as the Langmuir, Freundlich and B.E.T. etc. are derived, depending on the theoretical model used. (However, sometimes no analytical expression can adequately fit a complicated experimental adsorption isotherm.)

The adsorption isotherms are related to the Gibbs adsorption equation given in Section 3.3.1. The Gibbs equation shows that the amount of any adsorbate (gas or liquid) on a substrate is proportional to the variation of surface tension of the adsorbent with the change in the chemical potential of the adsorbate [$\Gamma_2 = -(d\gamma/d\mu_2)_T$], as given by Equation (226), where, Γ_2 is the *excess moles of the adsorbate* that are adsorbed per unit area of the interphase. We know that the chemical potential of the adsorbate is related to its activity, a_2, as given in Equation (169), in Section 3.2.1, so that we have ($\mu_2 = \mu_2^* + RT \ln a_2$), and complete differentiation gives ($d\mu_2 = d\mu_2^* + RT\, d\ln a_2$) where the standard chemical potential of component 2, μ_2^*, is a constant quantity giving, $d\mu_2^* = 0$. Then, we obtain the [$d\mu_2 = RT\, d\ln a_2$] expression, and the Gibbs adsorption equation becomes [$\Gamma_2 = -(a_2/RT)(d\gamma/da_2)_T$], as given in Equation (227). For the adsorption of gases (or volatile vapors) on solid adsorbents, the activities can be related to the partial gas pressures. If the gas (or vapor) obeys the ideal gas laws, Equation (227) may be written $[\Gamma_2 = -(P_2/RT)(d\gamma/dP_2)_T]$ as given by Equation (230), where P_2 is the gas (or vapor) partial pressure. For gas–solid adsorption, it is more convenient to use the n_2^S term which is the excess number of moles of adsorbate on the surface $(n_2^S \equiv \Gamma_2 A^S)$, as given in Equation (223), to eliminate the surface excess term, Γ. Then Equation (230) becomes

$$n_2^S = -\frac{A^S P_2}{RT}\left(\frac{d\gamma}{dP_2}\right)_T \tag{597}$$

After rearrangement, the variation of surface tension of the solid adsorbent by the partial pressure of the adsorptive gas at a constant temperature can be expressed as

$$-d\gamma = \frac{n_2^S RT}{A^S P_2}dP_2 \tag{598}$$

Since most of the solid adsorbents are porous materials having a large surface area per mass, it is convenient to relate the interfacial area, A^S, with the weight of the adsorbent, and we can define a new term, *specific area*, $A_{Spec.}$ $(m^2\,kg^{-1})$, so that, $(A^S = A_{Spec.}\,w)$ where w is the weight of the adsorbent. Then, Equation (598) becomes

$$-d\gamma = \frac{n_2^S RT}{A_{Spec.}w}d\ln P_2 \tag{599}$$

The term $A_{Spec.}$ is constant for a given adsorbent solid, and at equilibrium, the ratio of (n_2^S/w) is a mathematical function of the P_2 of the adsorbtive gas at a constant T. Thus, we can integrate Equation (599) in the form

$$-\int_{\gamma_o}^{\gamma}d\gamma = \frac{RT}{A_{Spec.}}\int_0^{P_2}\frac{n_2^S}{w}d\ln P_2 \tag{600}$$

After integration of the left side of Equation (600) we have

$$\gamma_o - \gamma = \pi = \frac{RT}{A_{Spec.}}\int_0^{P_2}\frac{n_2^S}{w}d\ln P_2 \tag{601}$$

If the mathematical relation between the (n_2^S/w) and P_2 terms is known for any adsorption process, then Equation (601) can be integrated analytically, and it is possible to calculate the two-dimensional surface pressure (or spreading pressure), π, of the adsorbate on the solid adsorbent as a function of the fractional coverage.

Alternatively, if no mathematical relationship is known, Equation (601) can be converted into a plot of *adsorbate mass versus partial gas pressure*, to apply a graphical integration. The excess number of moles of adsorbate at the surface, n_2^S, can be written as $(n_2^S = m_2^S/M_w)$, where m_2^S is the mass (excess) of adsorbate at the surface, and M_w is the adsorbate molecular mass. For simplicity, if we take the adsorbent mass as a constant value at $w = 1\,kg$, then Equation (601) becomes

$$\pi = \frac{RT}{A_{Spec.}M_w}\int_0^{P_2}m_2^S\,d\ln P_2 \tag{602}$$

Then, the two-dimensional surface pressure, π, can be calculated by graphical integration of a plot of m_2^S versus $\ln P_2$.

If the n_2^S or m_2^S parameters are determined experimentally, then the corresponding molar area, A_{molar}, $(m^2 kg\text{-}mol^{-1})$ occupied by a mole of adsorbate gas on the solid surface can be calculated as

$$A_{molar} = \frac{A_{Spec.}}{n_2^S} = \frac{A_{Spec.} M_w}{m_2^S} \qquad (603)$$

8.3.4 Experimental determination of adsorption isotherms

The extent of adsorption of gases onto solid surfaces can be determined experimentally using a wide variety of apparatus and techniques, and the literature on this subject is extensive. In general, measurements fall into one of two categories: either the volume of the gas adsorbed is determined manometrically, or gravimetric methods are used, where the mass adsorbed on the solid is determined directly.

Volumetric methods are more popular, their application at low temperatures greatly increases the physiosorption amounts, and the measurements become much easier under these conditions. Nitrogen gas adsorption at about $-196°C$ (77 K) is commonly applied in the surface area determination of adsorbents in the B.E.T. method ($-196°C$ is the boiling point of liquid N_2 at atmospheric pressure, which is generally used to cool the adsorbent sample). In the volumetric method, the adsorbent (usually in the form of a powder) is placed into a bulb, and the bulb is evacuated. Then, the *dead space* which is the volume of the bulb not occupied by the adsorbent is determined initially by introducing a non-adsorbing (or weakly adsorbing) gas, such as helium, into the bulb at constant pressure and temperature. Then the bulb is evacuated again to withdraw the helium gas. Next the adsorptive gas, such as nitrogen, is admitted into the bulb at constant temperature and pressure. When the system reaches equilibrium, the definitive pressure, P_2, is measured. The volume of the adsorptive gas minus the dead space gives the amount of gas adsorbed, and n_2^S is calculated from the volume difference by using the ideal gas law. Then, the pressure is further increased and the volume experiment is repeated to determine the points in the isotherm.

In the gravimetric method, the adsorbent (usually in the form of powder) is placed into a bulb, which is mounted on a sensitive balance and the bulb is then evacuated. Next, the weight increase of the adsorbent solid as a function of the absorptive gas pressure is monitored at constant temperature. More recently, the quartz crystal microbalance (QCM) technique has been applied; this is very sensitive to mass increases. Quartz is a piezoelectric material and the thin crystal can be excited to oscillate in a traverse shear mode at its resonance frequency when a.c. voltage is applied across the metal (usually gold) electrodes, which are layered on two faces of the crystal. When the mass on the crystal increases upon adsorption, its resonance frequency decreases. The increase in the mass is calculated from the reduction in resonance frequency. On the other hand, adsorption on single flat surfaces can also be measured by *ellipsometry*, which measures the film thickness of transparent films optically using the difference between light reflection from bare and adsorbed surfaces.

The total interfacial area, A^S, must also be known in order to obtain the Γ_2 values in adsorption isotherms. The value of A^S is generally calculated from the $(A^S = A_{Spec.} w)$ expression, when $A_{Spec.}$ is known. Since nearly every solid surface has a roughness and porosity, the evaluation of A_{Spec} is generally difficult and requires special techniques. In most practical applications, the volumetric B.E.T. adsorption isotherm in the range of $0.05 < P/P_o < 0.35$ is determined for this purpose, mainly using nitrogen or argon gases as the adsorptive. The B.E.T. adsorption equation, given in Section 8.3.8, is used to calculate the $A_{Spec.}$ of the adsorbent. The *Mercury intrusion porosimeter* instrument can be used to determine the pore distribution of the adsorbent if the pore sizes are larger than 3 nm and up to 360 μm (but practically, a pore size of larger than 100 nm gives better results). Liquid mercury does not wet most solid surfaces because of its very high surface tension of 486 mN m^{-1} giving contact angles larger than 90°, and it must be forced to enter a capillary. Mercury is applied to a bulb filled with the adsorbent powder, and the volume of mercury intruding into the sample as a function of the applied pressure is measured. Washburn developed an expression to relate the pore diameter with the mercury pressure, by assuming that the pore or capillary is cylindrical and the opening is circular in cross section so that only pores whose radius

$$r > \frac{2\gamma_{Hg}|\cos\theta|}{P} \tag{604}$$

are filled where, θ, is the contact angle of the mercury drop on the so-called *flat* solid. In other words, mercury under external pressure, P, can resist entry into pores smaller than r, but cannot resist entry into pores larger than r. Then, one can determine which pore sizes have been invaded by mercury and which sizes have not, for a particular pressure. The volume of mercury moving into the sample can be measured by attaching a capillary tube to the sample cup and allowing this tube to be the reservoir for the mercury during the experiment. When the external pressure changes, the variation in the length of the mercury column in the capillary indicates the volume passing into or out of the sample cup. However, electronic detection of the rise and fall of mercury within the capillary is much more sensitive, providing an even greater volume sensitivity down to less than a microliter. A typical mercury intrusion porosimetry test involves placing a sample into a container, evacuating the container to remove contaminant gases and vapors, and, while still evacuated, allowing mercury to fill the container. Next, pressure is increased towards room pressure while the volume of mercury entering larger openings in the sample bulk is monitored. When the pressure has returned to ambient, pores of a diameter down to about 12 nm have been filled. The sample container is then placed in a pressure vessel for the remainder of the test. A maximum pressure of about 400 MPa is typical for commercial instruments, which will force mercury into pores down to about 3 nm in diameter. Mercury intrusion porosimetry gives the total pore volume and density of the adsorbent. Since the reversible work of PdV is required to insert the mercury into a unit area of a solid surface, this is equal to the work necessary to wet the solid surface. Thus, the entire surface area wetted by mercury liquid can be expressed from surface thermodynamics as

$$A_{total} = -\frac{1}{\gamma_{Hg}\cos\theta}\int_0^P PdV \tag{605}$$

8.3.5 Types of adsorption isotherm

Physical and chemical adsorption and desorption isotherms are important in characterizing the overall adsorbent surface. The slightest change in the shape of the plotted isotherm is indicative of a particular surface feature. There are various forms of adsorption isotherm, which are normally plotted as the amount of adsorbate gas, m_2^S, or surface excess moles, Γ_2, or fractional coverage, θ_f, versus the equilibrium partial pressure of the adsorptive gas, P_2, or alternatively relative pressure (P_2/P_2^0), where P_2^0 is the saturated vapor pressure at constant temperature. In addition, m_2^S is sometimes expressed as the mass per gram adsorbent (m_2^S/w) or moles per gram of adsorbent (n_2^S/w). (For adsorption from liquid solutions, the solution concentration, c, is used in the abscissa, instead of P_2.) It should be noted that all the above adsorption isotherms can be plotted from the same experimental data and the choice of the abscissa and ordinate parameters is arbitrary.

In the scientific literature, tens of thousands of experimental adsorption isotherms have been published, but they all probably belong to one of the nine categories given in Figure 8.1. Type I isotherms rise sharply at low relative pressures and reach a plateau at $\theta_f = 1$. The shape is consistent with the formation of a monolayer upon which no further adsorption takes place. As we will see in Section 8.3.7, the type I isotherm corresponds to the Langmuir adsorption isotherm in which the plateau in the plot is characterized by saturation at high gas pressures. This saturation may be due to the formation of an adsorbate monolayer. However, the Langmuir isotherm very often results from the adsorbent being microporous with pores whose widths are less than 2 nm. In this case, the limiting value of the adsorption is due to the filling of micropores rather than completion of the monolayer. On the other hand, the Langmuir isotherm also describes the ideal chemisorption where gas molecules are adsorbed chemically until the adsorbent surface becomes saturated.

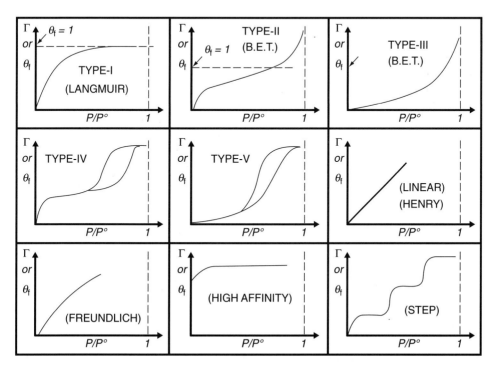

Figure 8.1 Typical adsorption isotherm plots.

When physisorption is considered in gas–solid systems, monolayer formation is not very common; instead, multi-layer adsorption is much more frequent, as shown in type II to V isotherms. Both types II and III are derived from experiments on multi-layer adsorption on non-porous solids (or solids with small macroporosity percentage), and both type IV and V isotherms, from multi-layer adsorption on appreciably porous solids. Type II is very frequently encountered as the B.E.T. multi-layer isotherm type for gas–solid adsorptions. The first concave part is attributed to the adsorption of a monolayer, and when the gas pressure is increased further, more layers adsorb on top of the first until many adsorbed layers are ultimately formed with this process. Thus, it is possible to form macroscopically thick layers in this type. Examples include N_2 gas adsorption on non-porous or macroporous powders such as carbons and oxides. Type III is a rare B.E.T. isotherm where the binding of the first monolayer to the adsorbent is weaker than the binding of molecules to already adsorbed molecules. Examples include H_2O adsorption on graphitized carbon or polyethylene polymer. The B.E.T. multi-layer adsorption isotherm equations will be given in Section 8.3.8. Type II is similar to type IV, except that the latter tends to level off at high relative pressures, and also the type III isotherm is similar to type V at low pressures, and the latter tends to level off at high pressures. The Type IV isotherm is important in the study of real catalysts and is used when adsorption hysteresis is present due to the different rates of adsorption and desorption. Desorption from a porous substrate is not as easy as adsorption as there is a high probability of readsorption during the desorption process. The characteristic *hysteresis loop* of the type IV isotherm is indicative of small mesopores which facilitate condensation. Industrial adsorbents and catalysts often show a type IV isotherm.

The most basic isotherm is the linear increase in the amount of adsorbate with an increase in adsorptive gas pressure, and this behavior is described by the *Henry's law limit* isotherm, which will be discussed in Section 8.3.6. The *Freundlich* type of adsorption isotherm, whose equations will be given in Section 8.3.9, is also very common for heterogeneous surfaces. It gives an exponential concave curve with respect to the abscissa. In a *high-affinity* adsorption isotherm, the molecules bind to the adsorbent very strongly so that all the adsorptive molecules present bind the solid surface as adsorbate. This type is often seen for polymer or protein adsorption from liquid solutions. The *Step* adsorption isotherm is characterized by porous adsorbents with large pores. At low gas pressure, a monolayer adsorbs similarly to the Langmuir adsorption; multi-layers then form at moderate pressures and the pores are filled. Finally, saturation is seen at high gas pressures due to the reduction of effective surface area after the pores have been filled by the adsorbate. Examples of step adsorption include noble-gas adsorption on well-defined uniform solids, such as highly oriented pyrolytic graphite.

8.3.6 *Ideal gas behavior: Henry's law limit*

For the simplest case, if (n_2^S/w) is directly proportional to the partial pressure of the adsorptive gas, P_2, then we can write

$$\frac{n_2^S}{w} = K_H P_2 \tag{606}$$

where K_H is a constant. Then, Equation (601) becomes

$$\pi = \frac{RTK_H}{A_{Spec.}} \int_0^{P_2} P_2 \, d\ln P_2 \qquad (607)$$

and upon integration we have

$$\pi = \frac{RTK_H}{A_{Spec.}} P_2 = \frac{RT}{A_{Spec.}} \left(\frac{n_2^S}{w}\right) = \frac{RTn_2^S}{A^S} = RT\Gamma_2 \qquad (608)$$

Equation (608) is exactly equal to Equation (433), given in Section 5.5.3 for the two-dimensional perfect gas for liquid solution surfaces. Equation (608) relates π to the surface excess and is called the *surface equation of state*. Similarly to Equation (436), we can write $[\pi A_{molecule}^S = kT]$ for gas–solid adsorption, where $A_{molecule}^S$ is the area available per adsorbate molecule in the monolayer, and k is the Boltzmann constant ($R = kN_A$). The adsorption isotherm given by Equations (607) and (608) corresponds to the so-called *Henry's law limit*, in analogy with the Henry's law equations that describe the vapor pressures of dilute solutions. Equation (606) predicts a linear relation between m_2^S (or fractional surface coverage, θ_f, and adsorbate gas pressure, P_2, as shown in the linear plot in Figure 8.1.

8.3.7 Langmuir adsorption isotherm

The *Langmuir* adsorption isotherm was developed by Irving Langmuir in 1916 from kinetic considerations to describe the dependence of the surface fractional coverage of an adsorbed gas on the pressure of the same gas above the adsorbent surface at a constant temperature. The Langmuir isotherm expression was re-derived thermodynamically by Volmer and statistically mechanically by Fowler. In his original treatment, Langmuir made several assumptions for his model:

1 Adsorption occurs on specific sites on the adsorbent and all sites are identical.
2 The interaction energy of adsorption is independent of how many of the surrounding sites are occupied.
3 Only one adsorbate occupies each site, and once a monolayer is deposited, occupying all the sites on the adsorbent, the adsorption ceases.

However, all of these assumptions were found to be considerably wrong following recent chemisorption studies performed on crystals and catalysts using advanced instruments. In addition, the Langmuir model does not consider the ordering of the adsorbate layer and its restructuring as it is deposited on the surface. However, Langmuir's model still provides a useful insight into the pressure dependence of the extent of surface adsorption, despite its limitations as a theory, because it contains all the required parameters to provide a very good first approximation.

The equilibrium that may exist between the gas adsorbed on a surface and the adsorptive molecules in the gas phase is a dynamic state, i.e. the equilibrium represents a state in which the rate of adsorption of molecules onto the surface is exactly counterbalanced by the rate of desorption of molecules back into the gas phase. For the derivation of his model, Langmuir used the fact that the rate of adsorption is equal to the rate of desorption during equilibrium, since otherwise the number of adsorbed molecules would change. Initially, an adsorbent has a total number of adsorption sites on the surface, ς_{total} (mol m^2). During the adsorption process, when the number of, ς_{occup}, sites are occupied with the adsorbate, then only ς_{vacant} sites are vacant ($\varsigma_{vacant} = \varsigma_{total} - \varsigma_{occup}$). Langmuir assumed that the adsorption rate is proportional to the number of vacant sites and the pressure of the adsorptive gas:

$$\text{Rate of adsorption} = k_{ad}P_2\varsigma_{vacant} \qquad (609)$$

where k_{ad} is the adsorption rate constant (s^{-1}Pa^{-1}). However, the desorption rate is only proportional to the number of occupied sites,

$$\text{Rate of desorption} = k_{des}\varsigma_{occup} \qquad (610)$$

where k_{des} is the adsorption rate constant (s^{-1}). When equilibrium is reached, the rate of adsorption is equal to the rate of desorption so that

$$k_{des}\varsigma_{occup} = k_{ad}P_2\varsigma_{vacant} = k_{ad}P_2(\varsigma_{total} - \varsigma_{occup}) \qquad (611)$$

After rearrangement, we have

$$\frac{\varsigma_{occup}}{\varsigma_{total}} = \frac{k_{ad}P_2}{k_{des} + k_{ad}P_2} \qquad (612)$$

However, ($\varsigma_{occup}/\varsigma_{total}$) is equal to θ_f, from the definition of fractional coverage given in Equation (596). If we denote a constant, b, as the ratio of the rate constants ($b = k_{ad}/k_{des}$), then Equation (612) becomes

$$\theta_f = \frac{bP_2}{1 + bP_2} \qquad (613)$$

where b is called the *Langmuir isotherm constant* (Pa^{-1}). Two extreme cases are possible:

1 At low pressures, ($bP_2 \ll 1$), and Equation (613) gives $\theta_f \approx bP_2$, which corresponds to Henry's law limit case, and the adsorption isotherm is linear.
2 At high pressures, ($1 \ll bP_2$), and Equation (613) reduces to $\theta_f \approx 1$, which corresponds to the asymptotic region of the type I curve in Figure 8.1.

Since we cannot directly measure the fraction of coverage experimentally, we need to calculate this term from the adsorbed amounts, or moles so that

$$\theta_f = \frac{n_2^S N_A \sigma^o}{A^S} = \frac{n_2^S N_A \sigma^o}{A_{Spec.}w} \qquad (614)$$

where σ^o is the surface area per adsorbate molecule and N_A is Avogadro's number. Alternatively, since ($\theta_f = \Gamma/\Gamma_{mono}$), Equation (613) can be expressed as

$$\Gamma = \frac{\Gamma_{mono}bP_2}{1 + bP_2} \qquad (615)$$

where Γ_{mono} is the number of adsorbed moles per gram or per unit area of the adsorbent when all binding sites are occupied by a monolayer.

If the experimental adsorption isotherm results are obtained from volumetric measurements, then θ_f is usually calculated from the ratio of the volume of the gas adsorbed, V, to the volume of the gas needed to produce a coverage of one monolayer, V_m ($\theta_f = V/V_m$). From Equation (613) we can write

$$V = \frac{V_m b P_2}{1 + b P_2} \tag{616}$$

After rearrangement, Equation (616) turns into a linear expression

$$\frac{P_2}{V} = \frac{P_2}{V_m} + \frac{1}{bV_m} \tag{617}$$

and a plot of (P_2/V) against the gas pressure, P_2, will give a straight line of slope $(1/V_m)$ and intercept $(1/bV_m)$. This will allow the determination of the b and V_m constants from the same isotherm.

The Langmuir isotherm equation can also be derived from the formal adsorption and desorption rate equations derived from chemical reaction kinetics. In Section 3.2.2, we see that the mass of molecules that strikes $1\,m^2$ in one second can be calculated using Equation (186), by applying the kinetic theory of gases as $[dm/dt = P_2 (M_w/2\pi RT)^{1/2}]$, where P_2 is the vapor pressure of the gas in (Pa), M_w is the molecular mass in (kg mol^{-1}), T is the absolute temperature in Kelvin, R is the gas constant 8.3144 (m^{-3} Pa mol-K^{-1}). If we consider the mass of a single molecule, m_w (kg molecule^{-1}), $(m = Nm_w)$, where N is the number of molecules, by considering the fact that $(R = kN_A)$, where k is the Boltzmann constant, and $(M_w = N_A m_w)$, we can calculate the molecular collision rate per unit area $(1\,m^2)$ from Equation (186) so that

$$\frac{dN}{dt} = \frac{P_2}{(2\pi m_w kT)^{1/2}} \tag{618}$$

If adsorption is assumed to be similar to a reversible chemical reaction, such as [X$_{gas}$ + Surface $\leftrightarrow$ X$_{ads}$-Surface], then from statistical mechanics, we can write for the rate of adsorption,

$$\text{Rate of adsorption} = \left(\frac{dN}{dt}\right)(1 - \theta_f)\exp\left(-\frac{U_{ads}}{RT}\right) \tag{619}$$

Since, $\lfloor(1 - \theta_f) = (\varsigma_{vacant}/\varsigma_{total})\rfloor$, this term in Equation (619) gives the probability of hitting a vacant site, and the final part of the equation is the standard Boltzmann factor in the rate equation, considering the adsorption activation energy, U_{ads}. Similarly, we can write for the rate of desorption

$$\text{Rate of desorption} = \theta_f N_S v_{des} \exp\left(-\frac{U_{des}}{RT}\right) \tag{620}$$

where N_S is the concentration of surface sites per unit area, v_{des} is the frequency factor, and U_{des} is the activation energy for desorption. If the rates of adsorption and desorption are equated and the expression is rearranged, we have

$$\theta_f = \frac{P_2(1 - \theta_f)}{(2\pi m_w kT)^{1/2} N_S v_{des}} \exp\left(\frac{U_{des} - U_{ads}}{RT}\right) \tag{621}$$

If we define the Langmuir isotherm constant, b, in kinetic terms as

$$b \equiv \frac{\exp\left(\dfrac{U_{des} - U_{ads}}{RT}\right)}{(2\pi m_w kT)^{1/2} N_S v_{des}} \tag{622}$$

then Equation (622) reduces to

$$\theta_f = bP_2(1 - \theta_f) \tag{623}$$

which simplifies to the Langmuir isotherm equation [Equation (613)] again. The value of the Langmuir constant, b, is increased by a reduction in the isotherm temperature, or an increase in the strength of adsorption, which can be expressed by ΔH_{ads}. In practice, a given equilibrium surface coverage may be attainable at various combinations of pressure and temperature so that, when the temperature is lowered, the pressure required to achieve a particular equilibrium surface coverage decreases.

Some molecules are dissociated upon adsorption, and this type of adsorption is called *dissociative adsorption*. The adsorption reaction can be shown as $[(XY)_{gas} + 2 \text{ Surface} \leftrightarrow X_{ads}\text{-Surface} + Y_{ads}\text{-Surface}]$, and the Langmuir adsorption expression must be modified because two sites on the adsorbent are consumed per adsorbate molecule. The probability of desorption is also different. When these differences are considered, the Langmuir adsorption equation for the dissociative adsorption becomes

$$\theta_f = \frac{\left(b' P_2\right)^{1/2}}{1 + \left(b' P_2\right)^{1/2}} \tag{624}$$

8.3.8 B.E.T. multi-layer adsorption isotherm

Physisorption arises from the van der Waals forces, and these forces also condense gas molecules into their liquid state. Thus, in principle, there is no reason to stop upon completion of a monolayer during physisorption. Indeed, the formation of multi-layers, which are basically liquid in nature, is very common in physisorption experiments. Brunauer, Emmett and Teller developed a theory in 1938 to describe physisorption, where the adsorbate thickness exceeds a monolayer, and this isotherm equation is known by the initials of the authors (B.E.T.). The original derivation of the B.E.T. equation is an extension of Langmuir's treatment of monolayer adsorption from kinetic arguments. Later, in 1946, Hill derived this equation from statistical mechanics. In the B.E.T. isotherm, it is assumed that:

1 Adsorbate molecules stay in their locations after adsorption.
2 Each first layer molecule acts as a potential adsorption site for a second layer molecule, which in turn acts as a site for a third layer molecule and so on.
3 Energy of adsorption is the same for all layers, other than the first.
4 The molecules in the first layer have different potential energies due to the effect of solid surface molecules, $U_1 = U_{ads} = -U_{des}$.
5 The molecules in the second and higher layers have the same potential energy as in the bulk liquid, $U_2 = U_n = U_{con} = U_{vap}$.
6 The layers do not interact with each other energetically.
7 A new layer can start before another is completely finished.
8 At equilibrium, the rates of condensation and evaporation are the same for each individual layer.

The general form of the B.E.T. equation is given as

$$\theta_f = \frac{N^S}{N^{tot}} = \frac{V}{V_m} = \frac{C\left(P_2/P_2^o\right)}{\left[1-\left(P_2/P_2^o\right)\right]\left[1+\left(P_2/P_2^o\right)(C-1)\right]} \tag{625}$$

where N^S is the number of adsorbate molecules per unit area on the solid substrate and N^{tot} is the total number of adsorbate molecules per unit area to produce one complete monolayer on the substrate surface, as given in Equation (596), and C is a constant:

$$C = \frac{f_1(T)}{f_{vap}(T)}\exp\left(\frac{\Delta U_{ads(1)}-\Delta U_{vap}}{RT}\right) \approx \exp\left(\frac{\Delta H_{ads(1)}-\Delta H_{vap}}{RT}\right) \tag{626}$$

Large values of C show a large difference in heat of adsorption between the first and subsequent layers. When C becomes large, the isotherm increasingly resembles the Langmuir adsorption isotherm. For volumetric adsorption measurements, Equation (625) can be rearranged in a linear form

$$\frac{P_2}{V\left(P_2^o - P_2\right)} = \frac{1}{V_m C} + \frac{(C-1)}{V_m C}\left(\frac{P_2}{P_2^o}\right) \tag{627}$$

and a plot of $\lfloor P_2/V(P_2^o - P_2)\rfloor$ against the reduced gas pressure (P_2/P_2^o) will give a straight line of slope $(C-1/V_m\, C)$ and intercept $(1/V_m\, C)$. This will allow the determination of the C and V_m constants from the same isotherm.

Some criticism can be made of the assumptions of the B.E.T. adsorption model. If the second and other layers are assumed to be in the liquid state, how can localized adsorption take place on these layers? Also, the assumption that the stacks of molecules do not interact energetically seems to be unrealistic. In spite of these theoretical weaknesses, the B.E.T. adsorption expression is very useful for qualitative application to type II and III isotherms. the B.E.T equation is very widely used in the estimation of specific surface areas of solids. The surface area of the adsorbent is estimated from the value of V_m. The most commonly used adsorbate in this method for area determination is nitrogen at 77 K. The knee in the type II isotherm is assumed to correspond to the completion of a monolayer. In the most strict sense, the cross-sectional area of an adsorption site, rather than that of the adsorbate molecule, ought to be used, but the former is an unknown quantity; however, this fact does not prevent the B.E.T. expression from being useful for the evaluation of surface areas of adsorbents.

Many adsorbents are porous, and porosity increases the surface area and provides spaces where adsorbate molecules may condense. However, the presence of porosity makes the treatment of the adsorption process more complex, because adsorption in small pores is preferred over that on plane surfaces. The reason is *capillary condensation*, which we have seen in Sections 4.7 and 4.8, whereby the pores are filled with the liquid at pressures less than P_2^o due to the strong attraction between condensed molecules. Pores having a width of less than 2 nm are classified as *micropores*, 2–50 nm as *mesopores*, and greater than 50 nm as *macropores*. It is generally realized that the adsorption isotherm is not coincident with the desorption isotherm, and this phenomenon is called *adsorption hysteresis*.

In the B.E.T. treatment, there is no restriction on the number of adsorbate layers that can be adsorbed on a flat surface. However, when pores are present, the number of adsorbed layers is clearly restricted. Brunauer modified the B.E.T. equation in 1944 so that

$$\frac{V}{V_m} = \frac{C\left(P_2/P_2^o\right)}{\left[1-\left(P_2/P_2^o\right)\right]} \frac{\left[1-(n+1)\left(P_2/P_2^o\right)^n + n\left(P_2/P_2^o\right)^{n+1}\right]}{\left[1+\left(P_2/P_2^o\right)(C-1)-C\left(P_2/P_2^o\right)^{n+1}\right]} \tag{628}$$

where n is the maximum number of adsorbate layers that can be accommodated on each of the walls of a hypothetical capillary having two parallel plane walls. When $n = 1$, Equation (628) reduces to the Langmuir equation, and when $n = \infty$, the standard B.E.T. equation results. Equation (628) gives a good fit for the $(P_2/P_2^o) > 0.35$ part of the isotherm, but it is not suitable for type IV and V isotherms because it does not consider the capillary condensation effects within the pores.

In spite of its limitations, the linear standard B.E.T. equation (Equation 627) is widely used to determine the surface areas of adsorbents. Specific surface area is often correlated with rates of dissolution and other rate-related phenomena, such as catalyst activity, electrostatic properties of powders, light scattering, opacity, sintering properties, glazing, moisture retention, shelf-life and many other properties that can influence the processing and behavior of powders and porous solids in industry. Therefore, surface area measurement is probably the most widely used means of characterizing porous materials. Since the surface area corresponds to the roughness of the particle exterior and its porous interior, gas adsorption is the preferred technique.

8.3.9   Other adsorption isotherms

There are many other types of isotherm, and some of the important types will be given in this section. On some surfaces, the Langmuir assumptions of independent and equal energy surface sites are not valid, and the isotherm deviates considerably from the Langmuir isotherm. Two or more sites having different adsorption energies may exist on such surfaces, and selective adsorption takes place where the energetically more favorable sites are occupied initially. Thus, heat of adsorption varies as a function of fractional coverage. Two important isotherms are applied for such cases:

Freundlich adsorption isotherm

This is one of the oldest empirical adsorption isotherms, developed by Freundlich in 1907. It is useful for adsorption from liquid solutions and also for chemisorption isotherms.

$$\theta_f = \frac{N^S}{N^{tot}} = \frac{V}{V_m} = K_F P_2^{1/n} \tag{629}$$

where K_F and n are two empirical constants and $n > 1$. The enthalpy of adsorption is assumed to vary logarithmically with fractional coverage, θ_f, for values in the range 0.2–0.8. In the Freundlich adsorption isotherm, no monolayer formation (complete coverage) occurs. The isotherm does not become linear at low pressures, but remains convex to the pressure axis. When we write Equation (629) in logarithmic form, we have

$$\ln V = \ln(V_m K_F) + \frac{1}{n}\ln P_2 \tag{630}$$

A plot of ln V versus ln P_2 gives a line with slope $1/n$ of the intensity of the adsorption, and the intercept $(V_m K_F)$ gives a measure of adsorbent capacity.

Temkin adsorption isotherm

This empirical adsorption isotherm is useful for the chemisorption experiments where a monolayer forms, and it considers that all sites are not energetically equivalent. The enthalpy of adsorption is assumed to vary linearly with the fractional coverage

$$\theta = K_T \ln(nP_2) \tag{631}$$

where K_T and n are two empirical constants. This is useful for fitting the middle region of chemisorption isotherms.

8.3.10   Heat of adsorption

The heat evolved during an exothermic adsorption process can be measured calorimetrically, under constant volume or pressure conditions, when a known amount of gas is allowed to adsorb onto a clean adsorbent. The heat of adsorption is made up of two contributions: (1) the heat arises from interaction of the adsorbate molecules with the adsorbent surface, and (2) the heat arises from lateral interactions between adsorbate molecules. The second heat source is small and is often neglected so that only adsorbate–adsorbent interactions are considered when the heat of adsorption of the homogenous and heterogeneous surfaces is investigated. A *homogenous surface* is energetically uniform and the heat of adsorption is independent of the fraction of the surface covered. A *heterogeneous surface* is energetically non-uniform and the heat of adsorption varies with the area fraction of surface covered. Liquid surfaces are fluid and consequently homogenous since any non-uniformity is very short-lived. Solid surface molecules have no fluidity and mostly give heterogeneous surfaces, and the slightest presence of homogeneity can cause considerable changes in adsorption properties.

Since, the heat of adsorption is dependent on the fractional coverage, its calculation from thermodynamics is a complex matter. In this section, we will only derive some basic parameters. When equilibrium is reached between the adsorbate and adsorptive gas molecules, the reversible free energy change of adsorbed gas molecules on the solid (adsorbate), dG_{ads}, is equal to the non-adsorbed gas molecules (adsorptive), dG_g in the medium. Then, we may write from Equation (126), given in Section 3.1.1, for a total number of moles of gas and unit area of adsorbent

$$-S_g dT + V_g dP_2 = -S_{ads} dT + V_{ad} \, dP_2 \tag{632}$$

After rearrangement we have

$$\left(\frac{\partial P_2}{\partial T} \right)_{n_2^S, A^S} = \frac{S_g - S_{ads}}{V_g - V_{ads}} \tag{633}$$

Since, $V_g \gg V_{ads}$, and if we assume ideal gas conditions, that is $[V_g = RT/P_2]$, then Equation (633) becomes

$$\left(\frac{\partial \ln P_2}{\partial T}\right)_{n_2^S, A^S} = \frac{S_g - S_{ads}}{RT} \tag{634}$$

When we write n_2^S is a constant in Equation (634), we mean a constant fractional coverage, θ_f, because the number of moles of adsorbed gas is constant. This is called *isosteric*, which means the *same coverage*. If the process is reversible, we may write the entropy difference from thermodynamics as

$$S_g - S_{ads} = \frac{Q_{st}}{T} \tag{635}$$

where Q_{st} is the *isosteric heat of adsorption*. Under constant pressure, $Q_{st} = -\Delta H_{st}$ where ΔH_{st} is the *isosteric enthalpy of adsorption*, and Equation (634) becomes

$$\left(\frac{\partial \ln P_2}{\partial T}\right)_{n_2^S, A^S} = -\frac{\Delta H_{st}}{RT^2} \tag{636}$$

Upon integration we have

$$\ln P_2 = \frac{\Delta H_{st}}{R}\left(\frac{1}{T}\right) + \text{Integral constant} \tag{637}$$

and the slope of the plot of $(\ln P_2)$ versus $(1/T)$ gives the isosteric enthalpy of adsorption, ΔH_{st}. For this purpose, adsorption isotherms taken at various temperatures are plotted on the same graph, and an arbitrarily chosen horizontal line representing the same surface fractional coverage in this graph cuts the isotherms at different pressures. Then, this isosteric temperature–pressure data is used to plot a new $(\ln P_2)$ versus $(1/T)$ graph to determine the ΔH_{st} parameter from Equation (637). This method is applicable only when the adsorption process is thermodynamically reversible.

On the other hand, two more heats of adsorption are also in use in adsorption science: the *integral heat of adsorption*, Q_i, is an experimentally found quantity from constant volume calorimeter measurements, using the simple expression

$$Q_i = \left(\frac{Q_{total}}{n_2^S}\right)_V \tag{638}$$

where Q_{total} is the total heat evolved during adsorption; Q_i, is also fractional coverage dependent. Lastly, the so-called *differential heat of adsorption*, Q_{dif}, is defined as

$$Q_{dif} = Q_{st} - RT \tag{639}$$

The experimental determination of the isosteric enthalpy of adsorption is important in distinguishing between physisorption and chemisorption on solids. As given above, ΔH_{ads} is low, typically in the region -10 to $-40 \, kJ \, mol^{-1}$, and is slightly larger than the latent heat of condensation, ΔH_{vap} of the adsorptive molecules in physisorption. On the other hand, the adsorption heats are much higher than the physisorption, and ΔH_{ads} is in the range -40 to $-1000 \, kJ \, mol^{-1}$ (typically in the range -100 to $-400 \, kJ \, mol^{-1}$) in chemisorption, due to chemical bond formation.

8.4   Catalytic Activity at Surfaces

In the chemical and petroleum industries, solid catalysts are used to promote and control a wide variety of chemical reactions. A *catalyst* is a substance that increases the rate of a chemical reaction without being consumed. A catalyst can increase the rate of a chemical reaction when the reaction approaches equilibrium, but it cannot induce a reaction that is not permissible under the laws of thermodynamics. Catalysts are also important in providing a useful distribution of products in complex chemical reactions. The adsorption of molecules (mostly chemisorption) on to a surface is required in any surface-mediated chemical process. As we have seen in Section 8.3, an adsorbate may rearrange, or fragment on the surface, or desorb. However, it is also possible for the adsorbate to react with other molecules that are co-adsorbed on the surface. When such a reaction happens, the adsorption characteristics of the product are also very important because, if it binds strongly to the surface, it cannot be desorbed, and thus, the catalyzed reaction stops. In general, a continuously cycling surface catalyzed reaction can be broken down into a short sequence of steps as follows. If we consider reactant molecules, A and B in the gaseous bulk above a solid surface that supports an ensemble of active sites S, there may be five steps:

1 diffusion of adsorptive reactants to the active site on the solid surface;
2 adsorption of one or more reactants (adsorbates) onto the surface: if molecule A is chemically adsorbed onto one of the active sites, a surface complex (S–A) is formed;
3 surface reaction: A reacts with B forming molecule (A + B);
4 desorption of products from the surface: (A + B) escapes the site, thus regenerating site S;
5 diffusion of products away from the surface.

The above scheme shows the importance of both adsorption and desorption processes. Adsorption of at least one of the reactant molecules is required for catalysis to occur. If the accelerated rate of reaction is simply due to the concentration of molecules at the surface, catalysis may result from physisorption of the reactants. On the other hand, chemisorption can be used primarily to quantitatively evaluate the number of surface active sites, which are likely to promote (catalyze) chemical reactions. Chemisorption analyses are applied to physically characterize a catalyst material, to determine a catalyst's relative efficiency in promoting a particular reaction, to study catalyst poisoning, and in monitoring the degradation of catalytic activity over time of use.

The performance of a catalyst depends on several variables. First, adsorption sites must be both numerous and available to the reactant molecules. If they are not available, they cannot be used as a catalyst. For example, some potential adsorption sites may be located deep within a micropore that is too narrow for the reactant molecule to enter, or for the reaction product to exit; in this case the surface site cannot be an active participant in chemisorption. Alternatively, a site could be located along a tortuous path that impedes the efficient flow of reactants towards the active site and products away from the site. Catalytic activity depends on how rapidly chemisorption occurs and the strength (energy) of the chemisorption bond. If the bond is too weak, the molecule may desorb prior to reacting; if it is too strong, the release of the product and regeneration of the site may be retarded. Isothermal chemisorption methods, as well as temperature-programmed chemisorption methods, can be used to study surface energy distribution. Another important design criterion is to maximize the number of active sites per unit of catalyst.

There is a problem in the investigation of catalysts. The chemisorption experiments carried out in laboratories are performed under high vacuum, whereas the real catalysts work under high pressure to catalyze chemical reactions; this may well result in different behavior. Recently, high-pressure adsorption cells and instruments using photons to probe species have been developed to examine catalysts in their original conditions.

Heterogeneous catalysts include metals, metal oxides and solid acids. Pure metals may be directly employed as solid catalysts, or alternatively they may be dispersed as small grains on the surface of supporting oxides such as TiO_2, ZrO_2, Al_2O_3, or SiO_2. One type of unsupported catalyst is composed of Raney metal catalysts, which are prepared by dissolving an aluminum–nickel alloy in a solution of sodium hydroxide, which dissolves the aluminum component. The end product is a highly porous active metal *sponge*, all other undesirable materials having been removed. Some metal oxides are fused or sintered with promoters to form a network of pores throughout the metal mass. For example, the iron catalyst, which is used in ammonia synthesis, is prepared in this way. Similarly, cobalt catalysts are used in the Fischer–Tropsch synthesis of hydrocarbons from CO and H_2; platinum catalysts are used in the hydrogenation of vegetable oils; platinum–palladium catalysts are used in the oxidation of hydrocarbons; and platinum–rhodium catalysts are used in the oxidation of NH_3 to HNO_3.

Zeolites are hydrated alumino-silicates and are used widely in the chemical and environmental industries. Their activity is influenced by the ratio of silica to aluminum. Amorphous silica–alumina catalysts have a lower activity than zeolitic catalysts and are used in mild hydrocracking reactions where acidic surfaces are required. The acidity is maintained by oxygen atoms attached to the aluminum atoms.

On the other hand, the preparation of a supported catalyst involves selecting precursors of the active components and any necessary promoters, and mixing them in a solvent. Then an inert carrier is coated with this mixture and the active metal or precursor is dispersed on the carrier. The product is dried, mixed with a binder then ground, pelletized, extruded, or otherwise shaped. Finally, the material is calcined and activated by oxidation, reduction, or other means.

The investigation of reaction performance and selectivity of catalysts is a very wide subject, both theoretically and experimentally, and is beyond the scope of this book.

References

1. Adam, N.K. (1968). *The Physics and Chemistry of Surfaces*. Dover, New York.
2. Aveyard, R. and Haydon, D.A. (1973). *An Introduction to the Principles of Surface Chemistry*. Cambridge University Press, Cambridge.
3. Adamson, A.W. and Gast, A.P. (1997). *Physical Chemistry of Surfaces* (6th edn). Wiley, New York.
4. Hiemenz, P.C. and Rajagopalan, R. (1997). *Principles of Colloid and Surface Chemistry* (3rd edn). Marcel Dekker, New York.

5. Butt, H.J., Graf, K. and Kappl, M. (2003). *Physics and Chemistry of Interfaces*. Wiley-VCH, Weinheim.
6. McCash, E. M. (2001). *Surface Chemistry*. Oxford University Press, Oxford.
7. Attard, G. and Barnes, C. (1998). *Surfaces*. Oxford Science Publications, Oxford.

Chapter 9
Contact Angle of Liquid Drops on Solids

9.1 Definition, Young's Equation and Use of Contact Angles in Industry

9.1.1 Theory of contact angles

The surface tension of solids, especially polymers with a low surface free energy cannot be measured directly because of the elastic and viscous restraints of the bulk phase, which necessitates the use of indirect methods. As we have seen in Chapter 8, the mobility of the surface molecules in a solid is exceedingly low when compared with any liquid, and a solid surface does not usually display those faces demanded by the macroscopic minimizing of surface free energy. Most solids are incapable of adjusting to such equilibrium conformations and in practice their surface structure will be largely a frozen-in record of an arbitrary past history, where some imperfections, humps and cracks are present. Thus, the laws of capillarity of liquids cannot be applied to solids. The only general method is the rather empirical method of estimating the solid surface tension from that of the contacting liquid. If we consider a liquid drop resting on a solid surface as shown in Figure 9.1, the drop is in equilibrium by balancing three forces, namely, the interfacial tensions between solid and liquid, SL; between solid and vapor, SV; and between liquid and vapor, LV. The *contact angle*, θ, is the angle formed by a liquid drop at the three-phase boundary where a liquid, gas and solid intersect, and it is included between the tangent plane to the surface of the liquid and the tangent plane to the surface of the solid, at the point of intersection. The contact angle is a quantitative measure of the wetting of a solid by a liquid. Unless it is very volatile, any liquid (including liquid metals such as mercury) having a low viscosity can be used as the liquid of the drop. Low values of θ indicate a strong liquid–solid interaction such that the liquid tends to spread on the solid, or wets well, while high θ values indicate weak interaction and poor wetting. If θ is less than 90°, then the liquid is said to *wet* (or sometimes *partially wet*) the solid. A zero contact angle represents *complete wetting*. If θ is greater than 90°, then it is said to be *non-wetting*. From a microscopic point of view, if the solid has a low-energy surface, it attracts the molecules of the liquid with less force than the liquid molecules attract one another. Therefore, the molecules in the liquid next to the surface have a weaker force field than in the liquid surface, so that the liquid molecules at the interface are pulled more strongly into the bulk of the liquid than they are by the solid. There is a tension in the layer adjacent to the solid, and the liquid molecules are somewhat

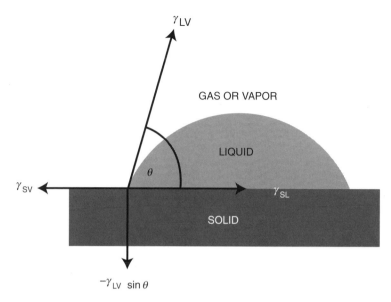

Figure 9.1 Vectorial equilibrium for a drop of a liquid resting on a solid surface to balance three forces, namely, the interfacial tensions, between solid and liquid, γ_{SL}, that between solid and vapor, γ_{SV}, and that between liquid and vapor, γ_{LV}, resulting in Young's equation: $(\gamma_{SV} = \gamma_{SL} + \gamma_{LV} \cos \theta)$, where θ is the *contact angle*. The down component of the vector forces $(-\gamma_{LV} \sin \theta)$ is also shown.

separated, owing to the one-sided force field. The situation is analogous to the behavior of a drop of one liquid on another immiscible liquid, the drop liquid having a higher surface tension than that of the lower liquid (but not equivalent because of the mobility of surface molecules at the interphase region between two immiscible liquids).

T. Young was the first to describe contact angle equilibrium, in 1805. The vectorial summation of forces at the three-phase intersection point (the so-called *three-phase contact point*) gives

$$\gamma_{SV} = \gamma_{SL} + \gamma_{LV} \cos \theta \tag{640}$$

where γ is the surface tension (or surface free energy) term. If $[\gamma_{SV} > (\gamma_{SL} + \gamma_{LV})]$ which shows the presence of a high surface energy solid, then Young's equation indicates ($\cos \theta = 1$), corresponding to ($\theta = 0$), which means the complete spreading of the liquid on this solid. On the other hand, in order to complete the resolution of vector forces about the three-phase contact point, an *up* component ($\gamma_{LV} \sin \theta$) should be present, as shown in Figure 9.1. This force is balanced with a ($-\gamma_{LV} \sin \theta$) force that corresponds to the strain field on the surface of the solid. It is shown that this strain causes the formation of micro-humps on the surface of some soft polymers when a strongly interacting liquid drop (giving low θ values) sits on them.

It is also possible to derive the Young's equation from thermodynamical considerations. The displacement of the contacting liquid is such that the change in area of a solid covered, ΔA, results in the change in surface free energy:

$$\Delta G = \Delta A(\gamma_{SL} - \gamma_{SV}) + \Delta A \gamma_{LV} \cos (\theta - \Delta \theta) \tag{641}$$

At equilibrium, when the interfacial area goes to zero at the limit

$$\lim\nolimits_{\Delta A \to 0} \frac{\Delta G}{\Delta A} = 0 \qquad (642)$$

The $(\Delta\theta/\Delta A)$ term behaves as a second-order differential and drops out in taking the limit of $\Delta A \to 0$, so one obtains from Equation (641)

$$\gamma_{SL} - \gamma_{SV} + \gamma_{LV} \cos \theta = 0 \qquad (643)$$

Equation (643) is identical to Young's equation, as given by Equation (640).

In Sections 5.6.1 and 7.5, we defined the *work of adhesion*, $[W_{12}^a = -\Delta G_{12}^a]$ as the reversible work at constant pressure and temperature conditions, per unit area, required to separate a column of two different liquids (or solids) at the interface creating two new equilibrium surfaces of two pure materials, and separating them to infinite distance. In terms of surface tension, the work of adhesion can be written as $[W_{12}^a = \gamma_1 + \gamma_2 - \gamma_{12}]$, as given by Equation (458). For a solid–liquid interaction, we may rewrite Equation (458) as

$$-\Delta G_{SL}^a = W_{SL}^a = \gamma_{SV} + \gamma_{LV} - \gamma_{SL} \qquad (644)$$

Dupré, in 1869, combined Equations (640) and (644) to give the Young–Dupré equation:

$$-\Delta G_{SL}^a = W_{SL}^a = \gamma_{SV}(1 + \cos \theta) \qquad (645)$$

Equation (645) shows that contact angle is a thermodynamic quantity, which can be related to the work of adhesion and interfacial free energy terms. When θ values are small, the work of adhesion is high and considerable energy must be spent to separate the solid from the liquid. If $\theta = 0°$, then $W_{SL}^a = 2\gamma_{LV}$; if $\theta = 90°$, then $W_{SL}^a = \gamma_{LV}$, and if $\theta = 180°$, then $W_{SL}^a = 0$, which means that no work needs to be done to separate a completely spherical mercury drop from a solid surface (or a water drop from a superhydrophobic polymer surface), and indeed these drops roll down very easily even with a 1° inclination angle of the flat substrate.

Later, in 1937, Bangham and Razouk showed that the γ_{SV} and γ_{LV} terms in Young's equation are related to the γ_S^o and γ_L^o of the pure solid and liquid phases (the components in contact with vacuum) at equilibrium, by the expression

$$\gamma_S^o - \gamma_{SV} \equiv \pi_{SV} \qquad (646)$$

$$\gamma_L^o - \gamma_{LV} \equiv \pi_{LV} \qquad (647)$$

where, π_{SV} is the *equilibrium film (or spreading) pressure* of the vapor (or gas) adsorbate on the solid, and π_{LV} is the *equilibrium film pressure* of the gas or vapor adsorbate on the liquid (π can only be zero or positive, see Section 5.5). If the solid is appreciably soluble in the drop liquid, then the π_{LV} term must be taken into account because any gas or vapor adsorption at the liquid–vapor interface is controlled thermodynamically by the activity of the solute. Nevertheless, since the solubility of a solid is generally very low in the contacting liquid drop, it is usual to ignore the π_{LV} term and assume that ($\gamma_L^o = \gamma_{LV}$). On the other hand, π_{SV} is the film pressure of the liquid vapor on the solid and it must be considered during contact angle measurements, especially on high-energy surfaces. The π_{SV} term arises from the liquid molecules evaporating and then condensing and adsorbing onto the solid next to the liquid drop during contact angle measurement, thus creating a tiny layer of liquid film of unknown thickness on the solid. Then, Young's equation can be modified as

$$\gamma_S^o - \pi_{SV} = \gamma_{SL} + \gamma_{lv} \cos\theta \qquad (648)$$

By rearrangement one obtains

$$\gamma_S^o - \gamma_{SL} = \gamma_{lv} \cos\theta + \pi_{SV} \qquad (649)$$

The $(\gamma_S^o - \gamma_{SL})$ term can be determined experimentally from contact angle and adsorption measurements, but unfortunately there is no straightforward method to determine the (γ_S^o) and (γ_{SL}) terms independently. There is some dispute about the presence of π_{SV} during contact angle measurements: Fowkes and Good proposed that π_{SV} is negligible in cases of a finite contact angle on low-surface-energy solids such as polymers. In contrast, some researchers have reported several experimental π_{SV} values for low-energy polymer surfaces. Generally, π_{SV} is assumed to be at monolayer or lower coverage for a liquid whose contact angle is less than 90° on high-surface-energy solids, and its value may be determined by the *Gibbs integral* method, in which the adsorption of the vapor of the liquid, Γ, is measured by ellipsometry (see Section 8) and using the expression,

$$\pi_{SV} = RT \int_0^{Past} \Gamma(P)\, d\ln P \qquad (650)$$

Equation (650) is equivalent to Equation (434), given in Section 5.5.3, and Equations (598) and (601) given in Section 8.3.3.

9.1.2 *Industrial applications of contact angles*

Measurement of contact angles provides a better understanding of the interactions between solids and liquids, or between immiscible liquids. Although it is a difficult task to measure the contact angle properly on solids, a large body of reliable data has been accumulated and a vast literature exists correlating contact angle data with surface tension of solids. The interactions between solids and liquids play a key role in understanding the chemical and physical processes in many industries. The adhesion between different composite structures (glass–metal, leather–fabric, wood–paper) and the wetting of adhesive on a substrate can be accessed by contact angle measurements. The determination of θ is very important in the paints and coatings industries. The motivation for new preparation methods is to obtain long-lasting adhesion between the coating and substrate surfaces (paper, metal, wood, plastic etc.), and in the automotive and building industries, there is the requirement to optimize the interfacial tension and to measure the strength of interaction, by the use of contact angles. The effectiveness of the coating formulation and the coating process, for example a car body coating, can be accessed by measuring the hydrophobicity (i.e. the contact angle) of the lacquer surface. The advent of new, environmentally friendly water-based coatings and inks started new research in the paper industry to improve ink performance. Adhesion of inks to polymeric food packaging film products also benefits from surface chemistry. As an example, all the materials involved in an offset printing process need to have a certain surface free energy in order to attain optimum printing quality, so contact angle measurements are required at many steps in the printing process. On the other hand, composite materials made of reinforcing fibers and polymeric (resin) matrix systems have replaced many of the traditional metals and other heavier and weaker materials, and have begun to be used in a wide range of products in the aerospace, automotive and sporting goods industries. Using contact angle measurements, it is possible to optimize the adhesion between the fiber and resin matrix system, and to find the right formulation of the resin matrix with proper wetting properties against the fiber. In the textiles industry, everything from carpet fibers to surgical gowns involves surface treatments such as anti-static or anti-stain coatings applied to the textile material to provide protection. The wettability of single fibers or fabrics as well as their hydrophobicity and washability can be checked by contact angle measurements.

The medical, pharmaceutical and cosmetic industries also use contact angle measurements in their research and quality control laboratories. Biocompatibility is an important issue in the medical and dental industries. Surface-modified biomaterials are being employed to create disposable contact lenses, catheters, dental prosthetics and body implants; however they must be biocompatible, that is, they should not be rejected by the human body. In the case of dental surgery, good adhesion between the tooth and embodiment is required. Thus, contact angle measurements are necessary in all wettability and biocompatibility studies. The effectiveness of the cleaning solution formulation for contact lenses can be improved by optimizing the surface free energy of the lens and the solution. On the other hand, applying special surface treatments can largely influence the distribution and dissolving behavior of a pharmaceutical powder. The dissolving behavior of an orally ingested pharmaceutical powder, tablet or capsule, or transdermally applied controlled-release drug product can be improved with the help of contact angle and surface tension measurements. In the cosmetics industry, the effectiveness of shampoos, cleaning solutions, suntans, body creams and lotions can be followed by measuring the contact angle. In order to improve absorbency and provide protection against wetness, superabsorbent personal hygiene products such as baby diapers have been developed with the help of contact angle measurements. The surface tension of pesticide or fertilizer formulations directly affects their spreading on plant leaves or in soil, which has an influence on environmental pollution. Similarly, oil-polluted sea and land can be treated using surfactant solutions, and the cleaning process can be monitored by contact angle measurements on the treated samples. Recently, contact angle methods have been used to assess the cleanliness of semiconductor surfaces in the electronics industry. Contact angle methods also have great potential for the newly developing nanotechnology field.

In comparison with other surface characterization techniques, contact angle methods are accepted as complementary techniques, providing supporting information to other more expensive surface analysis methods such as ESCA, SIMS, SAXS, Raman, IR etc. A researcher may be guided in the right direction by contact angle results before performing a more elaborate analysis using much more expensive surface characterization equipment.

9.2 Measurement of Static Contact Angles

Measurement of contact angles appears to be quite easy when first encountered, but this can be misleading and the accurate measurement of thermodynamically significant contact angles requires painstaking effort. If the substrate is not prepared properly, if very pure liquids are not used while forming drops and if some important practical issues during measurement, such as drop evaporation, the location of the needle in the drop, and maintaining a sharp image, are not considered, then incorrect and generally useless contact angle results can be obtained, which may be used as "evidence" for false thermodynamical conclusions. Unfortunately, many experiments, carried out improperly, have been reported in the scientific literature; this should be avoided for future studies.

Many different methods have been developed for the measurement of contact angles, but only a few are popular today. Two preferred approaches are: the measurement of the static contact angle of a sessile drop on a non-porous flat solid using a video camera or goniometer; and the dynamic contact angle measurement method using tensiometry, which involves measuring the forces of interaction, while a dynamic (moving) flat solid plate is immersed into or withdrawn from a test liquid (see Section 9.3). We will also present some other useful methods in Sections 9.2.2–9.2.5.

9.2.1 Direct measurement of static contact angle by video camera or goniometer

The most widely used method is to measure the angle of a sessile drop resting on a flat solid surface using a goniometer–microscope equipped with an angle-measuring eyepiece or, more recently, a video camera equipped with a suitable magnifying lens, interfaced to a computer with image-analysis software to determine the tangent value precisely on the captured image. A suitable cold light source and a sample stage whose elevation can be controlled to high precision are also required for the application of this technique. Formerly, the drop profile was photographed and the tangent of the sessile drop profile at the three-phase contact point drawn onto the photo-print to determine the value of the contact angle. In static θ measurement, the results are somewhat dependent on the experience of the operator.

The measurement of a single static contact angle to characterize the solid–liquid interaction is not adequate because, in practice, there is no single *equilibrium contact angle, θ_e*, on a solid surface. While deriving Young's equation, we assumed an ideal solid that is chemically homogeneous, rigid, and flat to an atomic scale. However, there is no such solid surface, because all solid surfaces have surface imperfections and are heterogeneous to a degree, as we saw in Chapter 1. Thus, there may be a range of static contact angles, depending on the location of the drop and on the application type of the measurement. Experimentally, only two types of contact angle measurement technique are standardized:

1 When a liquid drop is formed by injecting the liquid from a needle connected to a syringe onto a substrate surface, it is allowed to advance on the fresh solid surface and the measured angle is said to represent the *advancing contact angle, θ_a*. For each drop–solid system there is a maximum value of θ_a before the three-phase line is broken (it should be noted that the stainless steel needle must be kept in the middle of the drop during measurement of θ_a, and the needle may be coated with paraffin wax in order to prevent climbing of some strongly adhering liquids, such as water, on the metal needle surface; alternatively, plastic needles such as Teflon and polypropylene may be used with water).

2 The *receding contact angle, θ_r*, can be measured when a previously formed sessile drop on the substrate surface is contracted by applying a suction of the drop liquid through

the needle. Precise measurement of θ_r is very difficult (see Section 9.4.1 for drop evaporation effects).

These contact angles fall within a range where the advancing contact angles approach a maximum value and receding angles approach a minimum value ($\theta_a > \theta_r$). Alternately, both advanced and receded angles are measured when the stage on which the solid is held is tilted to the point of incipient motion of the drop (see Section 9.2.3).

Both θ_a and θ_r depend on the surface roughness (detailed shapes and configurations of the patches or strips) and also on the surface chemical heterogeneity. The direct determination of θ_a within ±2° is easy, but it is difficult to reduce the relative error to ±0.5°. This is because the direction of a liquid profile rapidly changes with the distance from the three-phase contact point. The difference between θ_a and θ_r gives the *contact angle hysteresis, H*, ($H \equiv \theta_a - \theta_r$), which can be quite large, around 5–20° in conventional measurements (or 20–50° in some exceptional cases). The reasons for contact angle hysteresis will be examined in Section 9.6.

There are differences between the techniques applied for the experimental measurement of static θ_a and θ_r. Historically, in the 1950s, Zisman and co-workers, who were the pioneers in contact angle standardization, used the tip of a fine platinum wire to bring a droplet to the surface and detach it from the wire. More liquid is added in successive droplets from the wire, and θ is measured after each addition by viewing through a goniometer microscope; the limiting value of θ is taken to be θ_a. The retreating angle was measured via stepwise removal of small increments of liquid by touching the tip of a fine glass capillary to the drop and withdrawing it to give θ_r. In the 1960s, Fowkes, Good and co-workers introduced the liquid drop by means of a micrometer syringe, which had a fine stainless steel needle up to a drop contact diameter of 4 mm. The liquid drop was held captive while additional liquid was added to the drop until a steady value of θ_a was obtained and the addition of the liquid stopped. The needle, having a diameter of less than 1 mm, must not be removed from the drop during measurement as this may cause mass and profile vibrations, which can decrease θ_a to some lower metastable state. Contact angles must be measured on both sides of the drop and reported separately. The liquid is withdrawn from the drop by means of the same micrometer syringe to measure θ_r. However, there is a strong influence on the θ_r value from the rate of liquid removal from the drop. Neumann and co-workers made a small hole in the flat substrate sample and first deposited a small drop on the substrate through a needle connected to this hole beneath the substrate. The size of the drop is then increased by feeding more liquid to the drop by means of this needle connected to a motorized syringe. This procedure prevents the drop oscillating and also destruction of the axisymmetry. By this means, they controlled the rate of advance or retreat of the symmetrical sessile drop on the substrate, to measure θ_a and θ_r precisely. They also developed a method to determine both the contact angle and surface tension of the liquid by applying a digital image analysis to drop profiles and a computation method named *axisymmetric drop shape analysis*, ADSA. In this method, an objective function is constructed which expresses the error between the physically observed profile and the theoretical Laplacian curve; the function is then minimized using an iterative procedure.

The inclusion of gravity correction into the Young–Laplace equation is feasible for large sessile liquid drops formed on solids. Determination of what is a *large* drop and what is *small* can be performed by simply comparing the contact radius of the sessile drop with the square root of the capillary constant, a of the liquid. If it is much smaller (say more than 10 times) than the square root of the capillary constant, then the influence of gravitation can be neglected. When volatile liquid drops are formed, measurements must be made in an enclosed chamber to prevent drop evaporation by permitting establishment of the equilibrium vapor pressure of the liquid. However, when high-boiling-point liquids, such as water, glycerol, and hexadecane, are used, and the measurements are carried out rapidly, there is no need for such a chamber. Some researchers do not measure θ_a or θ_r; they only measure the observed angle of the free-standing drop after removing the needle (an unknown period passes after the drop formation) calling it the *equilibrium contact angle*, θ_e. Values of θ_e are between θ_a and θ_r, but often nearer to θ_a. Since this angle does not represent the initial contact angle formed on the fresh surface, it is of a lower degree of scientific usefulness than is a true θ_a or θ_r. Some researchers use θ_e as the mean value of θ_a and θ_r, but this approach is thermodynamically wrong.

There are several advantages to static contact angle measurement. It can be used for almost any solid substrate, as long as it has a relatively flat portion and can be fitted on the stage of the instrument. Testing can be done using very small quantities of liquid. It is also easy to test high-temperature liquids such as polymer melts in heated chambers. However, there are several limitations to the method. First, conventional goniometry relies on the consistency of the operator in assignment of the tangent line. This may lead to subjective error, especially significant when the results of multiple users are compared. This problem can be reduced by computer analysis of the droplet shape and contact angle. The second problem is the variable rate of introducing the drop liquid through the needle during determination of θ_a and the variable rate of withdrawal of the liquid during determination of θ_r. The sessile drop method is not particularly well adapted to quantitative measurement of the dependence of contact angle on the rate of advance or retreat, because a linear rate of change in drop volume does not correspond to a linear rate of motion of the drop front. An appropriate rate is of the order of 0.01–0.10 mm min^{-1} linear advance or retreat by using a motor-driven syringe. Also, it is best to specify a constant time allowed before measuring the contact angle after the motion stops, e.g. 1–10 sec, to damp the drop oscillations formed in order to obtain more precise data. The third problem is distortion of the drop surface caused by the needle. If the needle enters the drop at a point very close to the solid, it may obscure the drop profile. It is best to keep the needle at the middle of the drop. If the needle passes through the upper surface of the drop, there will be some capillary rise of the liquid up the needle and distortion of the surface. (However, it has been claimed by some authors that this capillary rise does not perturb the liquid in the region of the contact line with the solid.) Removing the needle from the drop does not help, because that makes it impossible to study hysteresis. Finally, objects other than flat objects, such as cylindrical fibers, cannot be easily studied by the goniometry approach.

9.2.2 Captive bubble method

A captive air (or other gas) bubble is formed in the liquid contacting with the solid by means of an inverted micrometer syringe beneath the substrate which is kept in the test liquid. The contact angle is measured by means of a goniometer microscope or video camera. In this method, the solid–vapor interface is in equilibrium with the saturated vapor pressure of the liquid, which is present in the bubble. When more air is added to the bubble through the needle, the surrounding liquid front retreats and gives an angle which is equivalent to θ_r in the conventional sessile drop method in air. Withdrawing the air from the bubble causes the liquid front to advance, which is equivalent to θ_a. It is interesting to note that the θ_e measured in the captive bubble method is nearer to θ_r than to θ_a.

9.2.3 Sliding drop on an inclined plate method

A sessile drop is formed on a plate of solid substrate gripped at one end onto a motorized horizontal stage, which can be tilted to the point of incipient motion of the drop. When the plane of the solid surface reaches a critical slope, the drop starts to slide. The measured angle at the downhill edge of the drop approaches θ_a, and the angle at the uphill edge approaches θ_r, as shown in Figure 9.2. The angles should be measured immediately prior to the drop starting to slide. The tilt angle, ϕ_t, can also be used to derive thermodynamic conclusions; however, this method is not very reliable, because the determination of a clear and sharp drop image at the instance of sliding is difficult, and also it gives inconsistent results with rough substrates which show a strong pinning behavior with the liquid drop, so that no drop sliding occurs even at a tilt angle of $\phi_t = 90°$. In addition, some researchers cautioned against this method because it yields values of θ_a and θ_r that are strongly dependent on the drop size.

9.2.4 Drop dimensions method

The contact angle may be calculated indirectly from measurement of the dimensions of a sessile drop. In order to carry out such a calculation, the drop should be small enough so

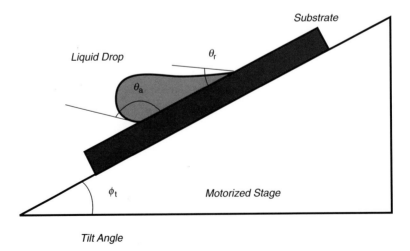

Figure 9.2 Determination of contact angle from the sliding drop on an inclined plate method. The advancing contact angle, θ_a, is the measured angle at the downhill edge of the drop, the receding contact angle, θ_r, is measured at the uphill edge, and ϕ_t is the tilt angle.

that its deviation from a spherical shape may be neglected. The height of the drop is $h = R(1 - \cos\theta)$, from plane trigonometry, R being the radius of the spherical segment, as shown in Figure 9.3. The contact radius of the liquid drop, r_b, is given as ($r_b = R\sin\theta$). Then, we may write

$$\frac{h}{r_b} = \frac{1-\cos\theta}{\sin\theta} = \tan\left(\frac{\theta}{2}\right) \tag{651}$$

If h and r_b are measured, θ can be calculated from Equation (651). For large drops, h and r_b can be so distorted by gravitation that Equation (651) cannot be used, and much more elaborate calculations are needed. On the other hand, the contact angle may be calculated if the drop volume can be experimentally measured by using advanced syringes. If the contact radius of the liquid drop, r_b of known volume, V is measured, then θ can be calculated from the dimensionless ratio obtained from three-dimensional spherical trigonometry

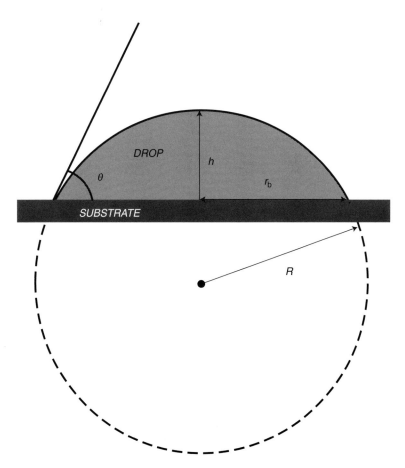

Figure 9.3 Determination of contact angle from the drop dimensions method: h is the height of the drop [$h = R (1 - \cos\theta)$], r_b is the contact radius of the liquid drop ($r_b = R\sin\theta$) and R is the radius of the circle of the spherical segment.

$$\frac{r_b^3}{V} = \frac{3\sin^3\theta}{\pi\left(2-3\cos\theta+\cos^3\theta\right)} \tag{652}$$

Alternatively, if both r_b and h can be measured, we can preferably use

$$\frac{r_b^2 h}{V} = \frac{3(1+\cos\theta)}{\pi(2+\cos\theta)} \tag{653}$$

Equations (651)–(653) imply that the base of the drop is an exact circle, but it is rarely circular due to the non-uniformity of most of the solid surfaces. This is the main source of error in this method.

9.2.5 Static Wilhelmy plate method

Previously, Guastalla had developed a method to measure the contact angles by using a balance and two hanging Wilhemy plates. These plates are immersed in the same liquid, each suspended from a balance. One of the plates is the (solid) surface under investigation and the other is a reference solid, which is completely wetted by the liquid (a metal plate for example). The slides are set up to have the same perimeter, and the ratio of the downward pull on the test specimen divided by the downward pull on the completely wet plate (both corrected for buoyancy) is ($\gamma_{lv}\cos\theta/\gamma_{lv}$) or ($\cos\theta$) only. Thus, the contact angle can be found from the differences in the work of adhesion of the sample and the reference solid within the same liquid.

Later, Neumann developed the *static Wilhelmy plate method* which depends on capillary rise on a vertical wall, to measure θ precisely. A Wilhelmy plate whose surface is coated with the solid substrate is partially immersed in the testing liquid, and the height of the meniscus due to the capillary rise at the wall of the vertical plate is measured precisely by means of a traveling microscope or cathetometer. If the surface tension or the capillary constant of the testing liquid is known, then the contact angle is calculated from the equation, which is derived from the Young–Laplace equation

$$\sin\theta = 1 - \frac{\Delta\rho g h^2}{2\gamma} = 1 - \left(\frac{h}{a}\right)^2 \tag{654}$$

where $\Delta\rho$ is the density difference between the liquid and the vapor, g is the acceleration due to gravity and a is the capillarity constant (see Section 4.4). However, if the surface is appreciably rough or chemically heterogeneous, contact angle results are not reproducible with this method.

9.3 Dynamic Contact Angle Measurement

Dynamic contact angles are the angles which can be measured if the three-phase boundary (liquid/solid/vapor) is in actual motion. A Wilhelmy plate is used in dynamic contact angle measurements, and this method is also called the *tensiometric contact angle method*. It has been extensively applied to solid–liquid contact angle determinations in recent years. In practice, a solid substrate is cut as a thin rectangular plate, otherwise a solid material is coated onto a Wilhelmy plate made of platinum or other metal by some means. The test plate is suspended from the beam of an electrobalance and is held vertically at a fixed position during the measurement, while the beaker containing the test liquid is raised and lowered via a motorized platform. Both the balance and platform assembly are interfaced to a computer for data acquisition and control. The plate is scanned in both advancing and receding directions at constant velocity, from which a force–distance plot is constructed. Before the experiment, the surface tension of the test liquid is measured using the same equipment, usually by using either a platinum Wilhelmy plate or a du Noüy ring. Later, the test plate is mounted vertically above the liquid. The plate is suspended with the bottom edge above the surface of the liquid to begin the measurement. When the beaker containing the test liquid is raised and touched to the plate, a force is detected on the balance. The location at which the plate contacts the liquid surface is called the *zero depth of immersion*. If the plate is immersed deeper into the liquid, then the balance detects a greater force, which is the sum of the wetting force, the weight of the probe and the buoyancy force. The forces on the balance are [F_{total} = *wetting force + weight of probe − buoyancy*], which can be given as

$$F_{total} = p\gamma_{lv} \cos\theta + mg - \rho_L g \varpi H d \qquad (655)$$

where F_{total} is the total measured force on the electrobalance, m is the mass of the plate, g is the acceleration of gravity, ρ_L is the liquid density, ϖ is the thickness of plate, H is the width of the plate, d is the immersion depth, p is the plate perimeter [$p = 2(H + \varpi)$], γ_{lv} is the liquid surface tension and θ is the contact angle at the liquid–solid–air three phase contact line. The weight of the plate probe (*m g*) can be measured beforehand and set to zero on the electrobalance during measurement, while the effect of buoyancy can also be removed by extrapolating the force back to the zero depth of immersion. Then, the remaining component force is only the wetting force

$$\text{wetting force} = F_{wet} = \gamma_{lv} p \cos\theta \qquad (656)$$

Thus, at any depth, the force data can be received and used to calculate contact angle. In practice, the force on the plate is measured as it is cycled slowly down and up. When the plate is immersed into the test liquid as the liquid level is raised, the contact angle thus obtained is called the *advancing contact angle*, θ_a. After the sample is immersed to a set depth the process is reversed, so that the plate is withdrawn from the test liquid back to its original position, at the zero depth of immersion. As the plate retreats from the liquid, the contact angle thus obtained is called the *receding contact angle*, θ_r. Dynamic contact angles may be assayed at various rates of speed. If low velocities are applied, then the contact angle results are similar to the properly measured static contact angles. However, dynamic Wilhelmy plate method results are usually found to be a little smaller than results obtained with other methods.

The progress of the experiment can be shown graphically as in Figure 9.4, with the x-axis representing the immersion depth and the y-axis the force over perimeter. The contact angles are calculated from the buoyancy versus depth information. As shown in Figure 9.4, the graph of force/wetted length versus depth of immersion appears as follows:

1 The sample is above the liquid and the force/length is zeroed.
2 The sample hits the liquid surface. If the sample has a contact angle <90°, then the liquid rises up, causing a positive force.

3 By raising the test liquid level, the sample is immersed, and the buoyant force increases to cause a decrease in the total force on the balance. Advancing contact angle is calculated from this force–immersion depth line.

4 After having reached the desired depth, the sample is withdrawn out of the test liquid. The receding contact angle is calculated from this force–immersion depth line.

There are many advantages of the dynamic tensiometric approach over conventional static sessile drop methods. Unlike static sessile drop methods, the accuracy of the dynamic Wilhelmy plate method can be better than 0.5° and is without human reading uncertainty. The value of θ_r can be measured as easily as θ_a with this method giving reliable hysteresis values. Contact angles on fibers can be measured, and it is possible to obtain reliable data from fibers with diameters below 10 μm. Another advantage is that the measurement is carried out for the entire perimeter of the immersed solid, giving an averaged value for the contact angle. (Some authors have suggested that this average value may have little significance if the surface tension is not highly uniform across a face.) The graphs produced by this technique are very useful in studying hysteresis. In addition, the variations generated

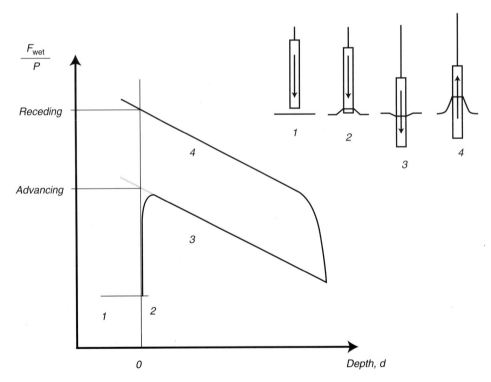

Figure 9.4   Dynamic contact angle measurement using the Wilhelmy plate method. Variation of the wetting force over wetted length versus depth of immersion with the stages: (1) the sample is above the liquid; (2) the sample hits the liquid surface; (3) the sample is immersed into the test liquid to give an advancing contact angle which can be calculated from this force–immersion depth line; (4) after having reached the desired depth, the sample is withdrawn out of the test liquid to give a receding contact angle from this line.

over multiple wetting/dewetting cycles can yield information on surface changes after the wetting (such as absorption or surface reorientation).

The dynamic tensiometric method has three major limitations. First, a large amount of very pure test liquid must be available in the beaker to immerse the solid in a single experiment. Second, the solid sample must be formed in a definite geometry such as a rectangular plate, cylindrical rod, or single fiber, so that it will have a constant perimeter over a portion of its length (immersion depth). In addition, the two sides of the sample plate must also be coated homogeneously with the same solid for proper dynamic contact angle measurement. Surfaces with a high roughness may give inconsistent results. Third, it is difficult to use this technique at high temperatures above 100°C.

9.4 Liquid Evaporation Effects During Contact Angle Measurement

9.4.1 *Receding contact angle determination from drop evaporation*

For sessile drops, liquid evaporation is inevitable unless the atmosphere in the immediate vicinity of the drop is completely saturated with the vapor of the liquid. In contact angle measurement, liquid evaporation is generally regarded as undesirable because it may cause the liquid front to retreat and the contact angle value to decrease, so usually a contact angle value which is intermediate between θ_a and θ_r is observed unintentionally. In practice, a closed chamber can be used to ensure saturation of the vapor of the liquid to minimize this effect. When water is used as the drop liquid, the humidity in the chamber must be controlled and be set to saturation conditions to prevent water evaporation. However, the use of closed chambers is not popular in general contact angle practice in either industry or academic circles, so incorrect contact angle measurements are common, if the measurement has not been made very rapidly when moderately volatile liquids are used as drop liquids.

It has been suggested that drop evaporation may also be used beneficially. The receding contact angles of water drops on several polymer surfaces can be determined by following time-dependent drop evaporation with a video-microscopy technique, in comparison with the static needle-syringe sessile drop method. For initial contact angles of water drops of less than 90°, evaporation proceeds with a decreasing contact angle and a pinned contact radius; this persists for much of the evaporation time, and the contact radius finally begins to recede. The contact angle at this point therefore represents a natural receding angle, θ_r. The contraction of the contact radius does not proceed smoothly but rather undergoes a stick–slip motion so giving rise to local peaks in the receding angle. It was concluded that the drop evaporation method allows a minimum rate of liquid withdrawal from the drop, which minimizes (or standardizes) the linear rate of retreat of the drop on the substrate, giving reliable receding contact angles, θ_r.

9.4.2 *Drop evaporation theory for spherical and ellipsoidal drops*

If we want to use a contact angle value in thermodynamic equations, measurement of the initial value of a contact angle, θ_i, at the instance of drop formation is required. Determination of θ_i is difficult in many instances, due to liquid evaporation effects. Thus, a more complete understanding of how liquid evaporation varies the contact angle of a sessile drop on a solid surface, in still air or in controlled atmospheric conditions, is important in wetting and surface characterization processes. If drop evaporation can be modeled adequately, then it is possible to predict precise initial contact angle values. There are two modes of drop evaporation: 1 at constant contact angle with diminishing contact area and 2 at constant contact area with diminishing contact angle. (A mixed type is also present especially at the last stages of drop evaporation where the mode would change from one to the other.) Since it is possible to monitor time-dependent drop evaporation with the advanced video-microscopy technique, it is possible to test diffusion-controlled evaporation models. When a liquid drop is sufficiently small and surface tension dominates over gravity, it is generally assumed that the drop forms a *spherical cap* shape which can be characterized by four different parameters: the drop height (h), the contact radius (r_b), the radius of the sphere forming the spherical cap (R), and the contact angle (θ), as shwon in Figure 9.3. If only r_b and θ are measured during an evaporation experiment, then the drop volume is given by Equation (652); if, however, all the r_b, θ and h parameters can be measured during an experiment, we can preferably use

$$V = \frac{\pi r_b^2 h}{3} \frac{(2+\cos\theta)}{(1+\cos\theta)} \tag{657}$$

for the drop volume calculation. The rate of volume decrease by diffusion-controlled (slow) evaporation with time, where the effect of convective air currents are neglected, is given from the diffusion theory as

$$-\left(\frac{dV}{dt}\right) = \frac{4\pi RD}{\rho_L}(c_S - c_\infty) \tag{658}$$

where t is the time (s), D is the diffusion coefficient ($cm^2 sec^{-1}$), c_S is the concentration of vapor at the sphere surface at R distance ($g\,cm^{-3}$), c_∞ is the concentration of the vapor at infinite distance (R_∞) ($g\,cm^{-3}$), and ρ_L is the density of the drop substance ($g\,cm^{-3}$). When $\theta <$ 90°, in the regime of constant contact radius, r_b, it has been proposed that the rate of mass loss, calculated from the drop profile by applying a model based on a two-parameter spherical cap geometry, is directly proportional to h and r_b, but not to the spherical radius, R. For a constant contact radius, the variation of contact angle with time is given as

$$-\left(\frac{d\theta}{dt}\right) = \frac{2D(c_\infty - c_S)\sin^3\theta}{\rho_L r_b^2(1-\cos\theta)} \tag{659}$$

After differentation, Equation (659) can be reduced to the angular function, $F(\theta)$ on one side and the diffusion parameters on the other side

$$F(\theta) = \ln[\tan(\theta/2)] + \frac{1-\cos\theta}{\sin^2\theta} = -\frac{4D(c_\infty - c_S)(t-t_o)}{\rho_L r_b^2} \tag{660}$$

The angular function fits a straight line in a $F(\theta)$ versus θ plot, over the range 30–90° according to the expression [$F(\theta) = -1.592 + 1.632\theta$], where θ is in radians.

Mathematical expressions for an ellipsoidal cap drop resting on a solid surface having three parameters (base radius, height and contact angle) were subsequently also derived, and this model fitted the experimental data better than those evaluated from the spherical cap model. The volume of the ellipsoidal cap can be evaluated by rotating the ellipse curve about the *y*-axis resulting in:

$$V = \frac{\pi r_b h}{3} \frac{(2r_b \tan\theta - h)}{\tan\theta} \tag{661}$$

and the ellipsoidal diffusion model becomes

$$-\frac{dV}{dt} = \frac{2D\pi h(c_\infty - c_S)}{\rho_L}\left[\frac{1}{2} + \frac{a}{2he}\left(\tan^{-1}\frac{ea}{b} - \tan^{-1}\frac{(b-h)ea}{b^2}\right)\right] \tag{662}$$

where *a* and *b* are the semi-axis lengths of the ellipse profile in the *x* and *y* directions, and *e* is the eccentricity, which is geometrically defined as $[e \equiv (\sqrt{b^2 - a^2})/b]$.

Next, a method was developed to determine the *initial peripheral contact angle*, θ_{ip}, of sessile drops on solid surfaces from the diffusion controlled rate of drop evaporation, for the constant drop contact radius mode. Application of this method requires use of the product of the vapor diffusion coefficient of the evaporating liquid, with its vapor pressure at the drop surface temperature ($D\Delta P_v$), which can be found directly from independent experiments following the evaporation of fully spherical liquid drops in the same chamber. It is then possible to calculate θ_{ip}, from

$$\sin\theta_{ip} = \frac{-(dV/dt)\rho_L RT r_b}{A_i(D\Delta P_v)M_w} \tag{663}$$

where M_w is the molecular weight, ρ_L is the density of the liquid, *R* is the gas constant, *T* is the absolute temperature, r_b is the contact radius between the solid–liquid interface, and A_i is the initial surface area of the liquid–air interface at time *t* = 0 of the evaporation, which can be found from ($A_i = 2\pi r_b^2/1 + \cos\theta_i$). The peripheral contact angle thus obtained may be regarded as the mean of all the various contact angles existing along the circumference of the drop. Thus, each determination yields an average result not unduly influenced by irregularities at a given point on the surface. In addition, the error in personal judgment involved in drawing the tangent to the curved drop profile at the point of contact is eliminated.

So far, we have not considered the presence of the flat substrate, which restricts the space into which vapor may diffuse and so reduces the evaporation rate. It has been predicted theoretically that a completely spherical small droplet possessing a 180° contact angle and sitting on a flat solid surface will have a rate of mass loss reduced by (ln 2), compared with an identical, completely spherical droplet far removed from any solid surface. When the effect of the presence of horizontal substrate is taken into account, we need to modify Equation (659) as

$$-\left(\frac{dV}{dt}\right) = \frac{4\pi RD}{\rho_L}(c_S - c_\infty)f(\theta) \tag{664}$$

where $f(\theta)$ is a function of the contact angle of the spherical cap. Picknett and Bexon gave two useful polynomial fits covering the angular ranges 0–10 and 10–180°

$$2f_{PB}(\theta) = \begin{cases} 0.6366\theta + 0.09591\theta^2 - 0.06144\theta^3 & (0 < \theta < 10°) \\ 0.00008957 + 0.6333\theta + 0.116\theta^2 - 0.08878\theta^3 + 0.01033\theta^4 & (10 < \theta < 180°) \end{cases} \tag{665}$$

where the contact angle θ is in radians. Equations (664) and (665) are tested with liquid drops having contact angles smaller than 90°, and also with angles larger than 110° and 150°, and very good fits with the experimental results are obtained. In general, drop evaporation experiments provide very good information on contact angle and also liquid–solid interaction, hysteresis, surface roughness, and energy barriers on the substrate (slip–stick motion). Unfortunately, since it takes considerable time to follow drop evaporation and also to carry out an image analyses in a single experiment, it has not been accepted as a popular method at present.

9.5 Contact Angle of Powders

In the case of porous solids, powders and fabrics, the measurement of contact angles is not an easy task. Any method is limited to the packing instability of the powders in the test medium, and all data should be viewed with caution, and whenever possible, the contact angles measured in compressed powder cakes should be confirmed by other independent measurements such as measurements on films of this substance prepared by solvent casting or dip coating. There are mainly two methods to measure the contact angle of solid powders:

1 *Wicking*, which is the measurement of the rate of capillary rise of the test liquid in a porous medium to determine the average pore radius, surface area and contact angle.
2 *Powder tensiometry*, where the powder solid is brought into contact with a testing liquid and the mass of liquid absorbed into the solid column is measured as a function of time.

In both methods, Washburn's equation is used, which was derived from the Poiseuille equation to measure viscosity in capillary viscometers. The rate of volume flow (V/t) through a capillary tube with radius, r_c, is given by the Poiseuille equation as

$$\frac{V}{t} = \frac{(\rho_L g h + \Delta P)\pi r_c^4}{8\eta h} \tag{666}$$

where t, η, ρ_L and γ denote the time, viscosity, density and surface tension of the liquid, respectively; g is the acceleration of gravity, h is the length of the capillary tube and ΔP is the pressure difference across the ends of the tube. Washburn assumed that, any liquid penetrating into a porous solid or a powder column flows through the cylindrical pores of radius, r_c, and, by neglecting the gravitational contribution, Equation (666) becomes

$$\frac{dV}{dt} \cong \frac{\Delta P \pi r_c^4}{8\eta h} \tag{667}$$

From simple geometry, the volume of a cylindrical pore can be written as

$$\frac{dV}{dt} = \pi r_c^2 \frac{dh}{dt} \tag{668}$$

After combining Equations (667) and (668) and simplifying, we obtain

$$\frac{dh}{dt} = \frac{\Delta P r_c^2}{8\eta h} \tag{669}$$

Since the pressure difference can be expressed as ($\Delta P = 2\gamma \cos\theta/r_c$), from Equations (327)–(330) in Section 4.4, and Equation (350) in Section 4.8, we can eliminate the ΔP term in Equation (669) so that

$$\frac{dh}{dt} = \frac{\gamma r_c \cos\theta}{4\eta h} \tag{670}$$

Equation (670) is known as the *Washburn equation*. The r_c term can also be eliminated from Equation (670) by using a completely wetting liquid giving $\theta = 0$, and measuring the increase of the liquid level in the powder column by time. In the practical wicking method, the powder is filled inside a glass capillary tube, the bottom of this tube is brought into contact with the test tube, and the increase of the liquid meniscus with time is followed by a cathetometer.

On the other hand, the *powder tensiometry* method can be applied with any Wilhelmy type tensiometer. A special thin glass tube, which has a porous plug at the bottom, is filled with the powder which will be tested and the powder column is then brought into contact with the testing liquid. The mass of liquid absorbed into the porous solid is measured as a function of time by means of an electrobalance. The amount absorbed is a function of the viscosity, density and surface tension of the liquid, the material constant of the solid, and the contact angle of the interaction, as given in the following relationship:

$$t = \left(\frac{\eta}{K \rho_L^2 \gamma \cos\theta} \right) m^2 \tag{671}$$

where t denotes the time after contact, m the absorbed mass, θ the contact angle and K a material constant. A graph of the absorbed mass squared versus time yields a straight line with slope ($\eta/K \rho_L^2 \gamma \cos\theta$). Since the viscosity, density and surface tension can be measured from independent experiments, or taken from tables, the material constant, K, can be determined by the use of a completely wetting test liquid giving $\theta = 0$. After evaluation of the K parameter, θ of the powder solid with the test liquid can be calculated.

9.6 Contact Angle Hysteresis and its Interpretation

As defined in Section 9.2, *contact angle hysteresis* is the difference between the advancing and receding contact angles:

$$H \equiv \theta_a - \theta_r \tag{672}$$

Hysteresis of the contact angle results from the system under investigation not meeting ideal conditions. In order to apply Young's equation, the solid should be ideal: it must be chemically homogeneous, rigid, flat at an atomic scale and not perturbed by chemical interaction or by vapor or liquid adsorption. If such an ideal solid surface is present, there would be a single, unique contact angle. On the contrary, it is common to find contact angle hysteresis on practical non-ideal surfaces, in the region of 10° or larger; and 50° or more of hysteresis has sometimes been observed.

In general, there appear to be five causes of contact angle hysteresis. Surface roughness and microscopic chemical heterogeneity of the solid surface are the most important ones, and the others such as drop size effect, molecular reorientation and deformation at the surface and the size of the liquid molecules to penetrate into the pores and crevices of the solid surface are less important. A great deal of research has gone into analysis of the significance of hysteresis. It has been used to characterize surface heterogeneity, roughness and mobility on surface groups.

9.6.1   Effect of surface roughness

If the surface of a substrate is rough, then the actual surface area is greater than the plan surface area and thus for a given drop volume, the total liquid–solid interaction is greater on the rough surface than on a flat surface. If the smooth material gives a contact angle greater than 90°, the presence of surface roughness increases this angle still further, but if θ is less than 90°, the surface roughness decreases the angle. Wenzel, in 1936, assumed that the drop liquid fills up the grooves on a rough surface and related the surface roughness with the contact angle by a simple expression

$$r_W = \frac{\cos\theta_e^r}{\cos\theta_e^s} = \frac{A^r}{A^s} \tag{673}$$

where r_W is the ratio of the actual surface area, A^r, to the apparent, macroscopic plan area, A^s, $\cos\theta_e^r$ is the equilibrium contact angle of the real solid, and $\cos\theta_e^s$ is the equilibrium contact angle on a flat, smooth surface. This relation is valid only for submicroscopic (small) roughening and cannot be applied to coarse roughening. When combined with Young's equation (Equation (640)), we can write for the contact angle on a rough surface

$$\cos\theta_e^r = r_W \cos\theta_e^s = r_W \left(\frac{\gamma_{SV} - \gamma_{SL}}{\gamma_{LV}} \right) \tag{674}$$

Wenzel's relation has been confirmed in terms of the first two laws of thermodynamics. Huh and Mason, in 1976, used a perturbation method for solving the Young–Laplace equation while applying Wenzel's equation to the surface texture. Their results can be reduced to Wenzel's equation for random roughness of small amplitude. They assume that hysteresis was caused by nonisotropic equilibrium positions of the three-phase contact line, and its movement was predicted to occur in jumps. On the other hand, in 1966, Timmons and Zisman attributed hysteresis to microporosity of solids, because they found that hysteresis was dependent on the size of the liquid molecules or associated cluster of molecules (like water behaves as an associated cluster of six molecules).

Quantification of surface roughness effects is difficult and somewhat controversial. The equilibrium contact angle cannot be measured directly due to the existence of metastable states. Thus, a rigorous thermodynamic relationship to describe wetting phenomena is difficult to apply because of the uncertainties in the basic parameter set. Recently, micro- or nano-patterned model surfaces have been prepared using a variety of modern techniques such as electron-beam etching, micro-contact printing, and photoresist structuring, and the validity of Wenzel's equation has been tested. Good fits are obtained for surfaces having micro-roughness, with liquids that wet the solid surface well. However, for non-wetting liquids, since these cannot penetrate into the pores of a surface, air pockets remain in these surface pores and Wenzel's equation cannot be applied due to the lack of solid–liquid interaction.

9.6.2 Effect of chemical heterogeneity

Contact angle hysteresis on a flat, atomically smooth solid surface is due to chemical heterogeneity of the surface. For heterogeneous surfaces, some domains exist on the surface that present barriers to the motion of the three-phase contact line. For the case of chemical heterogeneity, these domains represent areas with different contact angles. For example, when a water drop is formed on a heterogeneous surface, the hydrophobic domains will pin the motion of the contact line as the liquid advances, thus increasing the contact angles. When the water recedes, hydrophilic domains will hold back the draining motion of the contact line thus decreasing the contact angle. The movement of the wetting front can give rise to hysteresis, as demonstrated by θ_a and θ_r, and that actual contact line movement can appear as *stick-slip* behavior, resulting in the slow movement of the triple-line on a heterogeneous surface.

At a microscopic level, the non-uniformity of the solid surface allows many metastable configurations for the fluid interface and the energy barriers between them are the source of hysteresis. The range of hysteresis is dependent on the availability of energy to overcome such barriers. It has been proposed that the height of an energy barrier between successive metastable positions increases as the contact angle becomes closer to the stable equilibrium state. The problem is statistical in nature and the occurrence of the heterogenous regions in the triple-line zone can be estimated.

In 1944, Cassie and Baxter derived an equation describing contact angle hysteresis for composite smooth solid surfaces with varying degrees of heterogeneity:

$$\cos\theta_e^r = \sum f_i \cos\theta_i \tag{675}$$

where f_i is the fractional area of the surface with a contact angle of θ_i. For a two-component surface, the above equation can be expressed as:

$$\cos\theta_e^r = f_1 \cos\theta_1 + f_2\cos\theta_2 = f_1 \cos\theta_1 + (1-f_1)\cos\theta_2 \tag{676}$$

When air pockets are present on a rough surface, the Cassie–Baxter equation can also be applied to the contact angle estimate of a water drop on such a surface. Since the contact angle of water in air is equal to 180°, by taking $\theta_2 = 180°$, which corresponds to ($\cos\theta_2 = -1$), Equation (676) then becomes

$$\cos\theta_e^r = f_s \cos\theta_s - (1-f_s) = f_s(\cos\theta_s + 1) - 1 \tag{677}$$

where f_s is the fractional area of the solid surface with a contact angle of θ_s on the flat solid surface. The Cassie–Baxter equation was found to be useful in the analysis of heterogeneous surfaces, but it cannot explain the corrugation of the three-phase contact line and a wide scatter in contact angle data often observed for heterogeneous systems, which has been attributed to the drop size (see next section). In 1984, De Gennes and Joanny proposed a model for contact angle hysteresis based on contact line pinning at solid surface defects. They exploited the analogy between physically rough surfaces and chemically heterogeneous surfaces so that their conclusions were equally applicable to both cases. In conclusion, despite the simplicity of contact angle measurement, the contact angle hysteresis problem is very complex and far from being completely understood.

Hysteresis due to contamination

Solid surfaces are usually contaminated with foreign substances during their manufacture or formation. The liquids that are used in the drop formation may also contain foreign substances to decrease their surface tension. Thus rigorous cleaning of the solid surfaces and purification of the drop liquids before measurement of θ is essential. Solid surfaces are washed with solvents, which do not dissolve the solid but dissolve the possible contaminating materials. If such rigorous cleaning is not carried out and an oily contamination remains, this will result in a smaller water drop, a receding contact angle (because much of the oil would be spread on the water surface) and a different contact angle hysteresis value.

9.6.3   Other reasons for contact angle hysteresis

Hysteresis due to molecular orientation and deformation on solid surfaces

This type of hysteresis often occurs with polymer surfaces. Molecular reorientation in the polymer surface, under the influence of the contacting liquid phase, takes place especially if the polymer has polar or hydrogen-bonding chemical groups in its structure. The terms *surface reconstruction* or *surface reorientation* are also employed. In this process, the surface configuration of polymers (the spatial arrangement of atoms at the surface) changes in response to a change in the surrounding environment. As an example, the hydroxyl groups in a polymer backbone chain are buried away from the air phase for a polymer–air interface, but when a water sessile drop is formed on the polymer surface, the hydroxyl groups turn over to form hydrogen bonds with water. This movement results in reorientation of the surface under test and can be detected by the time-dependent change in contact angle. Such a change does not necessarily require long-range segmental motion but can be achieved by relatively simple short-range motion such as rotational motion of segments at the surface. It is obvious that the types of atoms or groups in a particular polymer chain in bulk do not determine the surface properties, but the types of atoms or groups actually existing at the top of the surface determine the surface-wetting properties. For such surfaces, after the initial θ_a and θ_r measurement, very different contact angle values may be measured after reorientation, and the hysteresis varies.

Hysteresis due to the drop size

The advancing and receding contact angle values may decrease with decreaseing drop size (or in the captive bubble method, with the size of the bubble). This decrease is more pronounced in θ_r values than in θ_a.

Since gravity effects are neglected for small drops, a possible explanation is the presence of the negative *line tension*. Gibbs was the first to postulate the line tension concept. He proposed that an additional free-energy component (line tension) for a three-phase system (solid/liquid/vapor) is needed to provide a more complete description of the system. The line tension results from an excess free energy for molecules located at or close to the three-phase contact line, and becomes increasingly important with decreasing drop size. By considering the line tension, Boruvka and Neumann modified Young's equation (Equation (640)) as:

$$\gamma_{SV} - \gamma_{SL} = \gamma_{LV} \cos\theta + \gamma_{SLV} K_{gs} \qquad (678)$$

where γ_{SLV} is the line tension and K_{gs} is the geodesic curvature of the three-phase contact line. Since K_{gs} is equal to the reciprocal of the drop base radius for a spherical drop sitting on a flat horizontal and homogeneous surface, Equation (678) can be expressed as follows:

$$\gamma_{SV} - \gamma_{SL} = \gamma_{LV} \cos\theta + \frac{\gamma_{SLV}}{r_1} \qquad (679)$$

where r_1 is the drop base radius. The line tension can be determined from the slope of a plot of $(\cos\theta)$ versus $(1/r_1)$ according to the dependence

$$\cos\theta = \cos\theta_\infty - \frac{\gamma_{SLV}}{r_1 \gamma_{LV}} \qquad (680)$$

where $\cos\theta_\infty = (\gamma_{SV} - \gamma_{SL}/\gamma_{LV})$ is assumed, and $\theta = \theta_\infty$ for $r_1 \to \infty$. However, experimental line tension values were found to be 10^5–10^6 times greater than the values predicted from theoretical calculations. The inconsistency between the theory and experiment is attributed to solid surface imperfections, heterogeneities and roughness. Line tension research is somewhat controversial, and a new parameter, pseudo-line tension $(\gamma_{SLV}{}^*)$, was also proposed, which includes the effects of surface imperfections to replace the thermodynamic line tension. Drelich and Miller derived a modified Cassie–Baxter equation containing the line tension contribution as:

$$\cos\theta = f_1 \cos\theta_1 + (1 - f_1)\cos\theta_2 - \frac{f_1 \gamma_{SLV_1}}{r_{l1} \gamma_{LV}} + \frac{(1 - f_1)\gamma_{SLV_2}}{r_{l2} \gamma_{LV}} \qquad (681)$$

The modified Cassie–Baxter equation was successfully used to interpret some of the contact angle data reported for heterogeneous surfaces.

Hysteresis due to liquid adsorption

When liquid molecules are adsorbed onto a solid surface, their surface concentration will be a function of the distance along the solid, and of time, in a band close to the three-phase contact line. Consequently, there will also be a corresponding gradient of surface free energy density, which will directly affect the horizontal component of force at the three-phase line resulting in contact angle hysteresis.

9.7 Temperature Dependence of Contact Angle

The surface tension of liquids, γ_{LV}, decreases with increasing temperature, and if we assume that the surface tension term related to a solid is much less affected by a temperature increase, and thus $(\gamma_{SV} - \gamma_{SL})$ is approximately constant, then the $(\cos\theta)$ term should increase from Young's equation, when the temperature rises. This corresponds to a decrease of θ with increase in temperature. Consequently, the temperature coefficient of the contact angles $(d\theta/dT)$ will be negative. The experimental results confirm this expectation. However the effect of temperature on contact angle is small; a common figure of about $-(0.1°/K)$ is observed. If the solid, such as a polymer, swells in the testing liquid drop, an increase in temperature further decreases the measured θ. This is because the higher temperatures favor intermixing, and θ would be smaller the greater the mutual interpenetration.

9.8 Solid Surface Tension Calculations from Contact Angle Results

Measurement of contact angles on solids yields data that reflect the thermodynamics of a liquid/solid interaction. These data can be used to estimate the surface tension of the solid. For this purpose, drops of a series of liquids are formed on the solid surface and their contact angles are measured. Calculations based on these measurements produce a parameter (critical surface tension, surface tension, surface free energy etc.), which quantifies the characteristic of the solid surface and its wettability.

9.8.1 Critical surface tension of solids (Zisman's method)

Zisman and co-workers introduced an empirical organization of contact angle data on solids (especially on polymers) in 1952. They measured θ for a series of liquids on the same solid sample and plotted $(\cos\theta)$ versus (γ_{lv}) for the test liquids; even for a variety of non-homologous liquids, the graphical points fell close to a straight line or collected around it in a narrow rectilinear band that approaches $\cos\theta = 1$ $(\theta = 0)$ at a given value of γ_{lv}. This value, called the *critical surface tension of solid*, γ_c, can be used to characterize the solid surface under test. It often represents the highest value of surface tension of the test liquid that will completely wet the solid surface. The linear expression fitting the $(\cos\theta)$ versus (γ_{lv}) is given as

$$\cos\theta = 1 - \beta(\gamma_{lv} - \gamma_c) \tag{682}$$

where the slope of the line gives, $-\beta$, the intercept gives $(\beta\gamma_c + 1)$, and both β and γ_c terms can be calculated from a single plot. The β value has been found to be approximately 0.03–0.04. This approach is most appropriate for low-energy surfaces that are being wetted by nonpolar liquids. Zisman warned that $(\gamma_c \neq \gamma_{SV})$ and γ_c is only an empirical value characteristic of a given solid; however γ_{SV} is a thermodynamic quantity. Binary solutions such as water plus methanol must not be used to construct Zisman plots, since one of the components is selectively adsorbed on the solid surface and causes deviations in the contact angle value. The γ_c results of some solids are given in Table 9.1.

However, there are objections to this method, because the value of γ_c is often uncertain since the extrapolation is quite long, and considerable curvature of the empirical line is present for solids on which a wide range of liquids form non-zero contact angles. It is gen-

Table 9.1 Critical surface tension, γ_c, dispersion component, γ_{SV}^d, and surface tension, γ_{SV}, values of polymeric solids. (Values compiled from standard references especially from: Kaelble, D.H. (1971) *Physical Chemistry of Adhesion*. Wiley-Interscience, New York. and Zisman, W.A. (1964) in *Contact Angle Wettability and Adhesion*, Adv. Chem. Ser. No: 43, American Chemical Society, Washington D.C.)

Solid	γ_c Critical surface tension (mN/m) from Zisman plot	$\gamma_{S^o}^d$ Dispersion Component (mN/m) from Fowkes method	γ_{S^o} Surface Tension (mN/m) from Fowkes method
Paraffin wax	22.0	23.2	23.7
Polyethylene	31.0	31.3	32.4
Polydimethylsiloxane	24.0	20.5	22.1
Polyhexafluoropropylene	16.2	11.7	12.4
Polytetrafluoroethylene	18.5	14.5	15.6
Polytrifluoroethylene	22.0	21.9	24.8
Polyvinylidenefluoride	25.0	26.2	32.3
Polyvinylfluoride	28.0	31.2	36.6
Polystyrene	33.0	38.4	40.6
Polyvinylalcohol	37.0	–	–
Polyvinylchloride	39.0	38.1	39.6
Polyvinylidenechloride	40.0	38.2	41.3
Polyethylene terephthalate	43.0	36.6	39.5
Nylon 66	46.0	33.6	41.3

erally believed that, when dealing with liquids where van der Waals forces are dominant, γ_c of the polymeric solid is independent of the nature of the liquid, and is a characteristic of the solid alone. However, when the polymer contains polar and hydrogen-bonding chemical groups, which contribute to the polymer/liquid interactions, the γ_c value may depend on both the nature of the liquids and the polymer. This has led to the concepts of surface and interfacial tension components theory, as given in Sections 9.8.2–9.8.5.

9.8.2 Geometric-mean approach (Fowkes' and later Owens and Wendt's method)

Fowkes proposed in 1964 that the work of cohesion, W_c, and the work of adhesion, W_a, can be separated into their dispersion, d, polar, p, induction, i, and hydrogen-bonding, h components:

$$W_c = W_c^d + W_c^p + W_c^i + W_c^h + \dots \tag{683}$$

$$W_a = W_a^d + W_a^p + W_a^i + W_a^h + \dots \tag{684}$$

The dispersion component of the work of adhesion between a solid and a liquid could be expressed as

$$W_a^d = \sqrt{(W_c^d)_S (W_c^d)_{LV}} = 2\sqrt{\gamma_S^d \gamma_{LV}^d} \tag{685}$$

and the interfacial tension for a solid–liquid system interacting by London dispersion forces alone can be given as

$$\gamma_{SL} = \gamma_S^o + \gamma_{LV} - 2\sqrt{\gamma_{S^o}^d \gamma_{LV}^d} \tag{686}$$

By combining his equation with the Young equation (Equation (648)), Fowkes obtained the Young–Fowkes equation

$$\gamma_{LV} \cos\theta = -\gamma_{LV} + 2\sqrt{\gamma_{S^o}^d \gamma_{LV}^d} - \gamma_{SV} \tag{687}$$

Since Fowkes assumed that $\pi_{SV} = 0$ for low-energy solid (polymer) surfaces, where ($\gamma_{LV} > \gamma_{SV}$) for finite contact angles of high-energy liquid drops on them, he then expressed the equilibrium ideal contact angle, θ_e as

$$\cos\theta_e = -1 + 2\sqrt{\gamma_{S^o}^d} \left(\frac{\sqrt{\gamma_{LV}^d}}{\gamma_{LV}} \right) \tag{688}$$

A plot of $(\cos\theta_e)$ versus $(\gamma_{LV}^d/\gamma_{LV})$ gives a straight line with an origin at $(\cos\theta_e = -1)$, and with a slope of $2\sqrt{\gamma_{S^o}^d}$. Fowkes assumed that $(\gamma_{LV}^d = \gamma_{LV})$ for all non-polar liquids, and calculated $(\gamma_{S^o}^d)$ values of some nonpolar polymers by using Equation (688). He later evaluated the γ_{LV}^d values for polar liquids as a fraction of their total γ_{LV}, initially for water by using water-immiscible hydrocarbon liquid interactions. He used an empirical equation,

$$\gamma_W^d = \frac{(\gamma_W + \gamma_O - \gamma_{WO})}{4\gamma_O^d} \tag{689}$$

where the subscript $(_W)$ denotes water and $(_O)$ hydrocarbon. Fowkes assumed that $(\gamma_O^d = \gamma_O)$ for all nonpolar hydrocarbons, and using data for eight hydrocarbons versus water, he found a value of $\gamma_W^d = 21.8\,\mathrm{mJ\,m^{-2}}$ for water, which is still in use today. Next, he used the contact angle data of polar liquids on nonpolar solids such as paraffin wax and polyethylene, and by applying Equation (688), he calculated (γ_{LV}^d) values for polar liquids, because it is impossible to determine these by any direct method. After using the liquid surface tension components, it is possible to calculate the surface tension of solids. In Table 9.1, the $\gamma_{S^o}^d$ and γ_{S^o} values of Fowkes are presented in comparison with the γ_c values of Zisman, for a number of polymers.

In 1969, based on the Fowkes equation, Owens and Wendt proposed a new expression by dividing the surface tension into two components, *dispersive*, γ_i^d, and *polar*, γ_i^p, using a geometric mean approach to combine their contributions. They assumed that the free energy of adhesion of a polymer in contact with a liquid can be represented by the equation

$$W_a = 2\left(\sqrt{\gamma_{SV}^d \gamma_{LV}^d} + \sqrt{\gamma_{SV}^p \gamma_{LV}^p} \right) \tag{690}$$

based on the assumptions

$$\gamma_i = \gamma_i^d + \gamma_i^p \tag{691}$$

and

$$\gamma_{SL} = \gamma_{SV} + \gamma_{LV} - 2\left(\sqrt{\gamma_{SV}^d \gamma_{LV}^d} + \sqrt{\gamma_{SV}^p \gamma_{LV}^p} \right) \tag{692}$$

by combining with the Young equation, one obtains

$$\gamma_{LV}(1 + \cos\theta_e) = 2\left(\sqrt{\gamma_{SV}^d \gamma_{LV}^d} + \sqrt{\gamma_{SV}^p \gamma_{LV}^p} \right) \tag{693}$$

Owens and Wendt applied only two liquids to form drops in their experimental surface tension determinations. They used $\gamma_{LV}^d = 21.8$ and $\gamma_{LV}^p = 51.0$ for water, and $\gamma_{LV}^d = 49.5$ and $\gamma_{LV}^p = 1.3 \, \text{mJ} \, \text{m}^{-2}$ for methylene iodide, in their calculations. After measuring the contact angles of these liquid drops on polymers, they solved Equation (693) simultaneously for two unknowns of γ_{SV}^d and γ_{SV}^p, so that it would then be easy to calculate the total surface tension of the polymer from the $(\gamma_{SV} = \gamma_{SV}^d + \gamma_{SV}^p)$ equation. Later, Kaelble extended this approach and applied determinant calculations to determine γ_{SV}^d and γ_{SV}^p. When the amount of contact angle data exceeded the number of equations, a non-linear programming method was introduced by Erbil and Meric in 1988.

The geometric-mean approach has been in constant use for more than three decades, even though many articles have been published proving it to be incorrect. The Owens and Wendt equation falsely predicts ethanol and acetone to be as immiscible in water as benzene! The main problem is the incorrect assumption that all polar materials interact with all other polar materials as a function of their internal polar cohesive forces. That is

$$W_{SL}^p \neq 2\sqrt{\gamma_{S0}^p \gamma_L^p} \tag{694}$$

as explained in Section 7.2. Once it is realized that polar interactions are mostly electron donor–acceptor (acid–base) interactions, and strong interfacial interactions occur only when one phase has basic sites and the other has acidic sites (otherwise there is no use for the polar surface tension components), then the use of geometric mean approximations for polar interactions is meaningless.

9.8.3 Harmonic-mean approach (Wu's method)

This method utilizes a similar approach to Owens and Wendt's method but uses a harmonic-mean equation to sum the dispersion and polar contributions. Wu reported that the Owens and Wendt equation was giving surface tension values for polymers in error by as much as 50–100% when compared with their melt values, particularly for polar polymers. He suggested that a harmonic (or reciprocal) mean approximation might be better for polar polymers as given:

$$\gamma_{LV}(1+\cos\theta) = \frac{4\gamma_{SV}^d \gamma_{LV}^d}{\gamma_{SV}^d + \gamma_{LV}^d} + \frac{4\gamma_{SV}^p \gamma_{LV}^p}{\gamma_{SV}^p + \gamma_{LV}^p} \tag{695}$$

It is obvious that the harmonic-mean approach has the same defects as the Owens and Wendt approach, and that internal cohesive polar interaction properties cannot determine the interfacial interaction energy between two different materials. This equation was also abandoned.

9.8.4 Equation of state approach (Neumann's method)

Neumann and coworkers proposed that the solid–liquid interfacial tension should be a function of the liquid and ideal solid surface tensions, $\gamma_{SL} = f(\gamma_{SL}, \gamma_{LV})$. They assumed the ideal solid surface to be smooth, homogeneous, rigid and non-deformable. Moreover, there is no dissolution of the solid in the liquid drop, nor is there any adsorption of any of the components from the liquid or gaseous phase by the solid. The semi-empirical equation of state approach was expressed as follows

$$\gamma_{SL} = \frac{\left(\sqrt{\gamma_{SV}} - \sqrt{\gamma_{LV}}\right)^2}{\left(1 - 0.015\sqrt{\gamma_{SV}\gamma_{LV}}\right)} \tag{696}$$

Neumann and co-workers demonstrated that the minimum γ_{SL} was zero and could not be negative. Later they modified Equation (696) as follows, to avoid the discontinuity as the denominator goes to zero

$$\gamma_{SL} = (\gamma_{SV} + \gamma_{LV}) - \left[2(\sqrt{\gamma_{SV}\gamma_{LV}})\exp{-\beta(\gamma_{LV} - \gamma_{SV})^2}\right] \tag{697}$$

where $\beta = 0.000115 \ (m^2 \, mJ^{-1})^2$. Combining Equation (697) with the Young equation will yield

$$\cos\theta = -1 + 2\sqrt{\frac{\gamma_{SV}}{\gamma_{LV}}} \ \exp{-\beta(\gamma_{LV} - \gamma_{SV})^2} \tag{698}$$

The equation of state approach is very controversial in many respects, and many papers have been published to invalidate this approach. First, it has been shown by Morrison to be based on erroneous thermodynamics. Second, it was shown that there are gross experimental disagreements between predictions from Equation (696) and observed interfacial tensions between water and organic liquids. Third, Neumann and co-workers have tended to ignore any chemical contributions such as hydrogen bonding or acid–base interactions to surface or interfacial tension calculations, treating all surface tensions similar to van der Waals interactions, even for water, where the contribution of hydrogen bonding to cohesive energy and surface tension is very large. Lee showed the limitations of this approach by stating that, without considering chemical interactions, this approach is incomplete and definitely not universal for interfacial tension calculations.

9.8.5 Acid–base approach (van Oss–Good method)

Based on the Lifshitz theory of attraction between macroscopic bodies, van Oss, Good and Chaudhury developed a more advanced approach after 1985 to estimate the free energy of adhesion between two condensed phases. They suggested that a solid surface consists of two terms: one the *Lifshitz–van der Waals interactions*, γ^{LW}, comprising *dispersion, dipolar* and *induction* interactions, and the other the *acid–base* interaction term, γ^{AB}, comprising all the electron donor–acceptor interactions, such as hydrogen bonding. They thought that the Lifshitz calculations yield γ^{LW}, that is the consequence of all the electromagnetic interactions taken together, whether due to oscillating temporary dipoles (γ^d), permanent dipoles (γ^p) or induced dipoles (γ^i). *LW* also includes the interactions of pairs, triplets, quadruplets etc. of molecules within each phase, in all the actual configurations that are taken on when they interact. Then, the corresponding components of work of adhesion are

$$-W_a = \Delta G_{SL} = \Delta G_{SL}^{LW} + \Delta G_{SL}^{AB} \tag{699}$$

and the combining rule for the *LW* component in Equation (699) is given as

$$\Delta G_{SL}^{LW} = \sqrt{\Delta G_S^{LW} \Delta G_L^{LW}} \tag{700}$$

Equation (700) is in concordance with the Fowkes approach for dispersion attractions (see Equation (685)). The term γ_{SL}^{LW} can now be written as

$$\gamma_{SL}^{LW} = \gamma_{S}^{LW} + \gamma_{L}^{LW} - 2\sqrt{\gamma_{S}^{LW}\gamma_{L}^{LW}} \tag{701}$$

or

$$\gamma_{SL}^{LW} = \left(\sqrt{\gamma_{S}^{LW}} - \sqrt{\gamma_{L}^{LW}}\right)^{2} \tag{702}$$

van Oss and Good did not apply a geometric-mean combining rule to acid–base (AB) interactions. Since hydrogen bonds are a sub-set of acid–base interactions, and surfaces of a number of liquids possess only electron donor properties and have no electron acceptor properties, or the reverse is true, one may consider the asymmetry for these interactions. Thus, van Oss and Good adopted Small's combining rule for acid–base interactions, which is not a geometric mean:

$$-\Delta G_{SL}^{AB} = 2\left(\sqrt{\gamma_{S}^{+}\gamma_{L}^{-}} + \sqrt{\gamma_{S}^{-}\gamma_{L}^{+}}\right) \tag{703}$$

where γ_{i}^{+} is the Lewis acid, and γ_{i}^{-} is the Lewis base parameter of surface tension. The term γ_{SL}^{AB} is now given as

$$\gamma_{SL}^{AB} = 2\left(\sqrt{\gamma_{S}^{+}\gamma_{S}^{-}} + \sqrt{\gamma_{L}^{+}\gamma_{L}^{-}} - \sqrt{\gamma_{S}^{+}\gamma_{L}^{-}} - \sqrt{\gamma_{S}^{-}\gamma_{L}^{+}}\right) \tag{704}$$

or

$$\gamma_{SL}^{AB} = 2\left(\sqrt{\gamma_{S}^{+}} - \sqrt{\gamma_{L}^{+}}\right)\left(\sqrt{\gamma_{S}^{-}} - \sqrt{\gamma_{L}^{-}}\right) \tag{705}$$

On the other hand, if the *LW* interfacial free energy is written in conjunction with the Young–Dupré equation, we have

$$-\Delta G_{SL}^{LW} = \gamma_{S}^{LW} + \gamma_{L}^{LW} - \gamma_{SL}^{LW} \tag{706}$$

By combining Equations (701) and (706), one obtains

$$-\Delta G_{SL}^{LW} = 2\sqrt{\gamma_{S}^{LW}\gamma_{L}^{LW}} \tag{707}$$

Later, by combining Equations (699), (703) and (707), one obtains the total interfacial free energy of adhesion as:

$$-\Delta G_{SL} = 2\left(\sqrt{\gamma_{S}^{LW}\gamma_{L}^{LW}} + \sqrt{\gamma_{S}^{+}\gamma_{L}^{-}} + \sqrt{\gamma_{S}^{-}\gamma_{L}^{+}}\right) \tag{708}$$

For negligible spreading pressure (π_{SV}), by combining Equation (708) with the Young–Dupré equation, the general contact angle equation is obtained:

$$\gamma_{LV}(1+\cos\theta) = 2\left(\sqrt{\gamma_{S}^{LW}\gamma_{L}^{LW}} + \sqrt{\gamma_{S}^{+}\gamma_{L}^{-}} + \sqrt{\gamma_{S}^{-}\gamma_{L}^{+}}\right) \tag{709}$$

In order to find the *AB* interactions of cohesion in a solid or liquid phase, Equation (703) is rewritten for a single phase:

$$-\Delta G_{i}^{AB} = 4\sqrt{\gamma_{i}^{+}\gamma_{i}^{-}} \tag{710}$$

since $(-\Delta G_{i}^{AB} = 2\gamma_{i}^{AB})$, and then Equation (710) becomes

$$\gamma_{i}^{AB} = 2\sqrt{\gamma_{i}^{+}\gamma_{i}^{-}} \tag{711}$$

If both γ_{i}^{+} and γ_{i}^{-} are present to interact, the substance is termed *bipolar*. If one of them is not present (equals zero), the substance is termed *monopolar*. If both γ_{i}^{+} and γ_{i}^{-} are absent, the substance is termed *nonpolar*. Therefore, $\gamma_i^{AB} = 0$ for nonpolar and monopolar substances, and γ_i^{AB} is present for only bipolar substances. The total interfacial tension can be obtained from the sum of Equations (702) and (705):

$$\gamma_{SL} = \left(\sqrt{\gamma_S^{LW}} - \sqrt{\gamma_L^{LW}}\right)^2 + 2\left(\sqrt{\gamma_S^+} - \sqrt{\gamma_L^+}\right)\left(\sqrt{\gamma_S^-} - \sqrt{\gamma_L^-}\right) \qquad (712)$$

The most important consequence of Equation (712) is the contribution of acid–base interaction results in negative total interfacial tension, in some circumstances. A solid–liquid system may be stable although it has a negative γ_{SL}. This occurs if $(\gamma_L^+ > \gamma_S^+)$ and $(\gamma_L^- < \gamma_S^-)$, or if $(\gamma_L^+ < \gamma_S^+)$ and $(\gamma_L^- > \gamma_S^-)$, and if $|\gamma_{SL}^{AB}| > |\gamma_{SL}^{LW}|$.

In order to apply Equation (709) to contact angle data, we need a set of values of γ_L^{LW}, γ_L^+ and γ_L^- for reference liquids. Since, $\gamma_{LV}^{LW} = \gamma_{LV}$ for nonpolar liquids, the problem is to determine a set of γ_L^+ and γ_L^- values for dipolar or monopolar liquids. van Oss and Good introduced an arbitrary relation for water. They assumed that $\gamma_W^+ = \gamma_W^-$ for water, and since $\gamma^{AB} = 51.0\,\text{mJ}\,\text{m}^{-2}$ is known, they calculated $\gamma_W^+ = \gamma_W^- = 25.5\,\text{mJ}\,\text{m}^{-2}$, from Equation (711). The values of all acid–base parameters derived therefrom are relative to those of water, and finally they suggest a set of liquid surface tension component data with these operational values; these are given in Table 9.2.

After finding the reference liquid surface tension component values, there are two methods to calculate the polymer surface values of γ_S^{LW}, γ_S^+ and γ_S^-. In the first method, three forms of Equation (709) are simultaneously solved using the contact angle data of three different liquids, two of them being polar. In the second method, γ_S^{LW} can be determined first by using a nonpolar liquid, then two other polar liquids are used to determine γ_S^+ and γ_S^-. Unfortunately, sometimes negative square roots of γ_S^+ and/or γ_S^- occur; this has not yet been definitively explained and causes much objection to this theory. It is recommended that if polar liquids are employed, water should always be used; otherwise if only two polar liquids other than water are used (e.g. ethylene glycol and formamide), highly variable γ_S^+ and γ_S^- values may be obtained.

van Oss–Good methodology has been successfully applied to the interpretation of immiscible liquid–liquid interactions. It is also somewhat successful in polymer solubility prediction in solvents, critical micelle concentration estimation of surfactants, polymer

Table 9.2 Values of surface tension components of test liquids: γ_{LV}^d of Fowkes; γ_{LV}^p of Owens and Wendt and γ_{LV}^{LW}, γ_{LV}^{AB}, γ_{LV}^+, γ_{LV}^- of van Oss–Good components. (Values compiled from Fowkes, F.M., McCarthy, D.C. and Mostafa, M.A. (1980) *J. Colloid. Interface Sci.*, **78**, 200; Good, R.J. (1993) *Contact Angle Wettability and Adhesion*, Mittal, K.L. (ed.). VSP, Utrecht)

Liquid	γ_{LV}	γ_{LV}^d	γ_{LV}^p	γ_{LV}^{LW}	γ_{LV}^{AB}	γ_{LV}^+	γ_{LV}^-
Water	72.8	21.8 ± 3	51.0	21.8	51.0	25.5	25.5
Glycerol	64.0	37.0 ± 4	27.0	34.0	30.0	3.92	57.4
Ethylene glycol	48.0	–	–	29.0	19.0	1.92	47.0
Formamide	58.0	39.5 ± 7	18.5	39.50	19.0	2.28	39.6
Methylene iodide	50.8	48.5 ± 9	2.3	50.8	0	0	0
Dimethyl sulfoxide	44.0	–	–	36.0	8.0	0.5	32.0
Chloroform	27.2	–	–	27.2	0	3.8	0
α-Bromonaphthalene	44.4	47.0 ± 7	0	47.0	0	0	0

phase separation, microemulsion formation in chemistry, and cell adhesion, cell–cell, antigen–antibody, lectin–carbohydrate, enzyme–substrate and ligand–receptor interactions in biology.

References

1. Adam, N.K. (1968). *The Physics and Chemistry of Surfaces.* Dover, New York.
2. Erbil, H.Y. (1997). Interfacial Interactions of Liquids. In Birdi, K.S. (ed). *Handbook of Surface and Colloid Chemistry.* CRC Press, Boca Raton.
3. Neumann A.W. and Spelt J.K. (1996). *Applied Surface Thermodynamics.* Surfactant Science Series Vol. 63. Marcel Dekker, New York.
4. Aveyard, R. and Haydon, D.A. (1973). *An Introduction to the Principles of Surface Chemistry.* Cambridge University Press, Cambridge.
5. Adamson, A.W. and Gast, A.P. (1997). *Physical Chemistry of Surfaces* (6th edn). Wiley, New York.
6. Hiemenz, P.C. and Rajagopalan, R. (1997). *Principles of Colloid and Surface Chemistry* (3rd edn). Marcel Dekker, New York.
7. Butt, H.J., Graf, K., Kappl, M. (2003). *Physics and Chemistry of Interfaces.* Wiley-VCH, Weinheim.
8. Rowan, S.M., Newton, M.I., McHale, G. (1995). Evaporation of Microdroplets and Wetting of Solid Surfaces. *J. Phys. Chem.*, **99**, 13268–13271.
9. Erbil, H.Y. and Meric, R.A. (1997). Evaporation of Sessile Drops on Polymer Surfaces: Ellipsoidal Cap Geometry. *J. Phys. Chem. B*, **101**, 6867–6873.
10. Meric, R.A. and Erbil, H.Y. (1998). Evaporation of Sessile Drops on Polymer Surfaces: Pseudo-Spherical Cap Geometry. *Langmuir*, **14**, 1915–1920.
11. Erbil, H.Y. (1998). Determination of Peripheral Contact Angle of Sessile Drops on Solids from the Rate of Drop Evaporation. *J. Phys. Chem. B*, **102**, 9234–9238.
12. Erbil, H.Y. and Dogan, M. (2000). Determination of Diffusion Coefficient – Vapor Pressure Product of some Liquids from Hanging Drop Evaporation. *Langmuir*, **16**, 9267–9272.
13. McHale, G., Erbil, H.Y., Newton, M.I. and Natterer, S. (2001). Analysis of Shape Distortions in Sessile Drops. *Langmuir*, **17**, 6995–6998.
14. Erbil, H.Y., McHale, G. and Newton, M.I. (2002). Drop Evaporation on Solid Surfaces: Constant Contact Angle Mode. *Langmuir*, **18**, 2636–2641.
15. Erbil, H.Y. and Avci, Y. (2002). Simultaneous Determination of Vapor Diffusion Coefficient from Thin Tube Evaporation and Sessile Drop Evaporation on Solid Surfaces. *Langmuir*, **18**, 5113–5119.
16. Erbil, H.Y. and Meric, R.A. (1988). Determination of surface free energy components of polymers from contact angle data using nonlinear programming methods. *Colloids and Surfaces*, **33**, 85–97.
17. Erbil, H.Y., Demirel, A.L., Avci, Y. and Mert, O. (2003). Transformation of a Simple Plastic into a Super-Hydrophobic Surface. *Science*, **299**, 1377–1380.

Chapter 10

Some Applications Involving Solid–Liquid Interfaces

10.1 Adsorption from Solution

When we introduce an insoluble solid into a solution, a change in composition of the solution usually occurs. This is as a result of preferential adsorption of one of the components on the adsorbent solid. Adsorption from solution is a broad subject including detergent, dye, ion, polymer and biological material adsorption on solids, and a huge amount of literature has been published in this field, since it is important to many industries. In this section, an introduction to the subject will be presented, but excluding ionic adsorption.

10.1.1 *Properties and experimental aspects*

The experimental investigation of adsorption from a solution is much simpler than that of gas adsorption. A known mass of adsorbent solid is shaken with a known volume of solution at a given temperature until equilibrium is reached, at which point there is no further change in the concentration of the supernatant solution. This concentration can be determined by a variety of methods involving wet chemical or radiochemical analysis, colorimetry, refractive index, etc. The experimental data are usually expressed in terms of an *individual adsorption isotherm* in which the amount of solute adsorbed at a given temperature per unit mass of adsorbent, which is calculated from the decrease (or increase) of solution concentration, is plotted against the equilibrium concentration. On the other hand, theoretical treatment of adsorption from solution is usually more complicated than that of gas adsorption, because it always involves competition between solute and solvent molecules, or between the components of a liquid mixture for the adsorption sites and also the heterogeneous nature of solid–solution interactions. Although there are several differences, the Gibbs adsorption isotherm given in Sections 3.2 and 3.3, and some basic concepts developed for gas adsorption on solids given in Chapter 8, can also be applied to adsorption from solution. The surface excess obtained using analytical methods is equal to the thermodynamic Gibbs surface excess in dilute solutions, since there is no problem in location of the Gibbs dividing plane. However, this is not true for concentrated solutions. There are large similarities between the adsorption from binary liquid mixtures at the liquid–vapor (which we examined in Chapter 5) and liquid–solid interfaces. In both processes, the fractional surface coverage, θ_{f}, is always complete and the compositions of the adsorbates on the surfaces change only when the bulk solution composition is changed. This is different from gas–solid adsorption where θ_t is a function of the adsorptive gas pressures. As we know, adsorption can be determined from surface tension data for the liquid–vapor interface; however, this is not possible for liquid–solid interfaces and adsorption can only be determined from the composition changes in the solution after the adsorbent is immersed, and the equilibrium is reached. If the proportion of one of the components at the solid surface is greater than its proportion in bulk solution, then that component is positively adsorbed and, consequently, the other component must be negatively adsorbed. For adsorption from binary solutions, it is clear that both components will be present at the liquid–solid interface, and the isotherm, which represents such an adsorption, is called the *composite adsorption isotherm*, as will be explained in Section 10.1.2. Individual adsorption isotherms for the adsorption of each component can be drawn, especially for the solutes present in dilute solutions (see Section 10.1.3).

In adsorption from solution, physisorption is far more common than chemisorption, although the latter is sometimes possible. Solute adsorption is usually restricted to a monomolecular layer, since the solid–solute interactions, although strong enough to compete successfully with solid–solvent interactions in the first adsorbed monolayer, do not do so in subsequent monolayers, because the interaction is screened by the solvent molecules. Thus, multilayer adsorption has only rarely been observed in a number of cases, and identified, when the number of adsorbate molecules exceeds the number of monolayer molecules possible on the total adsorbent surface area. However, this analysis cannot be applied to polymer adsorption, because it is generally impossible to determine the surface area of a monomolecular layer of a polymer adsorbed flat on the solid surface. This is because the adsorbed polymer can only be anchored to the surface at a few points, with the remainder of the polymer in the form of loops and ends moving more or less freely in the liquid phase.

Adsorption from solutions onto solid surfaces is important in many industrial practices, such as dye or organic contaminant removal, edible oil clarification by activated carbon, and ion exchange, where the adsorption of ions from electrolyte solutions is carried out. Adsorption from solution is also used in analytical chemistry in various *chromatography* applications. On the other hand, surfactant, polymer and biological material adsorption on solids, to modify the surface of solid particles in stabilizing dispersions, are also very important industrial fields.

10.1.2   Composite adsorption isotherms from binary liquid mixtures

A mass m_A of insoluble adsorbent is introduced into a binary solution, containing n_1^o moles of component (1) and n_2^o moles of component (2), before the adsorbent is immersed into a solution, where n_t^o is the total number of moles of solution before adsorption $(n_t^o = n_1^o + n_2^o)$, and then the solution is allowed to reach equilibrium. After the adsorption takes place, n_1^S is the number of moles of component (1) on the surface of unit mass of solid adsorbent, giving a total adsorption of $(m_A n_1^S)$ of component (1) whereas, n_2^S is the number of moles of component (2) on the surface of unit mass of solid adsorbent, giving a total adsorption of $(m_A n_2^S)$ of component (2). The number of moles of component (1) that remain in solution at the adsorption equilibrium is denoted by n_1; and the number of moles of component (2) in solution at adsorption equilibrium is denoted by n_2. Theoretically, the amounts of components (1) and (2) present in the system are the same before and after adsorption takes place. Thus, we can write

$$m_A n_1^S = n_1^o - n_1 \tag{713}$$

$$m_A n_2^S = n_2^o - n_2 \tag{714}$$

In order to use the mol fraction terms in the above expressions, from $(n_1/n_2 = x_1/x_2)$, we can write

$$m_A n_1^S = n_1^o - \frac{n_2 x_1}{x_2} \tag{715}$$

$$m_A n_2^S = n_2^o - \frac{n_1 x_2}{x_1} \tag{716}$$

which can be rearranged as

$$m_A n_1^S x_2 = n_1^o x_2 - n_2 x_1 \tag{717}$$

$$m_A n_2^S x_1 = n_2^o x_1 - n_1 x_2 \tag{718}$$

Since $(n_2 x_1 = n_1 x_2)$, subtracting Equation (718) from Equation (717) yields

$$m_A (n_1^S x_2 - n_2^S x_1) = n_1^o x_2 - n_2^o x_1 \tag{719}$$

Since $(x_2 = 1 - x_1)$, $(n_t^o = n_1^o + n_2^o)$ and $(n_1^o = n_t^o x_1^o)$ are all valid, Equation (719) can be rearranged as

$$\frac{n_t^o \Delta x_1}{m_A} = n_1^S x_2 - n_2^S x_1 \tag{720}$$

where $(\Delta x_1 = x_1^o - x_1)$. The term on the left-hand side of the above equation $(n_t^o \Delta x_1/m_A)$ can be determined experimentally, and Equation (720) gives the composite adsorption isotherm for binary liquid mixtures, because both n_1^S and n_2^S appear on the right-hand side of the equation. The data are usually drawn as plots of $(n_t^o \Delta x_1/m_A)$ against x_1. Positive values of $(n_t^o \Delta x_1/m_A)$ indicate the positive adsorption of component (1) $(x_1 < x_1^o)$ and correspondingly, $(n_t^o \Delta x_2/m_A)$ will be negative indicating the negative adsorption of component (2), which shows that component (2) is richer in the bulk phase than the solid–solution interface. The composite adsorption isotherms may be *U*-shaped or *S*-shaped, as given in Figure 10.1. If one of the components is preferentially (positively) adsorbed on the solid surface for all the bulk concentrations, then a *U*-shaped composite isotherm results. On the other hand, if one of the components is positively adsorbed over part of the concentration range, and negatively adsorbed over the remainder, then an *S*-shaped composite isotherm results. In general, the shape of composite isotherms is dependent on the nature of the adsorbent and the bulk solution properties. Homogeneous surfaces favor *U*-shaped isotherms, whereas heterogeneous surfaces, such as an oxygen complex containing active carbon, favor *S*-shaped isotherms.

In practice, the adsorption from solution behavior can often be predicted qualitatively in advance, in terms of the polar/non-polar nature of the solid and of the solution components. A non-polar adsorbent will tend to adsorb non-polar adsorbates strongly and polar adsorbates weakly, and vice versa. In addition, non-polar solutes will tend to be

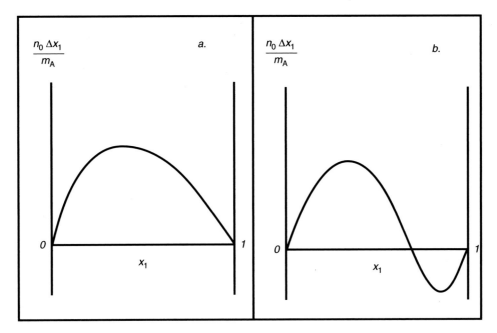

Figure 10.1 Composite adsorption isotherms for adsorption on solids from binary liquid mixtures: **a.** *U*-shaped isotherm. **b.** *S*-shaped isotherm.

adsorbed strongly from polar solvents (low solubility) and weakly from non-polar solvents (high solubility), and vice versa.

10.1.3 Individual adsorption isotherms from dilute solutions

Adsorption onto solids from dilute solutions is important in dyeing processes, detergent action and the purification of liquids by passing through adsorption columns. For a dilute solution, where solute (2) is dissolved in solvent (1), it is generally assumed that $(x_1 \cong 1)$ and $(x_2 \cong 0)$ in Equation (720), so we can write

$$\frac{n_t^o \Delta x_2}{m_A} \cong n_2^S \tag{721}$$

For this derivation, n_1^S is not assumed to be zero, but only $(n_1^S x_2)$ is very small in magnitude. The n_2^S term can be related to the excess concentration of the solute on the solid–solution surface

$$n_2^S = \Gamma_2^S A_{spec} \tag{722}$$

where A_{spec} is the specific surface area of the adsorbent. The determination of A_{spec} in solutions is difficult as a result of the ill-defined and heterogeneous nature of surfaces. Individual adsorption isotherms from dilute solutions are drawn by plotting n_2^S or Γ_2^S against x_2, having varying shapes. However, the most commonly obtained form is very similar to the type-I (Langmuir) isotherm of gas–solid adsorption. Although the Langmuir adsorption theory cannot be applied directly to solid–solution interfaces, because the adsorbed film contains solvent as well as adsorbate molecules, it is common to apply the Langmuir and/or Freundlich adsorption equations to adsorption from solution data. When the adsorption isotherm can be fitted using the Langmuir equation, the saturation adsorption of the solute can be obtained from the isotherm, and if the cross-sectional area of a solute molecule at the interface is known, A_{spec} of the solid can be calculated. A dye such as Methylene Blue is used for this purpose because of the ease of its concentration determination colorimetrically. For colorless adsorbates, the change in the refractive index of the solution can be determined. However, there are uncertainties in the cross-sectional areas of relatively large and asymmetric adsorbate molecules, such as moderately long-chain fatty acids and various dyestuffs, and it is necessary to make assumptions regarding their orientation and packing efficiency to calculate their effective surface coverage. In view of the uncertainties involved in such calculations, it is usually desirable to calibrate a particular adsorption from solution system with the aid of a surface area determination in air, which is measured by the BET method, using nitrogen as the probe gas. It is clear that adsorption from solution has the merit of being experimentally easier than gas adsorption; however, the problems in interpretation are far greater.

10.2   Detergency

Detergency is the removal of solid or liquid dirt particles from the adsorbed (or attached) solid surfaces by using surfactants in an aqueous solution. In this definition, the dissolution of the dirt by pure chemicals or mechanical cleaning is excluded in order to discriminate detergency from the *dry-cleaning* processes where organic solvents are used and from the mechanically operated special washing machines, which are in daily use at present. The sole detergent action can independently remove all oily substances and solid particles such as dust, soot etc. from a solid surface in a washing process; but it is usually supported by mechanical agitation in washing machines to speed up the cleaning process. The washing of textiles by the application of detergents accounts for the bulk of all surfactant usage, but detergent action is not limited to textiles. There is a wide variety of possible substrates to be cleaned by detergents, including hard materials such as plastic, metal, glass, ceramic; soft natural materials such as body skin and hair etc. On the other hand, the dirt materials are versatile; they may be liquid or solid (usually a combination of both), polar or non-polar, of small or large particle size, and chemically reactive or inert towards the substrate and/or the detergent. Thus, the development of a general theory of detergent cleaning mechanism is limited for such a wide variety of possible substrate–dirt systems. However, the basic principles of detergent action are nearly the same for every cleaning process, although their application is not possible for some substrate–dirt systems, especially where chemical reaction or chemisorption occurs.

10.2.1   Mechanisms of detergent action

It is clear that the solid or liquid dirt particles are adsorbed on a substrate, because it is energetically more favorable to adsorb. If we can reverse the process, so that it is energetically more favorable for dirt material to leave the substrate surface and to suspend or to dissolve in the solution, then the cleaning action will be accomplished spontaneously. In thermodynamic terms, when a dirt particle is adsorbed on a substrate surface, it forms a solid–dirt interface (*SD*) with an interfacial tension of γ_{SD}. When the dirt is removed from this solid surface in aqueous solution, two new surfaces will be created; one is the solid–water solution interface (*SWs*) having an interfacial tension of γ_{SWs}, and the other is the dirt–water solution interface (*DWs*) with an interfacial tension of γ_{DWs}; both have the same surface area with the former (*SD*) interface, in ideal cases. For solid dirt particles, the constancy of the interfacial area may be possible, whereas if the dirt is liquid, the removal of the dirt particle will accompany an area change, since the removed liquid dirt drop preferably takes a spherical shape in the solution due to surface free energy minimization. In this process, an increase in the temperature of the water solution will also help conversion into the spherical drop shape with the decrease in viscosity of the oily dirt.

The presence of solid or liquid dirt on a solid is in analogy with the formation of a sessile drop on a substrate, as shown in Figure 9.1. The dirt material is not necessarily in a spherical cap shape; it may be a rectangular block or in an irregular shape for solid dirt particles. For ease of schematic representation, if we replace the liquid drop with the dirt particle, we can write D instead of L ($L{\rightarrow}D$) and the vapor with the water solution ($V{\rightarrow}Ws$); then the work of adhesion between a dirt particle and a solid surface per unit area can be expressed from Equation (644) as

$$\frac{W_{SD}^a}{A} = -\frac{\Delta G_{SD}^a}{A} = \gamma_{SWs} + \gamma_{DWs} - \gamma_{SD} \tag{723}$$

We need to decrease the value of W_{SD}^a to separate the dirt particle from the surface. In thermodynamic terms, the quantity ΔG_{SD}^a must be negative for the separation to be spontaneous. When a detergent is dissolved in water, due to the positive surface excesses of detergent molecules at *DWs* and *SWs* interfaces, the values of γ_{DWs} and γ_{SWs} are reduced, so that the (W_{SD}^a/A) term decreases, from Equation (723). The reduction of W_{SD}^a increases the ease of removing the dirt particles with mechanical agitation.

On the other hand, if the dirt is liquid, it may form a layer or a spherical cap shape on the substrate surface, especially in warm solutions. Then, it is possible to apply the Young–Dupré equation (Equation (645)) to the detergent action

$$\frac{W_{SD}^a}{A} = -\frac{\Delta G_{SD}^a}{A} = \gamma_{DWs}(1+\cos\theta) \tag{724}$$

By combining Equations (723) and (724) we obtain

$$\gamma_{DWs}\cos\theta = \gamma_{SWs} - \gamma_{SD} \tag{725}$$

There may be two situations depending on the initial contact angle of the liquid dirt drop on the solid substrate. If ($\theta > 90°$), this corresponds to ($\gamma_{DS} > \gamma_{SWs}$), from Equation (725), and Equation (724) shows that W_{SD}^a decreases rapidly with the increase in contact angle by the detergent action, as shown in Figure 10.2 *a* (W_{SD}^a also decreases with decreasing contact interfacial area between D and S). If, however, ($\theta < 90°$), this corresponds to ($\gamma_{SWs} > \gamma_{DS}$), and Equation (724) shows that W_{SD}^a can decrease only with a decrease in the contact angle by the detergent action, as shown in Figure 10.2 *b*. However, the work of adhesion

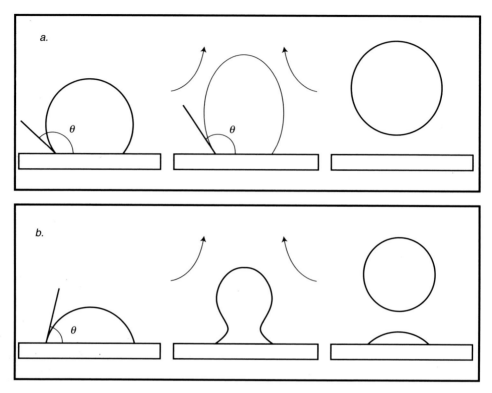

Figure 10.2 Schematic representation of the detachment of oily dirt material from a substrate surface: *a.* The sequences left to right illustrate the detachment of the dirt having an initial contact angle of $\theta > 90°$. *b.* The sequences left to right illustrate the detachment of the dirt having an initial contact angle of $\theta < 90°$. The *lift-off* hydraulic force detaches the spherical droplet at the end, but some drop remains on the surface for case *b*.

values are large for the latter case, only part of the oily dirt can be removed from the surface with the help of mechanical action, and some will remain adsorbed on the solid surface, as shown in Figure 10.2 *b*. On the other hand, W_{SD}^a is small for the ($\theta > 90°$) case and the oily dirt can be entirely removed with the help of mechanical action in a washing machine, as shown in Figure 10.2 *a*. The *lift-off* process arising from the hydraulic forces in the solution also helps to remove the dirt in both cases.

In microscopic terms, the water molecules with a high surface tension ($\approx 72.8 \, \text{mJ} \, \text{m}^{-2}$) cannot enter the dirt–solid interfacial region to dislodge the dirt in detergent-free conditions. However, when a suitable surfactant with detergency properties is dissolved in water, it decreases the surface tension of water near the surface tension of the dirt ($\approx 20–40 \, \text{mJ} \, \text{m}^{-2}$), and the hydrophobic parts of the detergent molecules line up both on the solid surface and on the dirt particles, thus increasing the surface excess of detergent molecules in these interfaces, and reducing the adhesion of the dirt to the solid. Afterwards, the dirt may be removed by mechanical action and will be held suspended in the aqueous solution due to detergent adsorption on the dirt particles. Meanwhile, several other detergent molecules form an adsorbed layer on the cleaned solid surface.

10.2.2 Properties of a good detergent

In summary, a good detergent must possess good wetting characteristics both with the dirt and with the solid substrate, so that the detergent can come into intimate contact with them so as to remove the dirt into the bulk of the liquid and disperse it, so that the dispersed dirt cannot re-deposit on to the cleaned surface again; otherwise, a washing procedure would only lead to a uniform distribution of the dirt. When a surfactant has ease of adsorption at the solid–water and dirt–water interfaces, it will then act as a good detergent. A detergent with a longer hydrocarbon chain in its structure can remove dirt better. On the other hand, the adsorption at the air–water interface, with the consequent lowering of surface tension down to ≈ 20–$40\,mJ\,m^{-2}$ is not necessarily an indication of detergent effectiveness. For example, non-ionic detergents usually have excellent detergent action, but they are poor surface tension reducers. In addition, the detergent action is dependent upon the concentration of unassociated surfactant and practically unaffected by the presence of micelles. The micelles may act as a reservoir to replenish the unassociated surfactant adsorbed from solution.

Re-deposition of dirt on solid surfaces can be prevented by stabilizing the dispersion of the detached dirt, by the adsorption of detergent molecules on the dirt particles in the aqueous solution. The adsorbed detergent molecules prevent the aggregation and flocculation of dirt particles by electrostatic repulsion, or by forming hydration barriers. Since the substrate and dirt surfaces are generally negatively charged, anionic detergents tend to be more effective than cationic detergents for this task. As a result of strong hydration of the poly (ethylene oxide) chains, non-ionic detergents are also effective for this purpose and mixed anionic plus non-ionic detergents are usually better than anionics alone. In textile cleaning, the rate of diffusion of a detergent into a porous fabric is very important, and the choice of a surfactant involves a compromise between a small hydrocarbon chain length for rapid diffusion, and a longer hydrocarbon chain length for better dirt removal and dispersion characteristics.

It is possible to test the action of a detergent in the laboratory: *launderometer* is a typical method. A white textile cloth is polluted with standard dirt mixtures comprising carbon black and grease such as vaseline, and cleaned with a standardized washing procedure (the temperature of washing solution, agitation rate and washing duration are kept constant); then the optical reflectivity is measured to evaluate the efficiency of the detergent. Recently, a *tergotometer* method has been developed to test particular washing applications, where special model dirt mixtures and paddle agitation are used. Radioactive labeling may also be used instead of optical reflectivity measurements to evaluate the cleanliness of the cloth. The microscopic determination of the shape and the contact angle of the droplet with a video camera is also an important tool to monitor detergent action.

10.2.3 Functions of detergent additives

The *builders*, such as silicates, pyrophosphates ($Na_4P_2O_7$) and tripolyphosphates, are generally incorporated into detergent formulations to improve their performance. They produce mildly alkaline solutions, which are favorable to detergent action. They also form soluble non-adsorbed complexes with the metal ions such as Ca^{2+} and Mg^{2+}, which contribute to water hardness, and act as deflocculating agents, thus helping to avoid dirt redeposition. Sodium carboxymethyl cellulose additive improves detergent performance in washing cotton fabrics by forming a protective hydrated adsorbed layer on the cleaned fabric; this helps to prevent dirt re-deposition. Optical brighteners are commonly incorporated into textile detergents to absorb ultraviolet light and emit blue light, in order to mask the yellow tint, which may develop in white fabrics.

References

1. Aveyard, R. and Haydon, D.A. (1973). *An Introduction to the Principles of Surface Chemistry.* Cambridge University Press, Cambridge.
2. Hiemenz, P.C. and Rajagopalan, R. (1997). *Principles of Colloid and Surface Chemistry* (3rd edn). Marcel Dekker, New York.
3. Erbil, H.Y. (1997). Surface Tension of Polymers. In Birdi, K.S. (ed.). *Handbook of Surface and Colloid Chemistry.* CRC Press, Boca Raton.
4. Butt, H.J., Graf, K. and Kappl, M. (2003). *Physics and Chemistry of Interfaces.* Wiley-VCH, Weinheim.
5. Shaw, D.J. (1996). *Introduction to Colloid & Surface Chemistry* (4th edn). Butterworth-Heinemann, Oxford.
6. Adamson, A.W. and Gast, A.P. (1997). *Physical Chemistry of Surfaces* (6th edn). Wiley, New York.

Index